CW01551768

The Pigment Compendium

Optical Microscopy of Historical Pigments

The Pigment Compendium

Optical Microscopy of Historical Pigments

Nicholas Eastaugh, Valentine Walsh, Tracey Chaplin, Ruth Siddall

ELSEVIER
BUTTERWORTH
HEINEMANN

AMSTERDAM BOSTON HEIDELBERG LONDON NEW YORK OXFORD
PARIS SAN DIEGO SAN FRANCISCO SINGAPORE SYDNEY TOKYO

Elsevier Butterworth-Heinemann
Linacre House, Jordan Hill, Oxford OX2 8DP
30 Corporate Drive, Burlington, MA 01803

First published 2004

British Library Cataloguing in Publication Data
A catalogue record for this book is available from the British Library

ISBN   0 7506 4553 9

For information on all Butterworth-Heinemann publications
visit our website at www.bh.com

Typeset by Charon Tec Pvt. Ltd, Chennai, India
www.charontec.com
Printed and bound in Italy

Acknowledgements

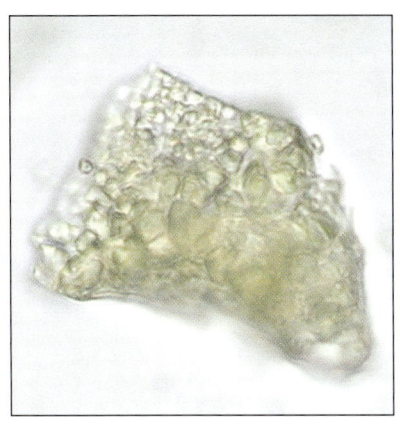

The authors are enormously grateful for the support and encouragement they have been given over the five years it has taken to take this book to move from the original concept of a small volume to the much larger and somewhat daunting volume that it has now become. Among the many who encouraged us, some must be singled out for special thanks.

First, thanks must go to our publishers and, especially, our editor Alex Hollingsworth for never wavering in their belief in the project and staying with it through the many changes in the scope of the work. The authors would not have persevered if they had not been certain that this book and its companion, *The Pigment Compendium: A Dictionary of Historical Pigments*, would ultimately be published.

This book would, of course, have been impossible without a large collection of pigments to analyse and illustrate. Many people have been extremely generous in giving samples for the collection and providing background information. First, several important historical collections were opened up to the authors. The Teylers Museum in Haarlem was very helpful in permitting access to the important Hafkenscheid Collection and the Molyn Collection of pigments that have become the core of the historical pigments in this book. The Courtauld Institute of Art in London also granted access to their historical collection and the Museo Archaeologico in Pompeii allowed a generous number of pigments to be analysed and included in this project. The Swiss Art Research Institute gave a sample of vermilion from the Terracotta Army and the Royal Ontario Museum, Toronto, gave samples of Han blue and Han purple from artefacts of the period. For all of these we are extremely grateful. The Geology Department of the National Museums and Galleries of Wales, Cardiff, donated some rare mineral samples. For these we are very grateful.

It appears that no work today on pigments is undertaken without at some time consulting Dr Georg Kremer and this book is no

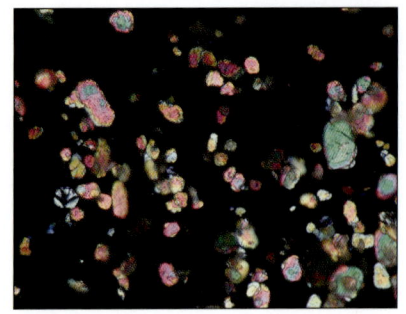

exception. Dr Kremer has been unfailingly generous with his pigments, time and advice over several visits to his factory and then on the occasion of many pleas for help and information over the years. Hans-Christophe von Imhoff has also graciously allowed access to his collection of pigments, one that has become established as a core of reference standards; he has also enthusiastically encouraged the authors throughout the project. De Kat in Amsterdam donated a number of pigments as well as giving advice.

Numerous other people have shared the results of their research, generously giving samples of pigments they have collected or made themselves. Wholehearted thanks therefore goes to: Dr Leslie Carlyle, Instituut Collectie Nederlands; Dr James Ovenstone, University of Greenwich; Dr David Scott, Getty Conservation Institute; Rowena Hill; Dr Ian Steele, University of Chicago; Dr John Winter, Freer Gallery of Art; Christophe Herm, Swiss Institute for Art Research; Ineke Pey; Dr Marie-Claude Corbeil, Canadian Conservation Institute; Susan Lake, Hirshorn Collection; Dr Claudio Seccaroni, ENEA Rome; Keith Edwards; Dr Janet Ambers, The British Museum; Carol Grissom, SCMRE; Gareth Hatton, RLAHA, University of Oxford; and Richard Tayler.

The analysis of this collection would not have been possible without the remarkable amount of help and support given by Dr Alison Crossley and Frank Cullen at the Department of Materials, University of Oxford. Frank has consistently impressed the authors with his extremely deep knowledge of X-ray diffraction analysis and interpretation. Gill Gibbs, of the Microstructural Studies Unit, University of Surrey, has also been very generous with her time and help with SEM-EDX analysis of the collection. Professor David Price and Dr Ian Wood, Department of Earth Sciences, University College London, have given advice concerning crystallography. Dr Paul Bown, Micropaleontology Unit, Department of Earth Sciences, University College London, has helped in the identification of coccoliths and geological stratigraphy for chalks.

Mrs Richard Walsh has contributed generously to help with the funding of the analysis and without that assistance the authors would not have managed to carry on. Heartfelt thanks go to her. Thanks must also go to Gordon and Eileen Eastaugh, whose significant financial gift supported in a number of important ways the work of the project.

Thanks must also go to Chris Collins who was initially part of the team and who contributed greatly in the initial phases, to Leica (Microsystems) UK Ltd and again to Dr Ian Wood, Department of

Earth Sciences, University College London, who have been generous in the loan of optical equipment.

Finally unstinting assistants have helped in untold ways to keep all strands of this project moving forward. They include Tanya Kielsich, Shona Broughton, Benedict Eastaugh, Maria Keller, Leonie Ainapore, and Sophie Godfraind and Helen Oglesby. Lastly, thanks must go to the authors' partners and families who have born the brunt of the highs and lows of this long project and without whose support and forbearance this book would long ago have been abandoned.

OPTICAL AND PHYSICAL PROPERTIES OF PIGMENT PARTICLES FROM POLARISED LIGHT MICROSCOPY

Those who have never used a polarized-light microscope will be surprised at the great beauty and vividness of the colours that arise as a result of the interference of the light waves passing through the specimen

(Robinson and Bradbury, 1992)

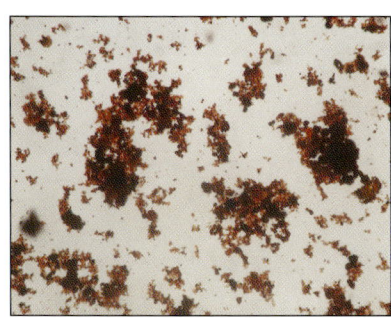

Introduction

Polarised light microscopy is perhaps the most widely applied analytical technique currently used for the identification of historical pigments. Despite the advent of newer methods, polarised light microscopy has maintained a place in this field as a routine and effective approach to the determination of the compounds to be found on painted artefacts. Consequently it may be surprising that no dedicated atlas of pigments has ever been published to support the researcher wishing to carry out such work.

In practice there were a number of important issues to resolve before such a project could be brought to completion, the major areas being:

1 *Naming conventions for pigments*: It was clearly absurd to illustrate pigments under the numerous common names that have been applied to them over the centuries. However, it was also realised that current naming was also often deficient. For example, 'cobalt violet' encompasses a wide range of pigments that may be the compounds cobalt phosphate, magnesium cobalt arsenate, lithium cobalt phosphate and so forth; many other cases could be given. As a consequence of this a more structured system was developed, based on the chemical nomenclature for the compounds found in pigments and their chemical groupings. The authors' systematic (taxonomic) listing of pigment compounds is presented in the companion volume to this, the *Dictionary of Historical Pigments*, along with extensive discussions of historical terminology. In essence this book presents the component compounds that occur within pigments, with some coverage of the more common pigment assemblages such as earth pigments; it is for the analyst to then determine the precise relationship between components found within a sample from an artefact. While the authors are aware that this departs significantly from much previous practice, it is to be hoped that both a simplification and some additional clarity will be brought to the subject.

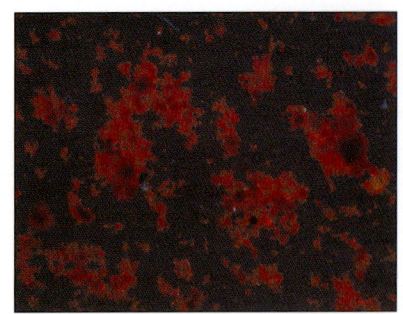

2 *Terminology for optical descriptions*: The majority of pigment compounds covered in this text are natural minerals or synthetic inorganic analogues. Therefore the methods and terminology used in their description relies heavily on those used routinely for the description of minerals and rocks under the microscope. Other pigment compounds are biologically derived, these being either biominerals, such as bone or shells, or compounds such as starch, sepia and other organic material or secretions. (Dye extracts are not usually appropriate for analysis by polarised light microscopy and are not discussed in detail here, though composite structures – 'lakes' – based on them are included.) Descriptive terminology for such materials may differ widely from that applied to natural minerals and their synthetic analogues. Information is given in the individual entries and general descriptive terms for biologically derived pigment forming materials is also outlined later in this section.

3 *Range of compounds*: Another difficult issue was to select the compounds to include. It was realised at an early stage that no handily complete list of pigment compounds existed and that one would need to be constructed. An extensive review of the pigment literature was carried out by the authors, the fruits of which led to the book on terminology mentioned above. This research also yielded a structured compound listing that has formed the basis of this work. The authors have also tried to take a moderately neutral view on inclusions and exclusions; while there might appear to be a core group of pigments that would be widely agreed as 'common', in fact as our knowledge of pigments and their analysis develops, so new compounds and contexts are being discovered. It was also decided that there were no clear geographical or historical boundaries that could be applied and that in fact a broader knowledge of practice was informative. For example, cobalt aluminium oxide ('cobalt blue') is generally known from its modern use, when in fact it was known to dynastic Egypt. Consequently this text does not discuss material context. None-the-less, the final selection of compounds represented covers, it is to be hoped, those pigments most likely to be encountered in a broad range of artefacts without too much undue emphasis on Western European easel painting.

4 *Illustrative material*: The samples accumulated for the illustrative purposes of this book are from varied sources and have been chosen with the aim of showing as wide a range in physical, chemical and optical properties as demonstrated by historical pigments as possible. A further requirement was that all samples should be well characterised and of known provenance. Therefore samples from artefacts were not used because,

although there might be an implied provenance, there is unlikely to be enough material to fully confirm all aspects of identity. Consequently samples are from a combination of sources including modern synthesis, collected minerals and established historical collections. (More information on the specific collections and samples is included in Appendix I.) The pigment collection has also been subject to analysis by multiple techniques, primarily scanning electron microscopy with energy dispersive X-ray spectrometry, X-ray diffraction and Raman spectroscopy, to ensure a clear and unambiguous identification in support of the polarised light microscope identifications. In excess of 1500 samples were collected, documented and analysed for this book, a major task. None-the-less this will not reflect the full range of variance and the authors have therefore drawn upon descriptions in the literature as well as their own experience to broaden the scope where necessary.

5 *Analytical methodology*: Those pigments included have been grouped only loosely by colour, a definition used broadly. The authors have also intentionally avoided the approach of producing a key or any other type of flowchart ('dichotomous schema') for the identification of pigment phases, in the belief that this method is too prescriptive and could (and does) lead to false identifications. It is of importance when applying polarised light microscopy that all optical properties are fully explored when examining pigment particles in dispersions – a single feature of a pigment, such as its colour or morphology, is not sufficiently diagnostic. The authors also recommend that analysts develop their own reference collections of characterised pigments, the better to observe features in direct comparison. Further, it should be appreciated that this technique, though powerful, cannot in all circumstances lead to a unique identification and that parallel analyses are often necessary (differentiation of mixtures of white pigments or the clear separation of indigo and Prussian blue come to mind). In summary, there is probably no adequate substitute for familiarity with a broad range of pigments and their appearance.

This book is not intended to act as an instruction manual for the practice of polarised light microscopy. However, some overview of the various phenomena exploited by the technique is appropriate.

CRYSTALLINE SUBSTANCES AND CRYSTAL SYSTEMS

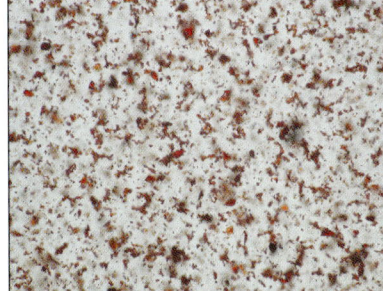

As many compounds encountered as pigments are crystalline, and because this underlies the phenomena observed in optical microscopy, a brief excursion examining the concept of crystallinity is included here. Every crystalline substance has a definite internal

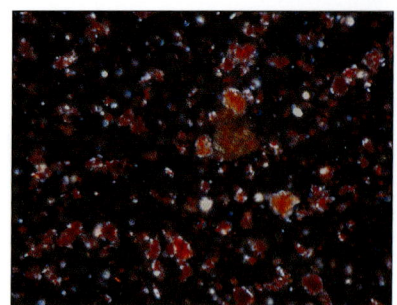

structure in which the atoms occur in specific proportions and are arranged in an ordered geometric pattern. This systematic arrangement of atoms is one of the most significant aspects of a compound in the crystalline state. It exists throughout the entire body of the material and, if crystallisation occurs under ideal conditions, this internal arrangement will be expressed in outwardly perfect crystal shapes. Not all crystalline compounds have perfect crystal faces or form because of restrictions in the environments in which they crystallise. However, compounds with poor or imperfect crystal faces do still possess a systematic internal structure, expressed by features such as *cleavage*, the fracturing along lines of particular weakness in the crystal structure.

All crystalline materials belong to one of seven crystallographic 'systems' based on the symmetry of the crystalline forms. On an atomic level the structure of all crystalline forms is periodic, with each repeating unit defining the crystal system. This smallest of crystallographic building blocks is called the *unit cell*. These are three-dimensional blocks, defined partly by the relative lengths of their three axes, *a*, *b* and *c*, and the angles between these axes, α, β and γ. For systems where the axes are of different lengths, *c* is conventionally the longest axis and *a* the shortest. Crystal systems are defined also by the rotational and bilateral symmetry displayed. The unit cell dimensions in terms of relative axial lengths, angles between them and the essential symmetries are outlined in the table below.

TABLE 1   The seven crystal systems (adapted from Putnis, 1992). The essential symmetry is in terms of a rotation axis.

Crystal system	Relative unit cell dimensions	Essential symmetry
Triclinic	$a \neq b \neq c$ $\alpha \neq \beta \neq \gamma$	None
Monoclinic	$a \neq b \neq c$ $\alpha = \beta = 90° \neq \gamma$	One 2-fold axis
Orthorhombic	$a \neq b \neq c$ $\alpha = \beta = \gamma = 90°$	Three perpendicular 2-fold axes
Tetragonal	$a = b \neq c$ $\alpha = \beta = \gamma = 90°$	One 4-fold axis
Cubic	$a = b = c$ $\alpha = \beta = \gamma = 90°$	Four 3-fold axes
*Trigonal	$a = b = c$ $120° > \alpha = \beta = \gamma \neq 90°$	One 3-fold axis
Hexagonal	$a = b \neq c$ $\alpha = \beta = 90°, \gamma = 120°$	One 6-fold axis

*Crystals in the trigonal system may be described by an hexagonal unit cell, even though they do not have a 6-fold rotation axis. Many texts therefore list trigonal as a subset of the hexagonal system.

Compounds may change their crystal system in response to variations in temperature, pressure or deformation and also after substitution of trace or major elements into the crystal structure. Examples here are distortions to the structure of lazurite and the other feldspathoids by substituting elements, and fluorapatite, which has a hexagonal structure but becomes monoclinic as hydroxyl groups substitute for the fluorine. Crystals are also never perfect, and are characterised by defects and dislocations (planar defects), the presence of defects actually facilitating crystal growth.

A *solid solution* series represents the variation in chemistry between two pure end-member crystalline phases whereby complete mixing can occur between the end-member compositions. For example, a hypothetical crystal with composition A and another of composition B can form solid solutions A_xB_{1-x}, where $0 \leqslant x \leqslant 1$. The end-members do not necessarily belong to the same crystal system, this property changing from one to the other at some intermediate point. A system with members in which mixing cannot occur is called a *eutectic*; the result of crystallisation of phases showing this behaviour is always a mixture of pure phases.

Materials that do not possess a crystalline structure are described as being *amorphous*. Amorphous substances that have undergone rapid solidification, which proceeds too fast for crystal nucleation to occur, are glasses; this term is applied irrespective of chemistry.

The processes of crystallisation and the relationship of the resultant morphologies to the conditions under which a substance was formed appears to be a relatively little studied aspect of historical pigments. For those interested in pursuing this aspect, a thorough review of the mechanisms of crystallisation can be found in Mullin (2001).

PHYSICAL PROPERTIES OF PIGMENTS

Colour

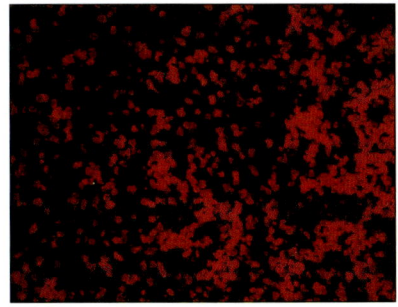

Colour is, naturally, a fundamental property of interest in pigments and most pigment phases appear strongly coloured under the microscope. However, the occurrence of colour is a far from simple concept. Observed colour arises from the transmission of some wavelengths of light and the absorption of others. Transmission of light – and therefore colour – is controlled by the crystal structure and chemical composition of the material. A full discussion of the production of colour on an atomic, molecular or bulk scale is far beyond the scope of this text and the interested reader is therefore referred to standard texts such as Nassau (1983), Putnis (1992) and Burns (1993) for further information.

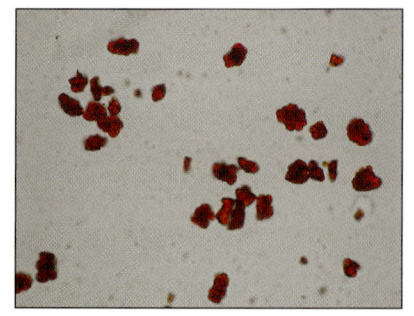

While the origin of colour in pigments is a field of interest in its own right, the documentation of colours is of more direct importance in the characterisation of these compounds. A number of systems for recording colour are available to the analyst, these falling into two broad groups – colour comparison charts and direct measurement. Discussion of these is beyond the scope of this text and the reader is referred, for example, to the discussion by Johnston-Feller (1986) regarding standard colour notation and the section on micro-spectrophotometry below. However, the formal specification of colour in particles seems to be generally more of a documentation ideal than a common practice. In this text the issue has been dealt with by giving broad colour categories only, or using qualifiers that are reasonably wide in their use and apprehension.

The analyst should though be aware of the external factors which can affect the appearance of a particle. Specifically, these are the illumination conditions of both the microscope itself and the general laboratory environment. All colour judgements for this book were made with a high-colour temperature microscope light source with a suitable dichroic filter to achieve illumination close to natural daylight and an ambient daylight (or equivalent fluorescent lighting) in the laboratory to minimise metameric effect during observation. Additionally, the impact of different lens qualities cannot be underestimated; as certain authors have noted elsewhere, the appearance of colour in particles under the microscope can shift significantly between lenses with and without substantial chromatic aberration. Optical systems in the microscopes employed for this book all had substantial levels of colour correction such that they provided highly accurate renditions.

Particle size and particle size distribution

Particle size can be described either relatively or absolutely. In simplest terms, particle size may be regarded relatively as very fine, fine, medium, large and coarse. Given the typically small particle size of pigments compared to, say, minerals observed in petrography, it is important that this classification be linked to known size ranges. The standard unit of measurement is the micrometre ($1\,\mu m = 10^{-6}$ metres, formerly known as a micron). Absolute measurement of particle size may be carried out using a stage micrometer, typically marked out in 1, 5 and $10\,\mu m$ divisions. Feller and Bayard (1986) have proposed a particle size classification, based partly on the Paint Research Station Classes (1956).

TABLE 2 Feller and Bayard's (1986) particle size classification for pigments.

Absolute particle size (μm)	Relative particle size
>40	Very coarse
10–40	Coarse
10–3	Large
3–1	Medium
1.0–0.3	Fine
<0.3	Very fine

To this scale has been added a 'very coarse' size class for particles greater than 40 μm. This classification is particularly applied to mineral-based pigments. The scale is based on a logarithmic increase in particle size, with the smallest optically distinguishable particles being around 0.3 μm, the diffraction limit for visible light. Consequently, the very fine category is unlikely to be used by the majority of analysts. Description of particle size may be more complex when describing particles with a high aspect ratio; in such cases, both length and width of the particle should be quoted. For particles of lower aspect ratio, at least one of the standard 'diameters' should be explicitly recorded – for example, the major axis, or Feret's diameter.

Further qualitative description of particle size distributions may be required when particle size is not uniform. This is particularly relevant to naturally occurring and traditionally prepared pigments, such as the ochres and other earth pigments. Where this occurs the range of particle sizes present should be given. It may be convenient to describe this in terms of a normal distribution curve, implying a mean sample size and deviations away from that mean. Distributions of grain size may therefore be described as being broad or narrow, the former implying a large range of grain sizes, the latter implying a very limited range in grain size, while allowing for the rare particle which falls outside the stated range.

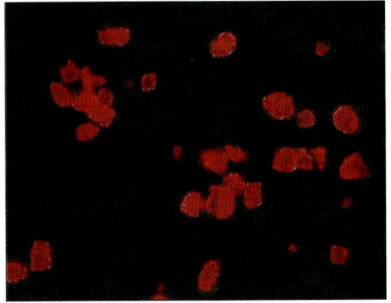

If there are two clear particle size ranges (say, one population of fine grained particles and another of large particles) then the sample may be described as having a bimodal particle size distribution. Clearly this can be extended to trimodal and so forth. The term 'seriate' is synonymous with a broad particle size distribution and refers to a sample population with all particle sizes between very fine and coarse.

Particle shape

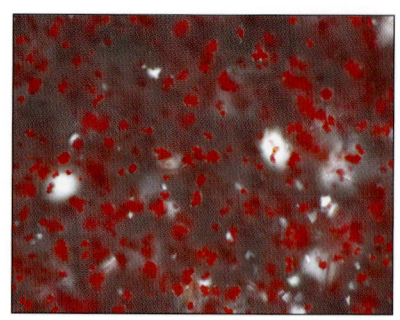

Particles in pigments may include mineral fragments, natural and synthetic organic and inorganic crystals, particles of glass or other amorphous substances, mineral fragments supporting organic lakes, fragments of shells, fossils and vegetable remains. These may have been prepared by crushing, grinding and milling larger particles and aggregates, as precipitates, or even simply by sieving naturally occurring substances. Therefore particle shape can be highly variable, but diagnostic of certain substances and certain preparation processes. Description of particle shape borrows largely from the fields of crystallography and petrography, where there is a well-defined terminology describing precisely mineral shapes; in these fields, particle shape is referred to as 'habit'. Some additional terms relevant to the identification of pigments are also added here.

All materials may form crystalline solids. Crystals, by definition, have an ordered, repeating and predictable structure. Despite the thousands of geometrical forms crystals may adopt, they may all be reduced in terms of their symmetry into the seven crystal systems: *cubic*, *tetragonal*, *orthorhombic*, *trigonal*, *hexagonal*, *monoclinic* and *triclinic*. Crystals belonging to these systems may well adopt the shape defined by their system. So, for example, pyrite and halite (common salt) both belong to the cubic system and adopt cubic habits. However, pyrite may also adopt polyhedral forms, such as the 'pyritohedron', which has 12 pentagonal crystal faces. Quartz and calcite on the other hand both belong to the trigonal system, but neither forms triangular prisms. Rather, it is the symmetry of these crystals as observed in three dimensions that will prove which system they belong to.

When crystals are allowed to grow in unrestricted environments, they will develop crystal faces and, as discussed above, these are a reflection of the symmetry of the underlying crystal system to which the phase belongs. Crystals observed under the microscope showing all crystal faces may be described as being *euhedral*. Those showing no crystal faces are *anhedral*. Intermediate forms showing some, but not all, faces are *subhedral*. Obviously these terms give no idea of the exact shape of the phase present although they are important qualifiers.

Terms that simply describe a geometrical figure may be used to describe euhedral crystals, such as 'rhombic', 'rectangular', 'hexagonal', 'cubic', 'triangular', 'polyhedral' and so forth. Additionally, some further useful and more specific terms are listed below:

Equant – any shape where length = width = height. A sphere is an extreme example of an equant shape. A cube is also equant and in fact the minerals adopting the cubic crystal system will form equant crystals. Members of other crystal systems can form crystals with a close approximation to equant forms.

Prismatic – forms of crystals having one axis greater than the other two. Terms applied to prismatic forms commonly include the following:

  Acicular – needle-shaped.

  Fibrous – an extreme case of acicular shape, with particles resembling textile fibres.

  Bladed – like a knife-blade; flattened but elongated.

  Lath-shaped – similar to bladed, particles appear as elongate rectangles with length much greater than width.

Platy – forming thin plates.

Lamellar – a stack of plates, like pages in a book.

Tabular – forming thicker, tablet-like plates.

Columnar – forming crystals with an aspect ratio similar to a column.

Dendritic – forming branching, tree-like shapes. These usually represent phases that have rapidly crystallised.

Pennate – (also 'pinnate'), refers to elongate particles having bilateral symmetry and the shape of a feather or leaf.

Subhedral and anhedral particles may occur naturally or as a result of pigment preparation. If some faces are apparent, then an indication of the crystal's ideal shape can be acquired. In the absence of any crystal shapes, anhedral particles may be described using the following criteria and terminology.

The degree of angularity of a particle's shape can be defined qualitatively, using the following series of terms: angular, subangular, subrounded and rounded, where particles range from sharp-edged to smooth-edged morphologies respectively (Powers, 1953). Highly angular particles, with morphology representing that of broken glass, may be called *shards*. Importantly, angular particles defined by cleavage (see below) may appear to be euhedral; they are in fact anhedral. The mineral calcite commonly adopts such forms, with rhombic crystals defined by the three cleavage planes.

The qualifier 'rounded', however, should not be confused with 'sphericity'. The concept of sphericity refers to the general aspect ratio of the particle. Particles with high sphericity are equant whereas particles with low sphericity are elongated. A ball, for example, has high sphericity and is rounded, whereas a cigar is also rounded but has low sphericity.

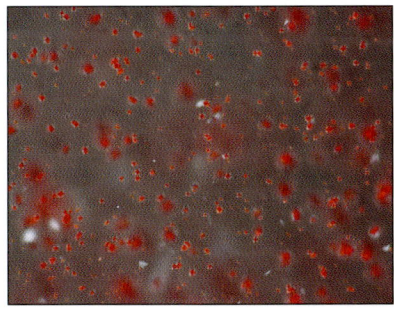

Many particles may show no evidence of crystal shape, nor do they fit with descriptive terms defined above. The terms *irregular, crumb-like*

or *ragged* may be used to define shapes with the appropriate morphologies. All imply embayed margins on various scales.

Very fine and fine-grained particles are at the limits of the resolution of the optical microscope. It is therefore impossible to define their exact shape without the aid of an electron microscope. All particles in this range often appear to be small spheres or slightly elongated, rounded, 'capsule'-like shapes. The term *bacterioid* is used here to refer to such habits.

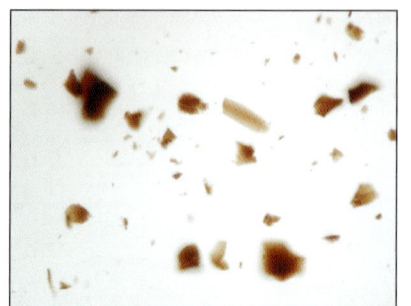

Some particle forms are directly biological in origin. If possible these should be identified and described using appropriate terminology. They will commonly be calcium carbonate-derived shell or fossil fragments, or intact calcium carbonate or siliceous microfossils. The former may typically include crushed shell (which has a lamellar structure), while the latter may include coccoliths (which are at the limit of optical resolution) and diatoms. For further information concerning such forms see Adams *et al.* (1991) and Tucker (1991). Terms for descriptions of these particles are discussed below and in the relevant entries.

Aggregation

Aggregates of several crystals of the same variety may form further diagnostic habits. An aggregate of euhedral crystals may be generally called a *glomerocryst*. However, where individual crystals are poorly defined or a particle appears to be composed of several crystals, then the term polycrystalline may be applied.

Elongate crystals often form glomerocrysts radiating from a single nucleation point. These give rise to *stellate* and *spherulitic* forms. Stellate forms are two-dimensional aggregations with the appearance of stars. Spherulites (or spherules) are three-dimensional and have the overall appearance of pom-poms. Fan- or cone-shaped aggregates of fibrous crystals are said to have *variolitic* texture. The latter are generally composed of fibrous or acicular crystals.

Fibrous or acicular crystals arranged in layers may develop what is known as *botryoidal* habit, referring to the surface on a macro-scale resembling a bunch of grapes. The mineral malachite frequently displays this form; however, the defining structure is generally lost when the mineral is crushed to make a pigment. Bundles of parallel fibres may be called *sheaves*.

Other aggregates of anhedral crystals may be very irregular and have the appearance of bread crumbs, the term *crumb-like* is simply used here. *Framboidal* aggregates are clusters of rounded, equant particles with the general appearance of (literally) raspberries.

Very fine aggregates of crystals such as those developed by so-called chalcedonic silica and calcite may be referred to as being *cryptocrystalline*. Calcite is a special case. Cryptocrystalline masses of this phase are described as *micrite*, whereas calcite that forms crystals visible down the microscope is called *sparite*. These terms should not be applied to other phases.

For some pigments, precipitates of one phase may be found adhering to a second phase and this relationship should be described as observed.

Particle surface

Using a binocular microscope a good impression of the surface texture of a particle can be obtained. This feature is highly influenced by the relief of the mineral, itself a function of refractive index (see below for definitions of both terms). Particle surfaces may straightforwardly be defined as being rough or smooth. Other descriptive qualifiers such as powdery, pitted, undulating, fractured and striated may also be applied to rough surfaces. Glass and some minerals show characteristic conchoidal fractures, which may be visible on crystal surfaces as scoop-shaped fracture surfaces with concentric pressure ridges (their appearance resembles a shell, hence the name). A surface showing several well-developed crystal faces may be described as being faceted (this phase will already have been described as euhedral). Particles with very high relief may appear domed (or convex).

Inclusions and other inter-particle relationships

Few naturally occurring and synthetic crystals are completely pure. They can contain inclusions which may be gas- and/or liquid-filled cavities (called fluid inclusions), or solid inclusions of another phase. Inclusions of both sorts may strongly influence the colour of a crystal both negatively and positively. The colour of some minerals is directly related to the presence of inclusions (for example, the purple and yellow colours of the Blue John variety of fluorite, which is thought to be influenced by sub-microscopic inclusions of hydrocarbons; see: Dunham, 1937). However, abundant inclusions of separate phases may also diminish the colour and attempts to remove the bulk of these will be part of the pigment processing.

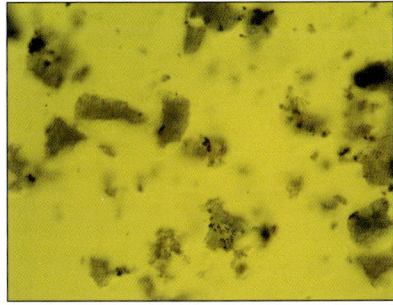

Fluid inclusions may contain both gas and liquid phases. In this case the gas will appear as a bubble in the liquid, which will start to move around as the light from the microscope heats the crystal. Fluid inclusions may be randomly orientated cavities and/or may have a crystallographically defined alignment.

Solid inclusions may well show a colour or relief contrast with the host particle and this should be noted. Where abundant solid inclusions occur, the particle may be described as being *poikilitic*. The host particle may be called the *oikocryst* and the included particles *chadocrysts*. Worm-like inclusions of one phase in another may be called vermiculate texture.

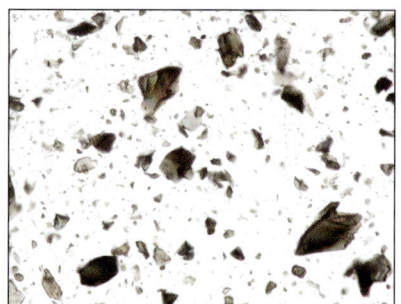

Other intergrowth textures, not strictly inclusions, may also occur. A *pseudomorph* is a crystal that grows replacing another crystal of different composition, but replicating its habit and, occasionally, some internal features such as cleavage. *Intergrowths* are where two crystal phases of different composition have grown simultaneously and interpenetrated each other as they do so. Complex, interfingering intergrowths between two phases may be called a *symplectite*. These are common in some minerals. Intergrowths between distinct phases all have specific, distinguishing names. For example, perthites are emulsions of one feldspar variety in another of slightly different chemistry. These relationships are rarely important in identifying pigment phases, particularly because they occur on scales greater than the ground pigment, and the interested reader is therefore referred to relevant mineralogical texts such as Hibbard (1995), MacKenzie and Guildford (1980), MacKenzie *et al.* (1984), Yardley *et al.* (1990) and Shelley (1992) for description and mode of formation of these textures.

Variations in chemistry within a single crystal (zoning) and the 'intergrowth' of two or more crystals of the same chemistry (twinning) are features generally only observable in cross-polarised light. These features are discussed in further detail below.

Fracture and cleavage

The way that crystals break may be a diagnostic feature. Many crystals possess a cleavage; that is, they will break along straight, parallel and equally spaced, crystallographically defined planes that represent weakness inherent in the crystal structure. Crystals may have one, two, three, four or even six sets of cleavage planes, or none. Because cleavage is crystallographically controlled, in crystals that display two or more cleavages the angle at which the cleavages intersect may be used to distinguish two otherwise similar crystals.

Crystals with strong, well-defined cleavage can be said to possess *perfect cleavage*, while a weak cleavage may be described as a *parting*. Some crystals (notably quartz) do not possess planes of weakness within their structure and therefore do not possess a cleavage.

Fracture refers to breaks within a crystal unrelated to cleavage. Fractures may be even, that is broadly planar, or uneven, that is leaving behind a rough surface. Conchoidal fracture is typical of glasses and some minerals including quartz. Conchoidal fracture forms curved, shell-like fracture surfaces, often with concentric pressure lines. Conchoidal cracks within a crystal are visible as curved lines.

As discussed above, the cleavage or fracture possessed by crystals can strongly influence their habit on crushing.

Diaphaneity

The degree of light transmittance through a substance, its diaphaneity, can be described in terms of transparency, translucency and opacity. Transparent phases do not impede the passage of light. Translucent phases transmit some, but not all, light and opaque phases do not transmit any light. The latter will appear black in dispersions.

Reflected light

Although much examination of optical properties is carried out using transmitted light ('dia-illumination'), dark-field epi-illumination is an effective method of determining the colour (as well as the texture) of phases that are opaque, or particles of such a pale colour that characterisation of the hue is difficult by transmitted light. It is common therefore to document the appearance of samples under these conditions. For this book, reflected light illumination was achieved using a polariser inserted in the epi-illumination path and a corresponding analyser. Using this approach it was possible to additionally observe the phenomenon of bireflectance (see below).

Red transmission

The degree to which certain, primarily blue and green, pigments transmit at the red end of the visible spectrum has been used as a means of distinguishing them. Most commonly this is via the application of the so-called Chelsea filter, a device inserted into the light train. However, the more general study of pigment transmission spectra across the visible spectrum is a field of recurring interest.

The Chelsea filter was originally developed by gemmologists to distinguish emeralds from other green stones. Emeralds have the unusual property of absorbing part of the yellow-green wavelengths of light and transmitting red light. Thus an emerald observed through the green glass of the Chelsea filter appears deep red, the filter absorbing the remaining green wavelengths (Liddicoat, 1993). However, other minerals have similar effects. For example, lazurite transmits a strong red light, despite the fact it is blue. Azurite, in contrast, appears blue-grey when observed using a Chelsea filter.

Though generally designed for use on items of jewellery, a Chelsea filter may be used for microscope analysis, simply inserted above the transmitted light source of a microscope (below the stage and polariser). Chelsea filters are available from suppliers to jewellers and gemmologists.

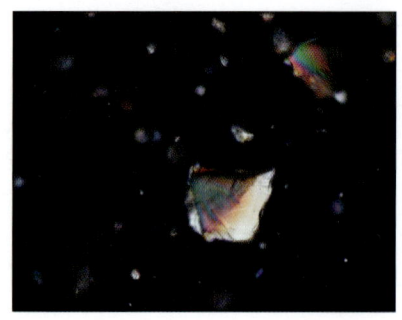

A more sophisticated approach (outside the scope of this text) is micro-spectrophotometry, whereby the transmission spectrum of particles under the microscope can be measured directly. This technique is potentially effective for such otherwise difficult applications as the differentiation of indigo and Prussian blue (Leona and Winter, 2001), detection of cobalt absorption bands in pigments such as cobalt blue and smalt (Bacci and Picollo, 1996), characterisation of red lake pigments (Kirby, 1977) and study of phthalocyanines (Talsky and Ristic-Solajic, 1987).

Fluorescence

Fluorescence is becoming an increasingly standard adjunct to paint microscopy due to its use in selective staining techniques as applied to media in cross-sections. However, it is a potentially useful tool in some cases for extending the examination of pigment dispersions.

Fluorescence is the phenomenon whereby a material absorbs light at one wavelength and then promptly re-emits it at another, longer, wavelength that is shifted towards the red end of the spectrum or beyond. Underlying this is absorption of energy from the light so that when it strikes the material, interacting with it, it is subject to a loss of energy. Since energy and wavelength are directly related, the wavelength (colour) of the light returned is therefore longer (redder). Fluorescence originates from the absorption of light photons by components within particular molecules, the absorption being associated with changes of energy levels of electrons in the molecular structure. Since such electron transitions are fairly precise, so the minimum energy (wavelength) of light required for fluorescence is also well defined. The shift in wavelength to the longer, redder, less energetic end of the electromagnetic spectrum is then related to gradual loss of energy of the electrons, generally through loss of heat, before the electrons fall back to lower molecular levels with the re-emission of another photon. This also affects the form of the range of emitted wavelengths, with the most photons being emitted at the higher end of the range. Fluorescence takes place rapidly, typically within around 10^{-9} to 10^{-8} seconds. The process of fluorescence can in fact take place at all wavelengths, not just in the visible light region, so that, depending on the substance, fluorescence can occur over a wide range of illumination (excitation) conditions.

In fluorescence microscopy it is important to define two values, the excitation (Ex) and emission (Em) wavelengths, with the excitation wavelength being that used to excite fluorescence, the emission wavelength being that which is given out by the material. Usually these values are effectively fixed by the microscope system, with standardised Ex/Em filter blocks targeted at specific fluorophores such as FITC and Rhodamine B.

A number of pigments have known fluorescence. Examples include zinc white, which has a strong yellowish fluorescence, cadmium sulfides, which fluoresce in the yellow to red part of the spectrum, certain lake pigments and some coals. For further information the papers by de la Rie (1982) and standard works on fluorescence microscopy such as Rost (1992, 1995) should be consulted.

OBSERVATIONS IN PLANE-POLARISED LIGHT

Polarised light

The technique of polarised light microscopy relies on the fact that light travels in waves. A ray of light from a light source travels out in every direction from that source. For microscopes not fitted with polarisers, this is the light observed and it is simply called transmitted light. However, on passing through a polariser, the light is forced to vibrate in a single plane; it is polarised and this is called plane-polarised light (PPL). The appearance of plane-polarised and transmitted light appear identical when observed by the human eye. When a second polariser (the analyser) is inserted, orientated at right angles to the first polariser, then the field of view is in cross-polarised light (XPL) and will be dark as the light is vibrating in the wrong plane and cannot pass through it. When a transparent or translucent crystalline material is placed between the polariser and the analyser, as light enters the crystal, two things will happen. First, the velocity of the light will be retarded (refracted) and second, the light beam will be split into two rays (of differing velocities), each vibrating perpendicular to the other. Therefore light can now be passed through the analyser and the crystal is said to be birefringent and interference colours will be observed.

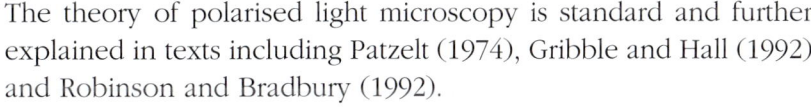

The theory of polarised light microscopy is standard and further explained in texts including Patzelt (1974), Gribble and Hall (1992) and Robinson and Bradbury (1992).

Pleochroism

The phenomenon of pleochroism is observed as a crystal changes colour or colour intensity as the stage is rotated. For uniaxial phases, crystals may be pleochroic (or strictly speaking 'dichroic')

in two colours. For biaxial phases, three colours may be observed (although all three will not be observed in the same particle). Each extreme in the two colours observed in a crystal will be observed twice in a 360° rotation. The phenomenon arises as a result of a difference in absorption of light in different crystallographic orientations. The more that light is absorbed, the darker the colour observed.

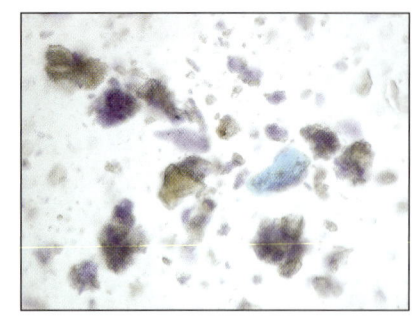

The intensity and colour changes observed should be noted. Pleochroism is a feature of all non-cubic phases. However, it may be very subtle or take place at wavelengths outside the visible region and therefore may not be apparent.

Refractive index and relief

As light passes from a vacuum into any substance, its velocity is reduced; it is refracted. The factor by which the light is slowed by the substance through which it is transmitted compared to its speed in a vacuum is known as the refractive index (RI) and may be expressed mathematically as:

$$RI = \frac{\text{Velocity of light in a vacuum}}{\text{Velocity of light in the substance}}$$

RI is a unit-less quantity. Substances that have the same refractive index in all directions are called *isotropic*. These include non-crystalline solids (such as glasses), liquids and gases as well as crystalline forms belonging to the cubic crystal system.

In *anisotropic* substances the refractive index varies with the orientation of light propagation through the crystal lattice. These crystals are classified as being either *uniaxial* or *biaxial*.

For a uniaxial crystal (belonging to the hexagonal, trigonal or tetragonal crystal systems) a beam of light entering it is split into two rays. One ray, the *ordinary* ray (with a refractive index n_ω) continues in a straight line through the crystal, whereas the second ray, the *extraordinary* ray (with a refractive index n_ε) is displaced. It is because of this that the phenomenon of double refraction occurs, which is particularly visible in the mineral calcite. For many uniaxial crystals, the RI of the extraordinary ray is greater than that of the ordinary ray and such materials are described as being optically positive (+). Less commonly, this is reversed and the ordinary ray has the greater RI. Such crystals are described as being optically negative (−). This is, in some cases, a useful diagnostic feature.

For a biaxial crystal (belonging to the orthorhombic, monoclinic or triclinic crystal systems) a beam of light entering it can experience three distinct RIs, these having values normally designated n_α, n_β and n_γ. In a similar manner to the uniaxial crystals, biaxial crystals may also be stated to be optically positive or negative. However, the designations refer in this case to the more complex relationship of whether $|n_\alpha - n_\beta|$ is greater or less than $|n_\beta - n_\gamma|$; such determinations are normally difficult, requiring a high level of expertise to carry out, especially on very small particles like those encountered in pigment analysis.

The difference in refractive indices of the rays, δ_n (or more usually just δ; calculated as $|n_\varepsilon - n_\omega|$, or $|n_\gamma - n_\alpha|$) is known as birefringence. Birefringence is responsible for the interference colours observed under crossed polars (see below); it is always zero for cubic crystals and amorphous substances.

Refractive index (RI) is an important and potentially diagnostic phenomenon. For each pigment in this book, numerical values for RI are given where known. A single refractive index is available for amorphous substances, two refractive indices are given for uniaxial minerals (for the ordinary ray and extraordinary ray n_ω and n_ε) and three refractive indices are given for biaxial minerals (for the minimum, intermediate and maximum, n_α, n_β and n_γ). Quantitative refractive index is difficult to determine for particles and involves immersing materials in a series of oils of different but known refractive index. However, a relative refractive index value compared to that of the medium in a dispersion (or adjacent particles in a cross-section) can be determined by observing the relief and using the Becke line test.

Relief is a three-dimensional optical effect that occurs as a direct function of the relative refractive indices of adjacent substances. The degree of relief observed increases the greater the difference between the two RIs. For example, if a pigment particle has an RI very similar to that of its mounting medium, it will have particle boundaries barely distinguishable from the medium. Alternatively a particle with an RI either substantially lower or higher than the medium will appear to stand out from the medium, with clear, well-defined particle boundaries.

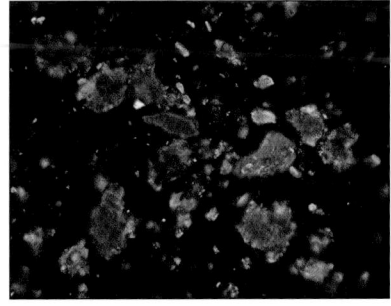

Relief is easy to qualify and the observed ability of the particle to 'stand out' from the medium can simply be stated as being low (particle boundaries almost indistinguishable from the medium), moderate and high (very distinct, thick, black lines defining particle boundaries).

Some phases, notably the carbonate minerals and their biogenic analogues, have variable relief. This is because the refractive index varies widely depending on the orientation of the crystal to the light source. For example, in calcite, RI varies from 1.658 to 1.486.

This is observed as the particle edges fading and then reappearing as the microscope stage is rotated.

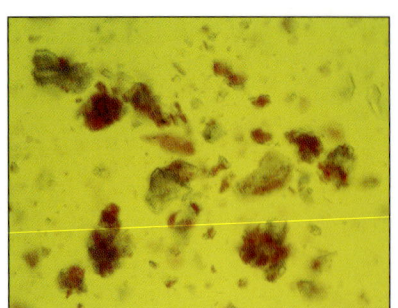

The relative refractive index of two adjacent substances may be determined by observation of the *Becke line*. The Becke line is a bright fringe formed by refraction and/or diffraction at the boundary between two substances of differing refractive index. Clearly this is of use if the refractive index of, say, the embedding medium of a microscope sample is known since one can then establish whether the unknown material has a higher or lower RI. When the focus of the microscope is adjusted the Becke line will move either into or out of the substance being examined, the direction being dependent upon whether the separation is increasing or decreasing, and the relative refractive index of the substances. As the distance between the sample and the objective lenses is increased, so the Becke line will move into that substance of *higher* refractive index. Consequently, it is simple to determine whether the particle has an RI greater than (>) or less than (<) that of the medium. For fine-grained particles the Becke line may be difficult to observe; however, for particles with refractive index greater than that of the medium, a bright spot appears in the particle centre as the stage is racked down.

As a consequence of this technique, it is important to know the RI of the medium being used, information that will be readily available from manufacturers. Meltmount™ is currently available in seven refractive indices, though that most commonly used for mounting pigments has an RI of 1.662. Canada Balsam, and materials of similar refractive index (RI ~ 1.54), are routinely used by mineralogists and petrologists. Therefore a mineral exhibiting moderate relief in Canada Balsam may have low relief in Meltmount™. Analysts should be aware of this when changing from one medium to another or, more particularly, when reading estimates of relief from texts intended primarily for petrologists (for example, Gribble and Hall, 1992). The RI of glass typically used in the manufacture of microscope slides is ~1.52, a value which may vary slightly with composition.

TABLE 3   Relationship of RI to relief for a medium with RI = 1.662.

RI	Description of relief
1.40–1.50	High
1.50–1.60	Moderate
1.60–1.70	Low
1.70–1.80	Moderate
1.80–2.0	High
>2.0	Very high

Dispersion

Refractive index also varies with wavelength of light, this being the phenomenon that underlies the splitting of light by a prism. Known as *dispersion*, blue light generally has a higher refractive index than red light. Dispersion also varies between substances, from a virtually negligible difference to marked levels. It should be noted that for some phases, where dispersion is high, interference figures might be difficult to obtain (Gribble and Hall, 1992; McCrone *et al.*, 1979). Dispersion may also strongly affect interference colours of minerals with low birefringence, leading to anomalous colours (see below). Dispersion has not been systematically measured for the samples used in this book; where available, data from the literature has been incorporated.

OBSERVATIONS IN CROSS-POLARISED LIGHT

Isotropism

Cubic and non-crystalline (amorphous) substances are isotropic – that is, the material behaves identically in all directions – and therefore has only one refractive index. Because the axial difference in refractive index is zero, isotropic phases do not transmit light through crossed polars and thus appear black. It is not usually possible to differentiate whether phases are cubic or amorphous using polarised light microscopy alone.

Anisotropism: birefringence and interference colours

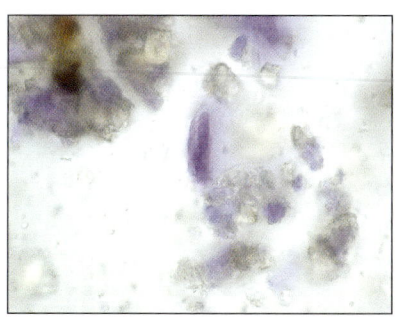

Light is transmitted through crossed polars when non-cubic crystalline structures with different refractive indices are placed between them. Such phases are anisotropic and birefringent. As discussed above, uniaxial minerals have two refractive indices, and biaxial minerals have three. Birefringence (δ) is a quantitative value, measured as the difference between the refractive indices of a crystal. As polarised light is transmitted through a crystal and then observed through another polariser, an array of colours is observed. These *interference colours* (also known as 'polarisation colours') are a function of the birefringence and the thickness of the sample through which the light is passing. A crystal exhibiting this property can be said to be *birefringent*. The colours that appear are graded according to Newton's scale, which is divided into orders of colours. *First order colours* are dark grey, through white, dull yellow and red. *Second order colours* are brighter and range through violet, green-blue, yellow and red. The jewel-like *third order colours* are bright blues, greens, yellows and pinks. *Fourth order* and above are increasingly subtle shades of pinks and greens. Above fourth order, the orders become difficult to distinguish by the eye. Because the colours observed depend upon sample thickness as well as birefringence, the thicker the sample, the higher the order of colours observed. For further discussion of this phenomenon, see Robinson and Bradbury (1992).

Mineralogical texts, such as Deer *et al.* (1992), MacKenzie and Guildford (1980) and Hibbard (1995), illustrate photomicrographs of minerals showing their 'standard' interference colours, which for petrographers are based on a mineral section 30 μm thick. As pigment particles are of variable particle size range (from less than 0.3 μm to substantially greater than 10 μm) then the order of interference colours observed may differ from the mineralogically expected norm, and in fact, will usually be lower. An estimation of interference colours based on particle size may be gleaned from the Michel-Lévy chart (see insert), where birefringence is shown as a series of rays plotted against particle thickness. Where the ray for the particular birefringence intersects the desired thickness, the position in Newton's scale is given. In this way it can be shown that all crystals can show orders of colours up to the maximum.

As a consequence of the refractive index variation of minerals with crystallographic orientation, the birefringence will also vary depending upon the section observed. Therefore orientations where the difference between refractive indices is small (generally the basal sections of crystals) will show lower interference colours than sections where the variation in refractive indices is large (that is, prismatic sections).

If, as in the case of many pigment-related phases, the crystal has a strong body colour, this will effectively 'mask' the interference colours. However, as a rule of thumb, the brighter the crystal appears under crossed polars, the higher the order of colours. In these cases it is possible to qualitatively describe the interference as low, moderate, high and very high, these categories broadly corresponding to first order, second order, third order and fourth to fifth order colours respectively.

Anomalous colours may form in many phases with strong body colours and, technically, low birefringence. True anomalous colours are blue, indigo, purple and brown, which do not equate to colours appearing on Newton's scale; they are thus abnormal colours. Chlorite and copper acetate compound F (a form of 'verdigris') are good examples of crystals exhibiting this phenomenon. Anomalous interference colours arise as a function of dispersion. Modification of interference colours in phases with strong body colours and high birefringence may also be termed as being anomalous, and this is the case for phases such as minium and cobalt phosphate.

For the entries in this book, only the standard value of birefringence, δ, is given. From this value, using the Michel-Lévy chart, the interference colours may be calculated using the variable of particle size.

Internal reflection

Some minerals which have very high relief and an RI much higher than that of the medium (greater than 2) often exhibit internal reflections, where light is trapped inside the particles and causes them to appear to glow. The same phenomenon is responsible for the 'fire' in diamonds. The reflections will be the same colour as the mineral. For extreme cases, the whole microscope slide will appear to glow in the respective colour. This phenomenon is usually more obvious under crossed polars and may be mistaken for birefringence, especially in fine-grained samples.

Extinction

In cross-polarised light, as the microscope stage is rotated through 360°, the observed crystal will be aligned such that it will not transmit light (that is, it will go black) at four positions, 90° apart, throughout the rotation. Uniaxial crystals possess 'straight' or 'parallel' extinction; they are extinct when a long axis of the crystal is parallel to the north-south or east-west cross-hairs at the 90°, 180°, 270° and 360° positions. Orthorhombic biaxial minerals also possess straight extinction (*ortho* meaning straight in Greek); however, other biaxial minerals possess inclined or oblique extinction – that is, they become extinct at an angle to the cross-hairs. The angle through which the crystal must be rotated from the extinction position to the nearest cross-hair is known as the extinction angle and this may be diagnostic of the chemical composition of some mineral phases. This feature is readily applied to the feldspar, amphibole and mica groups of minerals.

Many crystals show complete extinction; that is, as the extinction position is reached particles become completely dark. Variation in crystal growth mechanisms or subsequent deformation of the crystal may produce incomplete extinction phenomena. The terminology for description of these observations is given below.

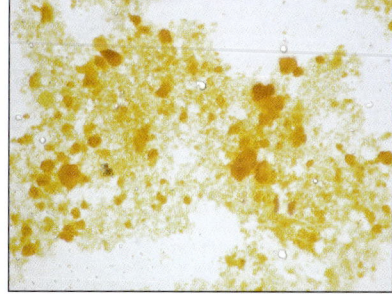

Undulose extinction occurs in particles that have undergone deformation. In this case, the entire crystal does not go black all at once, but 'undulates' across the crystal in patches.

Sweeping extinction is typical in aggregates of fibrous minerals and occurs because each crystal is orientated at a subtly different angle to its neighbours. In this way the extinction sweeps across the crystals. A special case occurs in spherulitic or stellate glomerocrysts, where crystals are orientated through the full 360° range. This means that even when the stage is rotated, there are always crystals orientated parallel to the extinction position. The resulting phenomenon is a cross-shaped extinction pattern that does not rotate, known as a fixed cross, standing extinction cross or 'Maltese' cross.

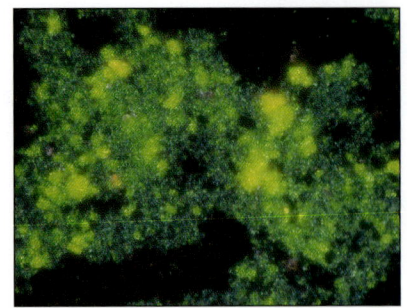

Mottled extinction is shown by carbonate minerals, micas and some clays. This is because sub-microscopic deformation of the crystal results in a very fine undulose extinction, which means that the crystal is never completely black when in the extinction position, but has a fine mottled appearance.

Cryptocrystalline materials are too fine to record accurate extinction positions. As the stage is rotated, the birefringent particles are clearly observed to 'flicker'. This phenomenon is described here as *twinkling extinction*.

Zoning and twinning

Zones in crystals develop as the crystal grows. These usually form concentrically to the particle margins and represent changes in composition. Some phases may develop zoning in sectors, like slices of a cake. When zoning is optically apparent, subtle changes of the colour of the crystals either in plane-polarised or cross-polarised light (or both) may be visible.

Twinning is another function of the mechanisms of crystal growth. Simple twinning is where a crystal and either its mirror image or a rotated crystal appear joined together. Multiple, lamellar, or polysynthetic twins are all terms applied to repeats of this phenomena. Each twin in a pair will have different optical properties. Cross-hatched twins are two sets of twins at different angles to each other.

Elongation

As discussed in the section concerning refractive index, non-cubic crystal structures have orientations at which the transmission of light is fastest or slowest, these directions corresponding to the principal crystallographic orientations of lowest and highest refractive index. In elongated crystals the major axis may correspond to either of these axes, leading to the designation of so-called 'length-fast' and 'length-slow' particles. (Length slow is also sometimes referred to as 'positive' elongation, length fast as 'negative' elongation.) This phenomenon can be diagnostic for some phases where other optical information is insufficient for differentiation. Determination of this property is usually achieved using the sensitive tint (or 'Red-1') retardation plate, an optical device that introduces an additional wavelength shift into the light train which thereby raises or lowers the birefringence in the aligned crystal axes.

Determination of elongation may be made for suitable particles with straight (or weakly inclined) extinction. It is important that the crystal be brightly illuminated when in the 45° position as this corresponds to the orientation of the accessory plate insertion slot in optical microscopes. Insertion of the accessory plate will lead to a rise or fall in the order of interference colours visible. A rise in colours will be observed if the slow direction of the plate corresponds with the slow direction of the particle or, correspondingly, when the fast direction plate corresponds to the fast direction of the particle; such phases are said to be length slow. Conversely, a fall in the order of the interference colours denotes that the particle is length fast. (The stage may also be rotated through 90° so that the colour change can be checked.) For phases with very high birefringence it may prove impossible to determine which direction is fast or slow, as the orders of colours are very similar; in such cases if smaller particles of the substance are present then these can be used for the determination.

OBSERVATIONS IN CONOSCOPIC LIGHT

Using conoscopic light (on insertion of the *Bertrand lens* or removal of the eyepiece) an interference figure may be observed on basal sections of crystals (these may be determined as those having the lowest birefringence or, where appropriate, no pleochroism). Determination of interference figures is impossible for phases with birefringence so low that they are to all intents and purposes isotropic.

Uniaxial minerals (hexagonal, trigonal and tetragonal crystal systems) have, as the name suggests, a single optic axis and in conoscopic light a cross-shaped interference figure, or *isogyre*, is produced; the optic axis lies at the centre of the cross. For biaxial crystals (those belonging to the orthorhombic, monoclinic and triclinic systems) there are two optic axes that intersect at their midpoints. Here the interference figure is of two separate curved isogyres. The optic axis corresponds to the hinge point of the isogyres. The smaller angle between the two axes is called the axial angle, or 2V. 2V can vary from almost zero to 90°. 2V may also be a diagnostic feature for some phases and may be estimated from the degree of curvature of the isogyres observed in conoscopic light. As 2V increases, the isogyres move further apart and become less curved. For 2V approximating zero, the isogyre has an almost right angle bend, whereas for 2V close to 90°, the isogyre is straight and revolves like a propeller in the field of view. Determination of optic sign and 2V ('2V' is zero for uniaxial phases) may be made using the conoscopic or Bertrand lens and a sensitive tint plate.

In practice, for particles of less than 50 μm in diameter it is almost impossible to observe interference figures. Some microscopes may be fitted with mask devices to aid with the conoscopic analysis of small particles.

MICROSCOPY OF ORGANIC PARTICLES

A small but significant group of pigments utilise organic materials as colouring agents, substrates or extenders. In addition, the humic earths and hydrocarbon-based pigments are derived from organic material and, although these are predominantly amorphous, they may also contain recognisable components with organic structures.

Description of properties observed using polarised light microscopy including relief, colour, birefringence, particle size and so forth broadly follows that outlined above for inorganic phases. Particle morphology, however, is likely to be biological in origin and therefore requires a separate terminology.

Fossils, shells and eggshell

Microfossils and fragments of macrofossils as well as unfossilised shells are not infrequent components of limestones of all varieties. Crushed shell has also been used as a pigment. Description of the morphologies of fossilised flora and fauna generally requires specialist language and reasonable coverage of this broad topic is beyond the scope of this book. However, simple descriptions of visual appearance and broad identification of organisms is straightforward enough. Entries are included in this book for avian eggshell, oyster shell and diatomites, and appropriate terminology is given therein. Additionally, the calcareous microfossils, coccoliths and associated microflora are described under the entry for chalk, a rock in which they are particularly abundant. Basic descriptive terminology for these components is given where appropriate. Further information general to the recognition of shells, fossilised or otherwise, is to be found in Adams *et al.* (1991) and in McCrone *et al.* (1973–80).

Bone

Bone in various stages of calcination may be encountered as pigments (bone blacks and bone ash white). Crushed particles of uncalcined bone have a composition of calcium hydroxylapatite and the optical properties of this phase are close to those for apatites of geological occurrence (see Deer *et al.*, 1993; McCrone *et al.* 1973–80). The biological structure of bone is often observable. Particles appear fibrous and have sweeping extinction. Some grains show the typical bone structure of flat plates (*lamellae*) and channels (*canaliculi*). Bone ash typically has irregularly shaped particles and very low birefringence. The authors and McCrone *et al.* (1973–80) have also observed pitted, bubble-rich particles associated with this phase. Fully coked bone and ivory is opaque and black.

Starch

Starch is a major component of turmeric rhizomes and starch from various other sources has been used as substrates for lakes. Although produced in leaves, fruits and barks, starch is only available in amounts suitable for commercial uses in tubers, roots and rhizomes of plants and in the seeds of legumes and cereals. This is so-called 'reserve starch'. Starches from different sources are differentiable using optical microscopy and are described in terms of their shape (globular, lenticular, ellipsoidal, ovoid, truncated or polygonal) degree of aggregation, size and *hilum*, or organic centre of the particles. The *hilum* is conspicuous in some starches, but difficult to observe (certainly in plane-polarised light) in others. *Hila* may be spots, small rings, elongated, or even dark clefts in various starches. *Hila* may also be situated centrally in the grain or off-centre. Under crossed polars, starch grains exhibit a distinctive extinction cross or 'Maltese cross', the centre of which is the location of the *hilum*. (It should be noted, however, that legumes have large, cleft-shaped *hila* have an extinction 'cross' with a central bar.) The appearance of extinction crosses may again be diagnostic and varies with the degree of crystallisation of the starch grains. Starch grains also frequently show concentric rings around the *hilum*. Again, these may be unobservable, faint or conspicuous. Rings are more obvious in oblique illumination. Observation of many starch grains with a sensitive tint or gypsum plate reveals a 'beautiful play of colours' (Winton, 1906) that may also aid in identification. The morphology of starches has additionally been described by Singh *et al.* (2003) and the optical properties applied in the determination of a variety of starches have been detailed by Winton (1906).

Gums and resins

Gums and resins, such as gamboge, are predominantly amorphous and therefore isotropic. Some examples may be characterised by the presence of air bubbles, thus having a spongy appearance.

Plant tissue

Plant tissue may be present in various states in organic derived pigments. Cell structure is often well preserved, and frequently even survives charring (Winter, 1983). Woody plant material (lignin) is also frequently well preserved in low-grade coals and lignites and a common component of impure humic earths.

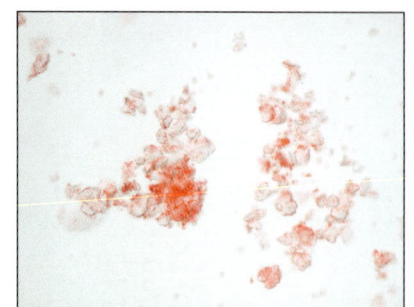

Cellular material from plants, roots and rhizomes is immediately recognisable under the microscope and has a polygonal network structure. The simplest form of plant tissue, with cellulose walls, is called *parenchyma*. The term *sclerenchyma* describes a variety of cellular plant tissues with noticeably thickened walls and composed of lignin. *Sclerenchyma* can contain rounded or polygonal cells, known as *stone cells*. These may occur singly or as dense groups; the latter case forms components such as peach and cherry stones. Sclerenchyma may also contain bundles of fibrous elements called *bast fibres*. The outer layers of plant stems, roots and rhizomes are formed of *cork cells*. These also have polygonal morphologies in surface view, but are elongate and arranged radially. These are particularly diagnostic of cork char blacks, especially apparent when observed using a scanning electron microscope. It should be noted that vegetable chars, while preserving tissue structure, are characterised by having substantially thinned cell walls, these structures being easily destroyed during grinding (Winter, 1983).

Crystallites of phases primarily including amorphous silica ('plant opal'), calcium oxalate (which can also occur in finely acicular habits known as *raphides*) and microconcretions of calcium carbonate may also be encountered. The microscopic identification of vegetable material has been described in some detail by Winton (1906). Further useful illustration of the appearance of plant tissue is given in McCrone (1973–80).

Pollen and spores are common constituents of dust and are remarkably resistant to degradation. Although not directly used as pigments, pollen may appear as unintended impurities. Again, as a first general reference, McCrone (1973–80) has an excellent section illustrating these particles.

Coals

As fossilised peat beds, coals may be regarded as rocks rather than organic deposits. The various components of coals are called macerals and these may be considered analogous to minerals. Identification is complex and based on the classification of the organic matter preserved. Coal macerals are divided into three main groups – *vitrinites*, *liptinites* ('exinites') and *inertinites* – based on their origin, reflectivity and UV fluorescence. Coal macerals are opaque and isotropic in polarised light. The *vitrinites* are derived from woody plant material and include the macerals telocollinite and textinite, both identifiable by their cellular, woody appearance as well as desmocollinite and eugelinite, which are amorphous, precipitated humic gels. Vitrinites have poor or no UV fluorescence and are moderately reflective. The liptinites (also 'exinites') are derived from fatty and waxy parts of plants; spores and pollen are preserved as sporinite, leaf cuticle wax as cutinite, resins, gums and bark oils as resinite. Liptinites are highly distinctive in that they have very low reflectance but strong green or yellow UV fluorescence. Finally the *inertinites* (most commonly the maceral fusinite) are natural charcoals, having been oxidised or aromatised during the coalification process. Inertinites are of strikingly high reflectance and exhibit no fluorescence under UV light.

UV fluorescence of coal macerals has been described by Rost (1995). General texts on coal microscopy are uncommon. However, Seyler (1929) provides a thorough and well-illustrated introduction, but understandably does not employ modern terminology.

VARIANTS AND COMPOSITE PIGMENTS

Two practical aspects of the examination of samples should be stressed. These are, first, the differentiation of variants – pigments of the same composition, but arising (for example) from different manufacturing processes or sources – and, second, composite pigments, where a number of different compounds are intermixed as a result of their origin or use. The analyst should consequently be aware of the range of normal variation within a pigment, other compounds found commonly or uncommonly in association with it and the implications such observable differences may have on the interpretation.

Sample homogeneity should normally be noted in a specimen, and should be expressed in terms of percentage estimates of each phase present. Inhomogeneity may, of course, result from intentional mixing of pigment phases or from the presence of naturally occurring impurities or contaminants. All phases present should be described.

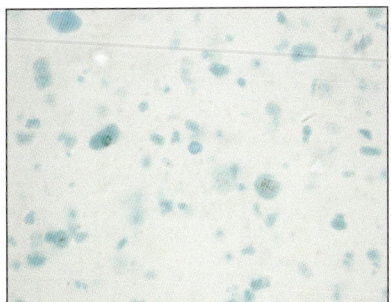

FURTHER READING

The aim of this book is to provide a first and encyclopaedic description of the optical properties of pigment-related materials. Pigment analysis is mainly concerned with the examination of naturally occurring minerals, synthetic crystalline compounds and amorphous substances, including the substrates of lake pigments. As such, pigment analysis borrows largely from the terminology devised for petrography with a few notable, but nevertheless important, differences. Consequently there are additional works already widely available that deal with the optical properties of materials encountered in other disciplines, but which overlap with those of our own. As further reading the authors therefore recommend the works below.

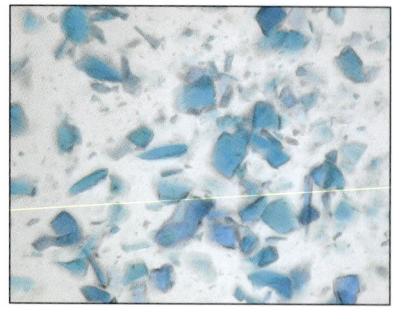

Examination of particles using the polarising microscope has been standard practice in many disciplines for over 100 years though the technique has perhaps found most routine employment in the fields of crystallography and petrography (that is, the description of rocks and their constituent minerals). Listings and descriptions of the optical properties of minerals are readily accessible in standard texts including Gribble and Hall (1992), Deer *et al.* (1992), and Hibbard (1995). Colour atlases of photomicrographs of rocks, minerals and fossils are also useful *aide mémoires* (see MacKenzie and Guilford, 1980; MacKenzie *et al.*, 1984; Adams *et al.*, 1991; Yardley *et al.*, 1990). For heavy minerals in dispersions ('grain mounts') see Mange and Maurer (1991). For the discussion and identification of strongly coloured minerals used as gems, decorative stones and their synthetic (or even fake!) analogues, including a great many varieties used as pigments, see Liddicoat (1993) and O'Donoghue and Joyner (2003). The technique also has applications in engineering, biology and material sciences. McCrone *et al.* (1979, 1973–80) give a detailed overview of the scope of applications. More recently, polarised light microscopy has been applied to other fields as a diagnostic tool including the analyses of archaeological ceramics (see Whitbread, 1989; Freestone, 1995) and, of course, pigments. The *Artists' Pigments* series of volumes edited by Feller (1986), Roy (1993), FitzHugh (1997) and Berrie (forthcoming) deals with a selection of the more prevalent pigment phases, including discussions of physical, chemical and optical properties.

STRUCTURE OF ENTRIES

A variety of optical properties may be observed in crystalline and non-crystalline solids. Key features are described below in the text accompanying each entry in this book. In addition a banner that gives standard optical properties for each phase accompanies each entry. This includes the refractive indices, and birefringence, 2V and optic sign where relevant. The 'crystallinity' of each phase is also stated. For normally crystalline substances the commonly adopted crystal system is given. Similarly the term amorphous is applied to those compounds with no crystal structure. For pigments in which more than one phase is present, crystallinity is given as 'composite'. Other terms are used as appropriate and further discussed in the related text.

All samples were examined using two microscope systems: a Leitz Orthoplan, and a Leica DMRX. Photographs were prepared using an Olympus C-3030 zoom 3.3Mpixel digital camera and Adobe Photoshop version 7.0. Analyses were primarily made using 10× (0.3NA), 40× (0.75NA) and 100× (0.6NA) objectives, the latter with immersion oil; the 10× and 100× objectives were plan fluotar lenses (low fluorescence designed for UV studies), the 40× objective a plan-apochromat. Illumination was with 100 W tungsten-halogen (transmitted/reflected light) or 100 W mercury vapour (fluorescence). Filters used were a Schott dichroic filter to achieve daylight balance of the normal illumination, a Chelsea filter, and Leica A and I3 fluorescence filter cubes. Elongation measurements were performed using a length-slow sensitive tint plate inserted into the Leica DMRX in a NE-SW orientation.

The feasibility of using an entirely digital image capture system became possible during the lifetime of the project and allowed the authors to capture the huge number of photographs required for this publication – far more than are reproduced here. As reported elsewhere (Eastaugh, 2002), the system used proved surprisingly effective for the purpose despite a number of technical compromises. In particular, the issue of capturing images with low light levels was never fully resolved, difficulty being encountered with dynamic range and sensor noise; the former made high-contrast images of birefringent particles under crossed polars especially difficult while the latter has led to residual background speckle being visible in some images. However, colour reproduction proved excellent, this being checked throughout the process by back-reference to the original samples. System components such as computer monitors were also colour calibrated to ensure consistency.

Information on the conditions under which each image was taken is given within the relevant caption. This appears as a code along with the sample collection data and any specific comments regarding the illustrative content of the image. The code is structured in the following manner:

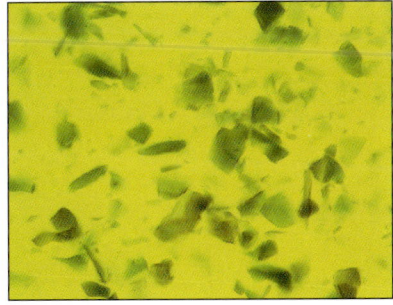

P*xxxx*	Pigment sample reference number (see Appendix I for details)
PPL	Plane-polarised light
XPL	Cross-polarised light
~XPL	Partially cross-polarised light
RPL	Reflected polarised light

CF	Chelsea filter
STP	Sensitive tint plate ('quarter-wave plate')
10×, 40×, 100×	Objective magnification (refer to scale provided to determine absolute magnification as printed)
●	The substage aperture was closed down, increasing effective depth-of-field and image contrast
H	Sample is from a historical collection

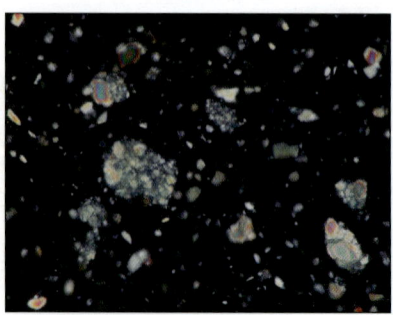

OPTICAL AND PHYSICAL PROPERTIES OF PIGMENT PARTICLES FROM POLARISED LIGHT MICROSCOPY

VIVIANITE

Fe$_3$(PO$_4$)$_2$.8H$_2$O

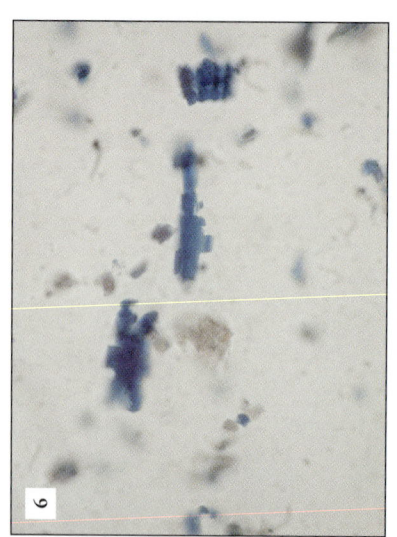

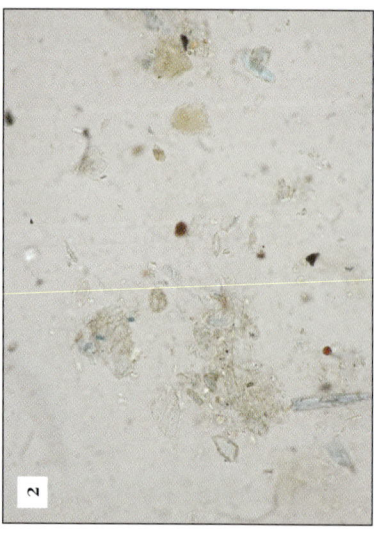

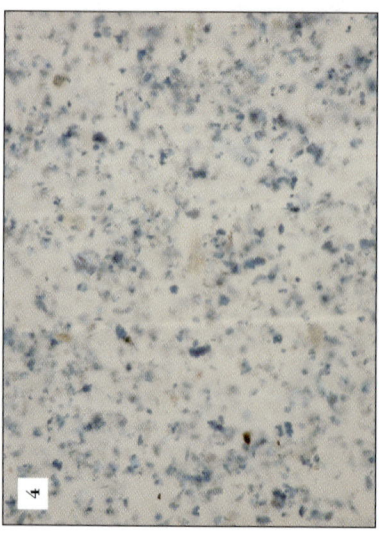

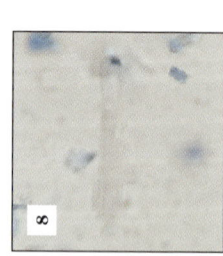

Key:

1. P0007: PPL/100×/ **O**. Green-blue vivianite plus other mineral impurities.

2. P0007: PPL/100×/ **O**. The field of view in Fig. 1 rotated through 90° to show pleochroism.

3. P0007: XPL/100×/ **O**. The field of view in Fig. 2 in crossed polars.

4. P1353: PPL/40× . Blue vivianite particles and other yellow impurities.

5. P1353: XPL/40×/ **O**. The field of view in Fig. 4 under crossed polars.

6. P1353: PPL/100×/ **O**. Bladed particles of vivianite.

7. P1353: PPL/ **O**. An enlarged view of the particle at the centre of Fig. 6.

8. P1353: PPL/ **O**. The particle in Fig. 7 rotated through 90° to show intense pleochroism, the particle is now colourless.

Crystallinity	Optic sign	2V	n_α	n_β	n_γ	δ
Monoclinic	+ve	81°–84°	1.579–1.582	1.602–1.604	1.629–1.635	0.053
					(data from Tröger, 1952)	

In plane-polarised light, the violet-coloured iron phosphate mineral vivianite is seen to form as euhedral bladed, lath-shaped crystals, and subhedral fragments of these. These may be up to 60 μm long in some cases. It may also form radiating aggregates of crystals or granular, anhedral masses. However, the most apparent diagnostic feature of vivianite is its intense pleochroism from colourless to a strong lilac-blue. In some samples, the blue has a greenish tinge. It has moderate to low relief and perfect cleavage, parallel to the long axis of the crystal.

Under crossed polars, interference colours may be masked by the body colour. However, it is strongly birefringent and shows up to fourth order colours. Anomalous pink and bronze colours may be observed. Vivianite normally has straight extinction, though sometimes it is sweeping. Multiple twins form parallel to the long axis of the crystals. Vivianite is length fast.

Vivianite appears a green-blue colour when using the Chelsea filter.

Examples of vivianite have been discussed by several authors including Howard (1995, 2003) and Hill (2001). Howard in particular has noted the apparent conversion to metavivianite, visible in samples as yellow coloured particles.

Lit.: Hill (2001); Howard (1995); Howard (2003)

The Pigment Compendium

FLUORITE

CaF

Key:

1. P1192: PPL/40×. Particles appear colourless at high magnification; note octahedral particle near centre.

2. P1193: PPL/40×. Colourless angular grains.

3. P0177: PPL/100×. Strongly banded crystal of antozonite and colourless fluorite crystals.

4. P0177: PPL/100×. Purple banded antozonite crystal.

5. P0177: PPL/100×. Banded and coloured crystals of antozonite.

6. P0177: PPL/100×. Banded and colourless crystals of antozonite.

7. P0177: PPL. Purple banded antozonite crystal.

8. P0177: PPL. Detail of banding in antozonite crystal.

9. P0177: PPL. A zoned antozonite crystal.

10. P0177: PPL. Shows a purple banded crystal and a fully coloured crystal.

Crystallinity
Cubic

n
1.433–1.435

δ
Isotropic
(data from Deer et al., 1992)

Under plane-polarised light, fluorite is usually colourless, although it may be very weakly coloured and, in some cases, strongly coloured purple (sometimes almost black) as in the variety known as antozonite. The strongly coloured purple and blue-purple particles of this variety sometimes show banding even after grinding, though colourless particles are also abundant. Such purple forms of fluorite have recognised use as a pigment (see: Richter et al., 2001; Spring, 2000). Fluorite may also exist in a wide range of other colours from yellow, green and blue, but these are usually colourless in thin particles. The origin of the colour in fluorites has been much studied and the causes seem various; however, dark purple fluorites have been reported to be relatively rich in strontium (Deer et al., 1992).

Fluorite has high relief, with RI much less than that of the medium. The dispersion is weak. Fluorite commonly forms cubic crystals and has 'octahedral' fracture, which leads to a range of square to octahedral, cubic, triangular or tabular forms. However, crushed particles will normally adopt the habit of angular, subhedral and anhedral shards. Richter et al. examined a number of samples of purple fluorite from paintings and observed that individual particles tended to appear as flattish, angular, broken fragments, often with conchoidal fracture. Particle size was found to be widely variable, typically in the range 10–50 μm, but with coarser material that could reach 80 μm or more.

Under crossed polars, fluorite is isotropic. Fluorescence is moderate.

Richter et al. note that calcite and quartz were frequently present in samples of purple fluorite they examined; other minerals typically found with fluorite include dolomite, gypsum, baryte, galena, sphalerite, cassiterite and apatite.

Lit.: Deer et al. (1992), 672–675; Richter et al. (2001); Spring (2000)

ERYTHRITE

$Co_3(AsO_4)_2 \cdot 8H_2O$

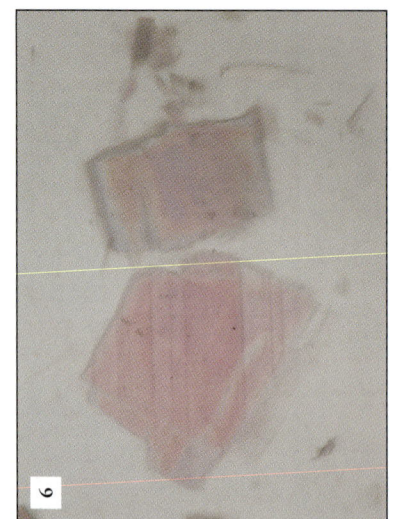

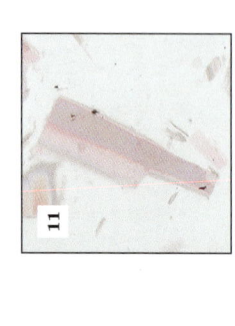

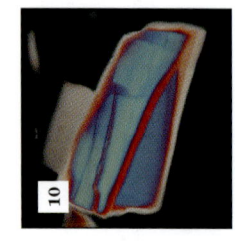

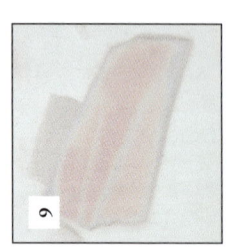

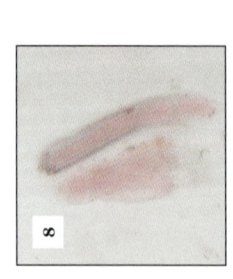

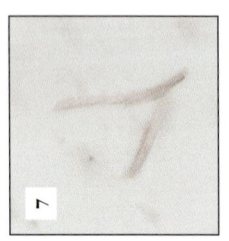

Key:

1. P1190: PPL/40×. General low relief and pleochroism; flexible bladed crystals.

2. P0862: PPL/10×. Subhedral to euhedral bladed crystals with colour variation as a result of pleochroism and twinning.

3. P0862: PPL/40×. Colour variation due to pleochroism.

4. P1190: XPL/40×. Same field of view as Fig. 1 under crossed polars.

5. P0862: XPL/10×. Same field of view as Fig. 2 showing third order interference colours.

6. P0862: PPL/40×. The same field of view on rotation of 90°.

7. P1190: PPL. Shows flexible crystals.

8. P1190: PPL. Shows flexible crystal.

9. P0862: PPL. Subhedral bladed crystal.

10. P0862: XPL. The crystal in Fig. 9 under crossed polars.

11. P0862: PPL. Bladed crystal showing twinning.

12. P0862: XPL. The crystal seen in Fig. 11 in crossed polars.

Crystallinity	Optic sign	2V	n_α	n_β	n_γ	δ
Monoclinic	+ve	85°–90°	1.626–1.629	1.662–1.663	1.699–1.701	0.0720–0.0730
						(data from Ford, 1932; cf. Webmineral, 2003)

Erythrite is a naturally occurring cobalt arsenate mineral of violet colour related to the cobalt arsenate violets.

In plane-polarised light erythrite has variable low to very low relief such that it is indistinguishable from the mounting medium in certain positions. It has a very pale purple body colour and is weakly pleochroic. Typically, erythrite forms euhedral acicular to elongate bladed crystals. Particles may be rich in fluid inclusions. Other habits may be earthy or granular. Erythrite has a perfect cleavage, parallel to the long axes of the crystals.

Under crossed polars erythrite has high birefringence, which appears as bright pale-coloured fourth order down to third order interference colours.

The crystals show inclined, sometimes sweeping extinction; crystals of erythrite are flexible and therefore prone to strain, which is generally responsible for the sweeping extinction. Lamellar twins are present, orientated parallel to the long axes of the crystals. Particles are length slow.

The colour of erythrite is unaffected by use of the Chelsea filter.

Lit.: Corbeil *et al.* (2002)

The Pigment Compendium

COBALT PHOSPHATE

$Co_3(PO_4)_2$

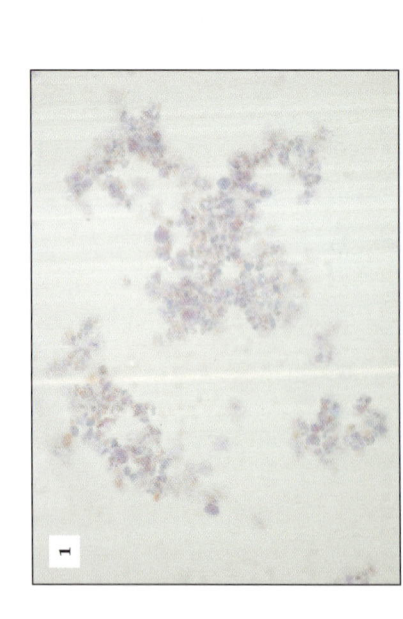

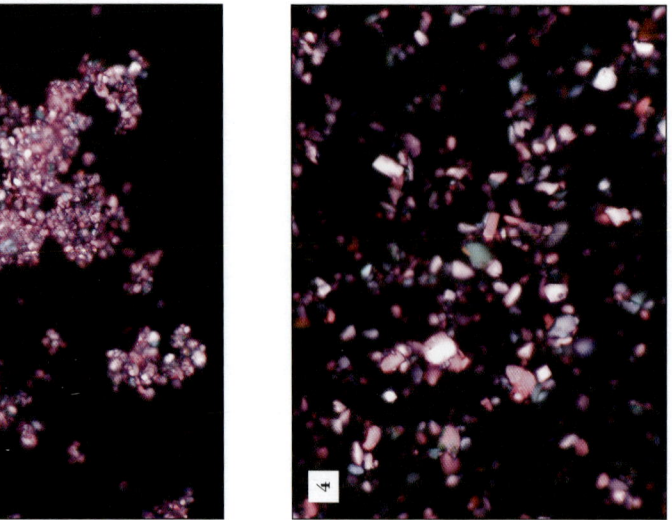

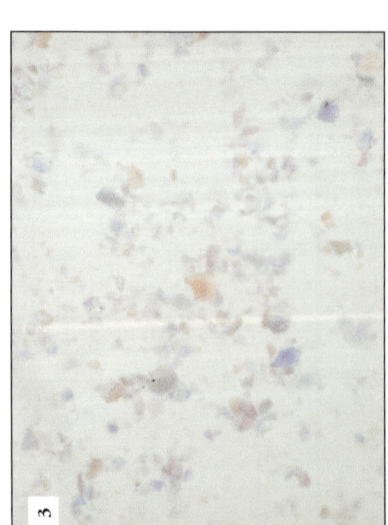

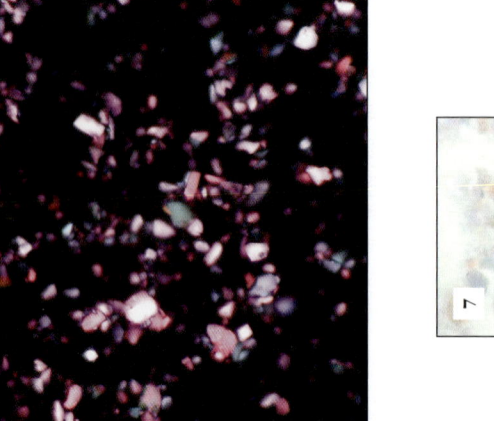

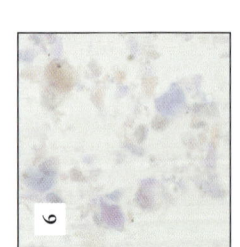

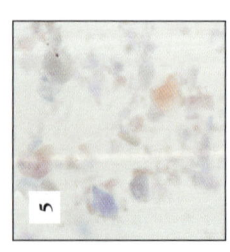

Key:

1. P0113: PPL/100×. Subrounded and angular particles.

2. P0113: ~XPL/100×. High birefringence masked by the particle body colour.

3. P0196: PPL/100×. Subangular particles showing suite of pleochroic colours.

4. P0196: XPL/100×. High birefringence masked by the particle body colour, with anomalous blue-green interference colours shown by some crystals.

5. P0196: PPL. Subangular particles.

6. P0196: PPL. Cobalt phosphate particles in Fig. 5 with polariser rotated through 90° clockwise to show change in crystal body colour.

7. P0113: PPL. Shows range of pleochroism.

Crystallinity	**Optic sign**	**2V**	$n_\alpha/n_\beta/n_\gamma$	δ
Monoclinic	[unknown]	[unknown]	≥1.662	High

(data from Corbeil *et al.*, 2002 and authors)

In plane-polarised light, cobalt phosphate (a type of 'cobalt violet') is composed of medium- to coarse-grained, subrounded plates or angular shards; Corbeil *et al.* (2002) have noted sizes in the range 1–100 μm for samples they examined. Characteristically, these particles are strongly pleochroic, with colours from pink to lilac to orange-yellow visible. Particles have variable moderate relief with an RI varying from equal to that of the medium to greater than that of the medium. Some larger particles may show fractures.

Under crossed polars, this compound exhibits clear diagnostic features. It has high birefringence, but the interference colours are strongly masked by the body colour and consequently appear bright pink. Some particles show distinctive anomalous green colours. Cobalt phosphate has a sweeping extinction.

Using the Chelsea filter, cobalt phosphate appears magenta in colour.

The optical properties of the various pigments sold under the name of cobalt violet have been recently described by Corbeil *et al.* (2002). Careful examination by optical microscopy should permit distinction within this group of compounds.

Lit.: Corbeil *et al.* (2002)

The Pigment Compendium

COBALT PHOSPHATE HYDRATE

$Co_3(PO_4)_2 \cdot 8H_2O$

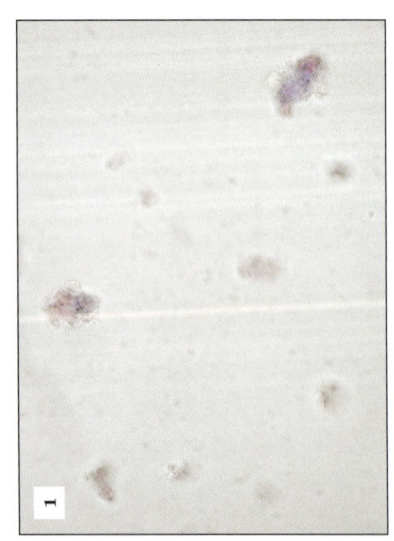

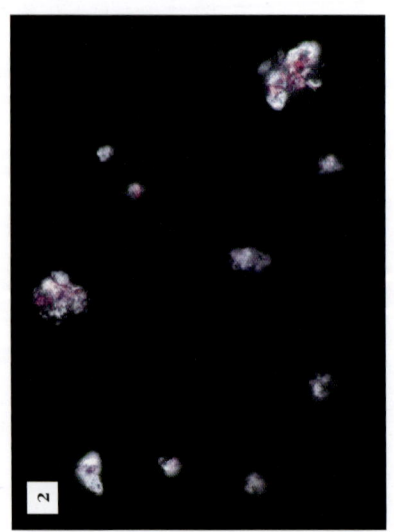

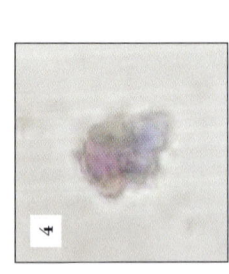

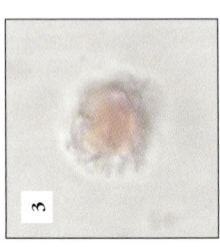

Key:

1. P1478: XPL/100×. Polycrystalline grains of cobalt phosphate hydrate.

2. P1478: XPL/100×/ **o** . The field of view in Fig. 1 under crossed polars.

3. P1478: PPL. Pleochroism in cobalt phosphate hydrate; note the orange coloration of this particle.

4. P1478: PPL. A similar particle showing the opposite end of the pleochroic spectrum.

Crystallinity	Optic sign	2V	$n_\alpha/n_\beta/n_\gamma$	δ
Monoclinic	[unknown]	[unknown]	≥1.662	High

(data from Corbeil *et al.*, 2002)

Under plane-polarised light cobalt phosphate hydrate (a type of 'cobalt violet') can be seen to form translucent pale purple particles. The particles are pleochroic, from purple to pink-orange. Relief is moderate and the RI is greater than that of the medium. Particles are irregular anhedral plates with uneven surfaces. Particle size ranges from fine to medium; Corbeil *et al.* (2002) note typical values of 1–15 μm.

Under crossed polars, cobalt phosphate hydrate has strong birefringence, but interference colours are strongly masked by the body colour and therefore bright pinks and purples are observed. Extinction is mottled to complete.

Cobalt phosphate hydrate is optically similar to magnesium cobalt arsenate (*q.v.*) except that the latter exhibits anomalous interference colours.

Characterisation of various cobalt violet pigments has been recently carried out by Corbeil *et al.* (2002). Careful examination by optical microscopy should permit distinction within this group of compounds.

Lit.: Corbeil *et al.* (2002)

AMMONIUM COBALT PHOSPHATE HYDRATE

$CoNH_4PO_4.H_2O$

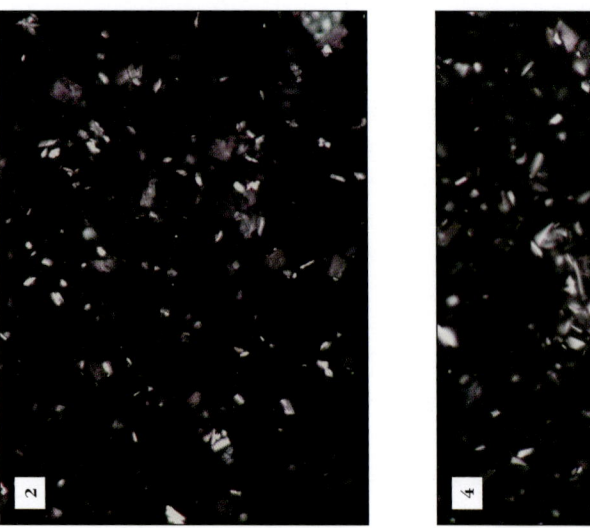

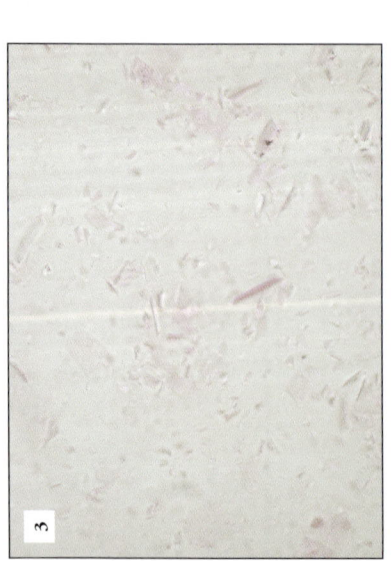

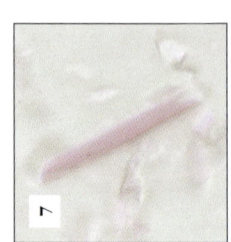

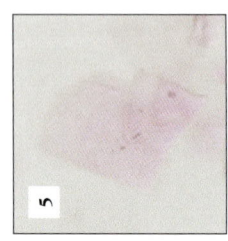

Key:

1. P0114: PPL/40×. Platy pink rectangular crystals with inclusions in some cases.

2. P0114: XPL/40×/ ●. Same view as in Fig. 1 in crossed polars.

3. P1470: PPL/100×. Acicular crystals and platy crystals with inclusions.

4. P1470: XPL/100×/ ●. Low first order colours. Same field of view as in Fig. 3 in crossed polars.

5. P0114: PPL. Subhedral low relief platy crystal.

6. P1470: PPL. Plate-likehabit of two overlapping crystals.

7. P1470: PPL. Acicular crystal.

8. P1470: XPL/ ●. Sheaf of fibrous crystals in crossed polars.

Crystallinity	Optic sign	2V	$n_\alpha/n_\beta/n_\gamma$	δ
Orthorhombic	[unknown]	[unknown]	≤1.662	Low

(data from Corbeil et al., 2002)

In plane-polarised light, ammonium cobalt phosphate hydrate (a type of 'cobalt violet') appears either as very coarse, euhedral crystals with a distinctive, rectangular platy habit or as medium-grained fibrous, variolitic aggregates that Corbeil et al. (2002) describe as 'rosettes and bundles'; these authors also indicate a size range of 1–100 μm for samples they examined. In both morphological cases particles have moderate relief with RI less than or equal to that of the mounting medium, in contrast to other cobalt violet pigments where the RI is commonly above that of the mounting medium. The platy crystals show three sets of cleavage at right angles to each other, allowing crystals to fracture into rectangular fragments. Some crystals contain fluid inclusions. The crystals have a pale pink body colour, but do not exhibit pleochroism.

Under crossed polars, ammonium cobalt phosphate hydrate with the platy habit has low, first order grey interference colours that are partially masked by the pink body colour. These are basal sections because the bundles of needles exhibit stronger interference colours. The crystals have straight extinction and elongation in the fibrous particles can be seen to be length slow. Evidence of concentric zoning is apparent in some sections.

The optical properties of the various pigments sold under the name of cobalt violet have been recently described by Corbeil et al. (2002). Careful examination by optical microscopy should permit distinction within this group of compounds.

Lit.: Corbeil et al. (2002)

The Pigment Compendium

LITHIUM COBALT PHOSPHATE

LiCoPO$_4$

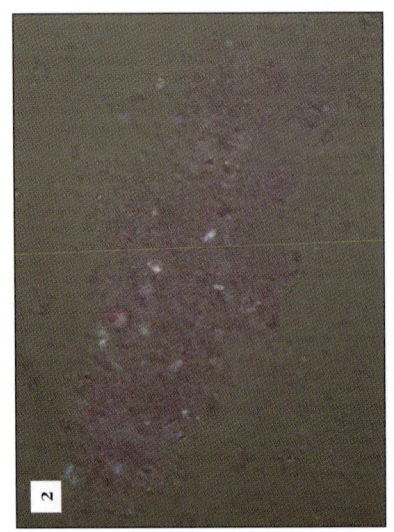

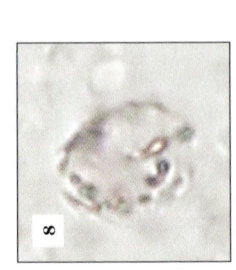

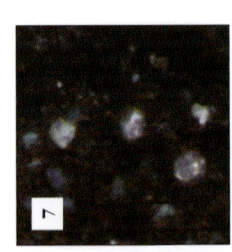

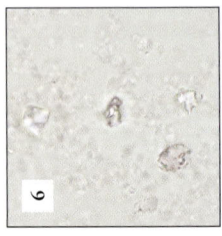

Key:

1. P1480: PPL/100×. Strongly coloured pink-purple fine-grained particles of lithium cobalt phosphate.

2. P1480: ~XPL/100×. The same field of view as Fig. 1 under partially crossed polars.

3. P0112: PPL/100×. Pale pink grains of lithium cobalt phosphate showing variation in grain size.

4. P0112: PPL/100×/◯. Pale pink particles of lithium cobalt phosphate showing variation in grain size plus impurities.

5. P0112: XPL/100×. Field of view as in Fig. 4 under crossed polars.

6. P0112: PPL/◯. Enlargement of particles in Fig. 4.

7. P0112: XPL. The particles in Fig. 6 in crossed polars showing anomalous blue-green interference colours.

8. P0112: PPL/◯. Enlargement of grain from lower left corner of Fig. 6 showing green inclusions.

Crystallinity	Optic sign	2V	$n_\alpha/n_\beta/n_\gamma$	δ
Orthorhombic	[unknown]	[unknown]	≥1.662	Low

(data from Corbeil et al., 2002)

In plane-polarised light lithium cobalt phosphate ('a type of 'cobalt violet') appears as having low to moderate relief with RI greater than or equal to that of the mounting medium. The individual particles are rounded and particle size is typically medium; Corbeil et al. (2002) note that in their samples this value was ≤5 μm, in contrast to other cobalt violets they examined where size ranges could be up to 100 μm. It is also noticeable in the sample shown here that there is a bimodal distribution in particle size, with a medium-grained fraction and less abundant, coarse-grained fraction. The particles have a weak, pale pink body colour. Pleochroism is not apparent.

Under crossed polars, this compound is weakly anisotropic and particles show low first order interference colours; some particles do show anomalous blues. Where discernible, particles appear to have sweeping extinction.

The optical properties of the various pigments sold under the name of cobalt violet have been recently described by Corbeil et al. (2002). Careful examination by optical microscopy should permit distinction within this group of compounds.

Lit.: Corbeil et al. (2002)

The Pigment Compendium

MAGNESIUM COBALT ARSENATE

$Mg_2Co(AsO_4)_2$

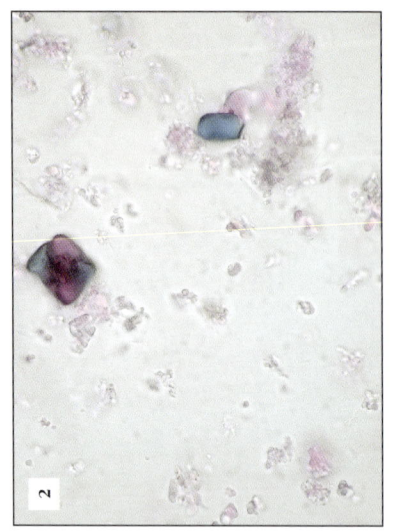

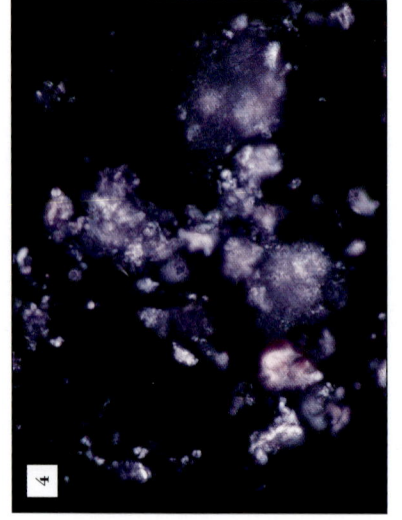

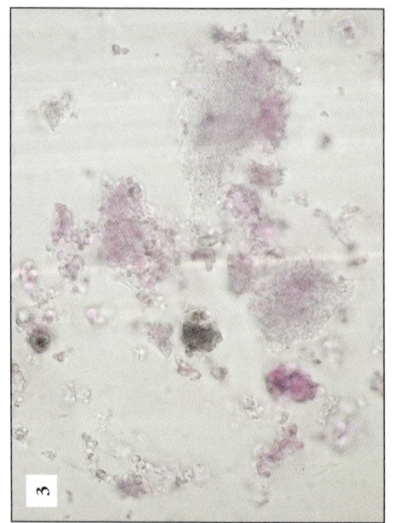

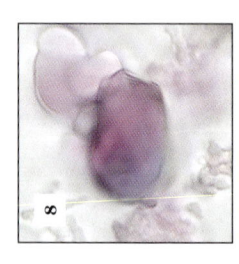

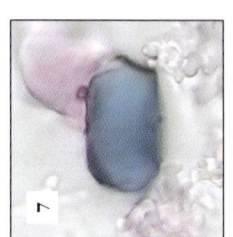

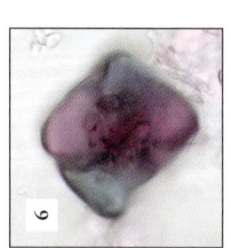

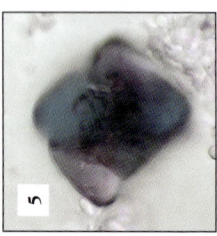

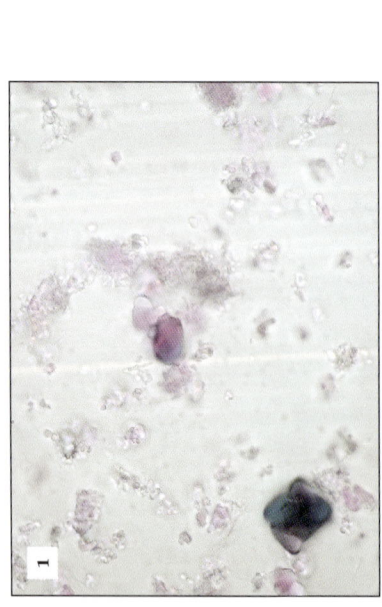

Key:

1. P1476: PPL/100×. Example of variety of particle size, morphology and colour.

2. P1476: PPL/100×. The same field of view as in Fig. 1 rotated through 90°.

3. P1476: PPL/100×. Example of variety of particle size, morphology and colour.

4. P1476: XPL/100 × ●. The same field of view as in Fig. 3 under crossed polars.

5. P1476: PPL. Enlargement of large grain from Fig. 1.

6. P1476: PPL. The grain from Fig. 5 rotated through 90°.

7. P1476: PPL. Subhedral bladed particle from Fig. 1.

8. P1476: PPL. The particle in Fig. 7 rotated through 90°.

Crystallinity Anisotropic	Optic sign [unknown]	2V [unknown]	$n_\alpha/n_\beta/n_\gamma$ ≥1.662	δ High
				(data from Corbeil et al., 2002)

Under plane-polarised light magnesium cobalt arsenate (a type of 'cobalt violet') can be seen to form translucent pale purple particles. The particles are pleochroic, from pink-purples to a dull blue or in some cases even a dull green colour. Particle shape is variable, ranging from individual, anhedral, subrounded plates to polycrystalline aggregates of particles. Particle size ranges from fine to medium; Corbeil et al. (2002) note typical values of 1–15 μm. Particle surface texture varies with crystal morphology. The platy crystals have smooth surfaces, whereas aggregates of particles have increasingly rough and pitted surfaces with a corresponding decrease in size of the constituent particles. Relief is moderate, with RI greater than that of the medium.

Under crossed polars magnesium cobalt arsenate has high birefringence colours, these being masked by the body colour. Interference colours typically appear as white and purplish-grey, although some particles have striking anomalous orange colours. Extinction is sweeping to complete on the individual plates and twinkling in the particle aggregates.

The pigment appears a weak purple colour in reflected light. Appearance of cobalt violet is also largely unaffected when viewed through the Chelsea filter and no UV fluorescence was noted by the authors.

Magnesium cobalt arsenate is optically similar to cobalt phosphate hydrate (q.v.) except that the latter does not exhibit the anomalous interference colours.

Characterisation of various cobalt violet pigments has been recently carried out by Corbeil et al. (2002). Careful examination by optical microscopy should permit distinction within this group of compounds.

Lit.: Corbeil et al. (2002)

MANGANESE PHOSPHATE

MnPO₄

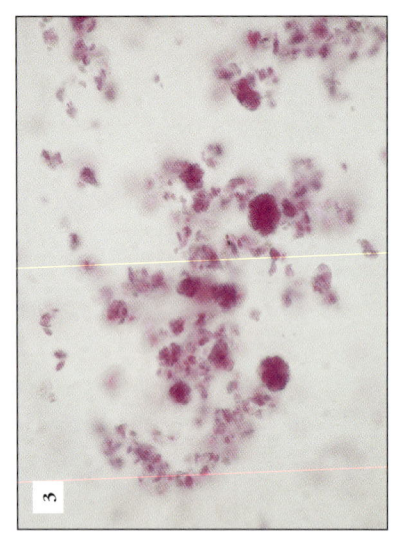

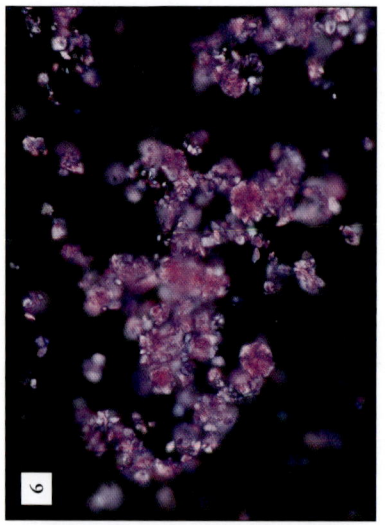

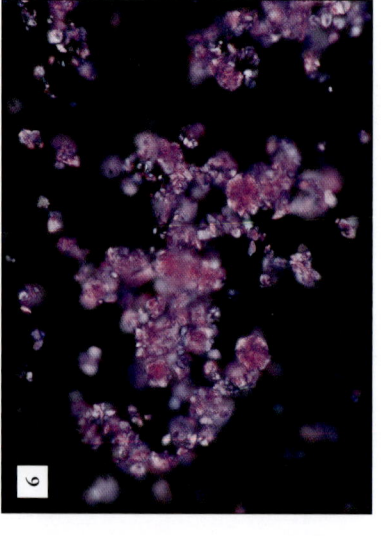

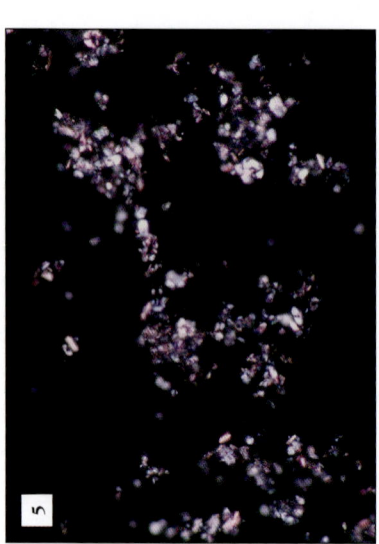

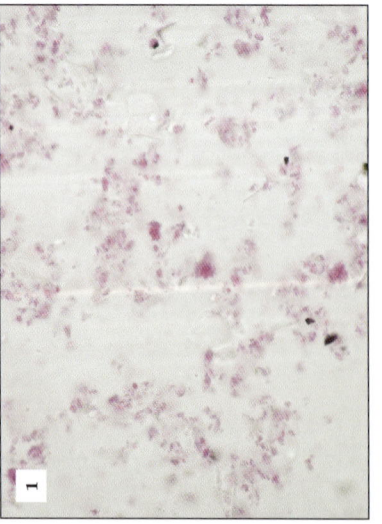

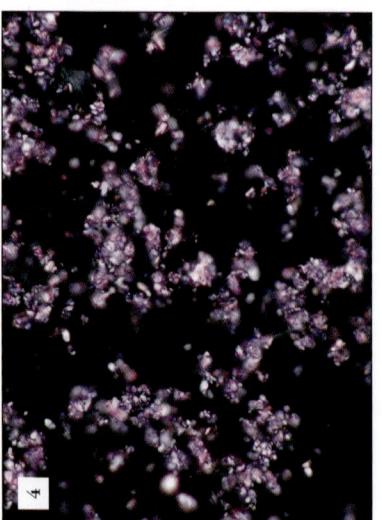

Key:

1. P0103: PPL/100×. Shows fine-grained purple and purple-red particles with larger colourless anhedral impurities.

2. P0192: PPL/100×. Fine-grained purple-pink particles with colourless impurities also present.

3. P1351: PPL/100×/H. Larger pink spherulitic aggregates and smaller single pinnate crystals of manganese violet.

4. P0103: XPL/100×/●. Birefringent particles with interference colours masked by the body colour. Same field of view as Fig. 1.

5. P0192: XPL/100×/●. Birefringent particles with interference colours masked by the purple body colour. Same field of view as Fig. 2.

6. P1351: XPL/100×/●/H. Interference colours of moderately birefringent particles masked by pink-violet body colour. Same field of view as Fig. 3.

7. P1351: PPL/H. An enlarged image of the spherulites seen in Fig. 3.

8. P1351: XPL/●/H. Particles of Fig. 7 in crossed polars with aggregate showing weak dark purple standing cross and pinnate crystals in top left corner.

Crystallinity	Optic sign	2V	n_α	n_β	n_γ	δ
Orthorhombic	+ve	38	1.850	1.912	1.920	0.700

(data for mineral purpurite from Webmineral, 2003)

Two forms of manganese phosphate were discerned from the examination of samples. Under plane-polarised light 'type I' forms fine-grained crystals with moderate to low relief, with RI greater than that of the medium. The samples shown here are impure, containing strongly pleochroic red-purple to purple phases and more abundant pinkish purple particles. Other colourless, transparent phases are present and a few particles are spherulitic. Further particle shapes vary from irregular, angular anhedral shards to euhedral, pennate forms. 'Type II' on the other hand forms fine-grained spherulites and pennate individual crystals. Particles are translucent, but strongly coloured a pinkish violet and pleochroic from light to darker shades. Relief is moderate with RI greater than that of the medium. Particle surfaces of the pennate forms are smooth; the fibrous crystals forming the spherulites are visible.

Under crossed polars both forms are moderately birefringent with the birefringence masked by the body colour such that particles appear as bright, purplish crystals. Both have inclined extinction. Additionally, the spherulitic particles of the type II morphology exhibit a poorly defined standing extinction cross. The fibrous crystals can be seen to exhibit length-fast elongation.

Manganese phosphate is a weak pink colour in reflected light and transmits a dull red when viewed through the Chelsea filter.

PURPURITE

MnPO$_4$

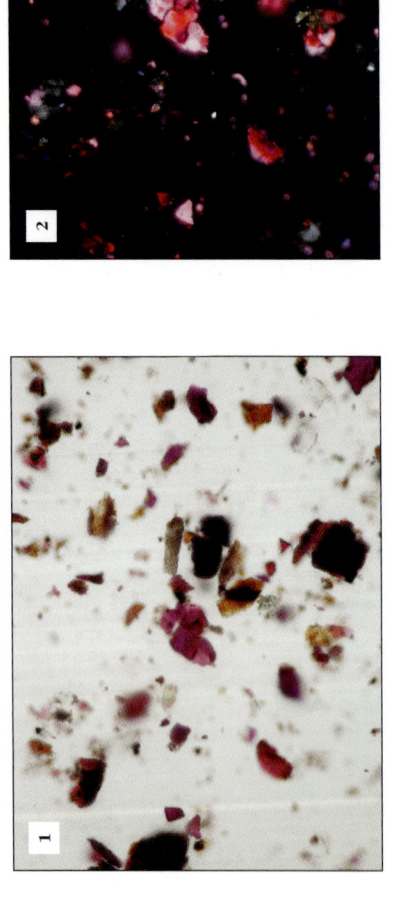

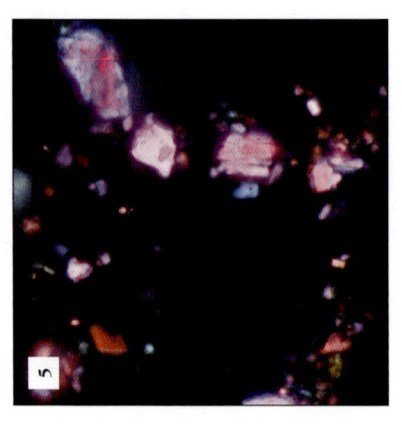

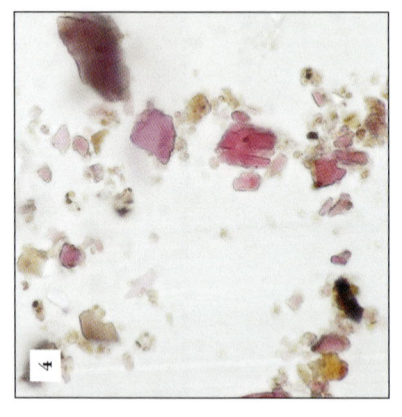

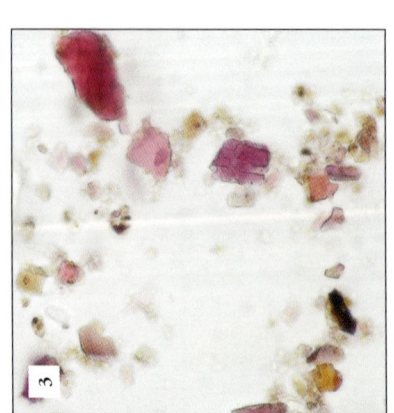

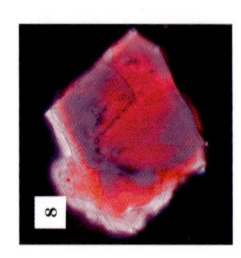

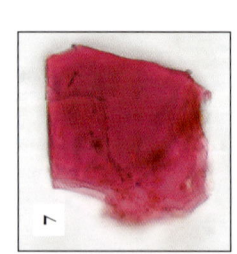

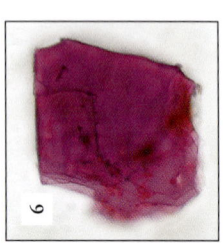

Key:

1. P0015: PPL/40×. General assemblage of particles found in purpurite.

2. P0015: ~XPL/40×. Highly birefringent purpurite crystal with interference colours masked by high body colour giving bright pinks, oranges and violets, and turquoise in paler particles.

3. P0015: PPL/100×. Anhedral purpurite crystals showing variation in body colour caused by the pleochroic scheme.

4. P0015: PPL/100×. Anhedral purpurite crystals showing colour change after the polariser has been rotated through 90° clockwise.

5. P0015: XPL/100×. Strongly birefringent particles as Fig. 4 with interference colours masked by the body colour.

6. P0015: PPL. Purpurite crystal showing purple colour.

7. P0015: PPL. The same crystal as Fig. 6 on rotation of the polarizer by 90° showing pleochroic change to pink.

8. P0015: XPL. The same crystal as Fig. 6 under crossed polars at maximum illumination.

9. P0015: PPL. Shows strong pink body colour.

10. P0015: PPL. The same crystal as Fig. 9 showing pleochroic change to grey on rotation of the stage.

Crystallinity	Optic sign	2V	n_α	n_β	n_γ	δ
Orthorhombic	+ve	38	1.850	1.912	1.920	0.700

(data from Webmineral, 2003)

Under plane-polarised light, the manganese phosphate mineral purpurite is distinctive. Particles are both strongly coloured and strongly pleochroic, with the colour changing from grey to orange-pink, or from orange-pink to a deep magenta (sometimes also described as 'deep blood-red'). In reflected light particles appear dull pink to pink in colour. It has high relief, with the RI greater than that of the medium. Dispersion is relatively strong.

Particles are rarely euhedral, but when they occur, they form slightly elongate hexagons; more frequently, crushing and grinding produce angular shards. Particle shapes may be influenced by the two perfect cleavages possessed by the mineral and shape may be irregular. Cleavage planes are visible on many particles. Examples of the crushed mineral show it as coarse grained, with a broad particle size distribution.

Under crossed polars purpurite is spectacular. The birefringence is high, with third order interference colours typical. However, these are masked by the body colour, resulting in bright pinks, oranges, and violets. Small, less strongly coloured, particles may appear turquoise or lime green. Purpurite has straight extinction.

Purpurite transmits a dull red when viewed through the Chelsea filter and it is not fluorescent under UV.

Impurities are common with the naturally occurring mineral, which is found in massive crusts associated with lithiophyllite (an orange-brown translucent phase with lower birefringence), psilomelane, bixbyite and pyrolusite (all of which are opaque), and quartz.

BARIUM COPPER SILICATE, BLUE TYPE AND EFFENBERGERITE

BaCuSi$_4$O$_{10}$

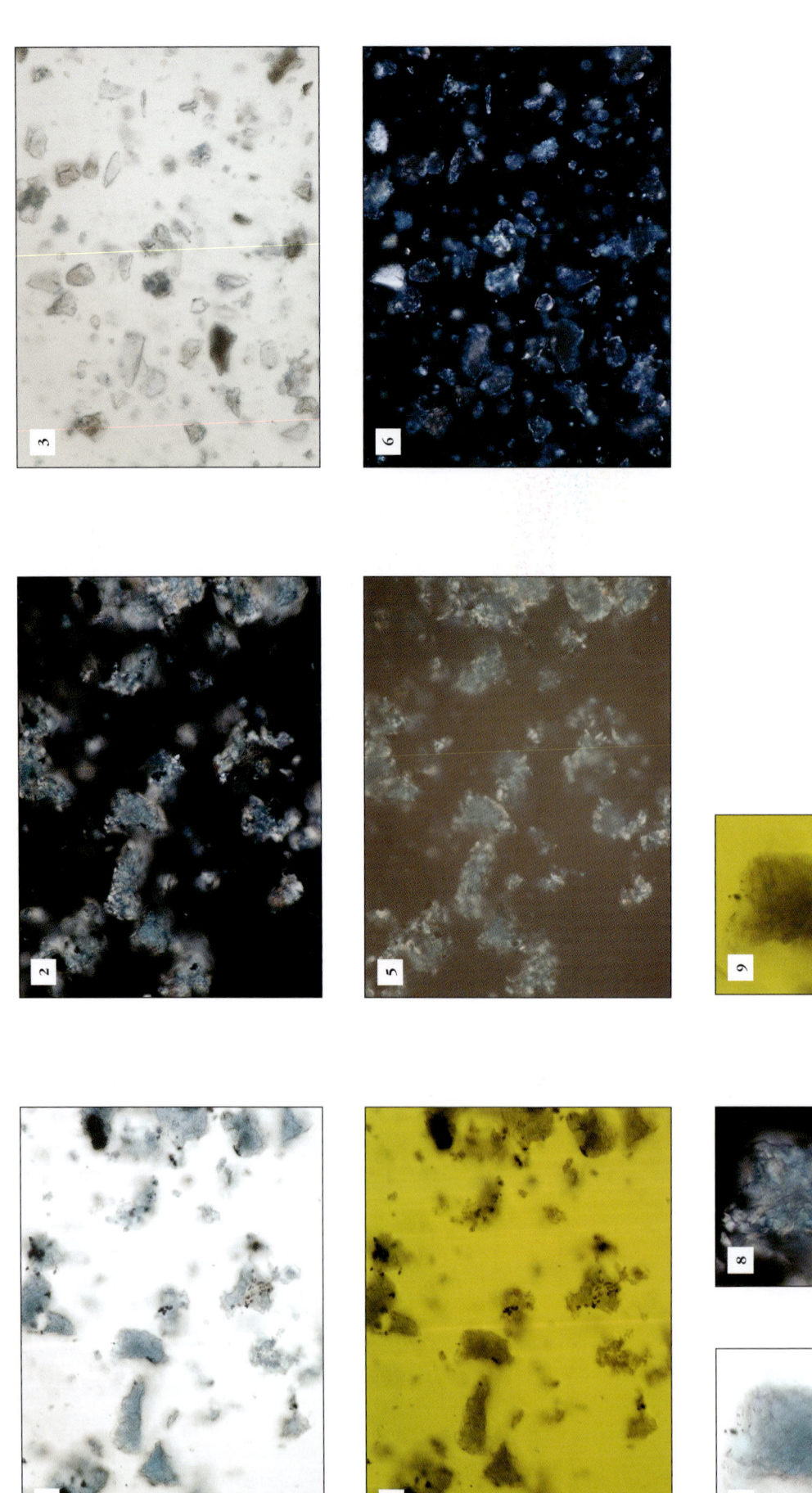

1. P1111: PPL/40×. Deep blue particles with low relief; opaque grains are also visible.

2. P1111: XPL/40×. Strongly birefringent grains with interference colours masked by the body colour in most areas; limited areas of extinction are visible due to the polycrystalline nature of the sample. Same view as Fig. 1.

3. P1114: PPL/40×. Colourless and pale blue particles with a range of morphologies; also present are other colourless and purple impurities.

4. P1111: PPL/CF/40×. Weak red transmission through the Chelsea filter. Same view as Fig. 1.

5. P1111: RPL/40×. Blue reflective surfaces as seen in reflected light. Same view as Fig. 1.

6. P1114: XPL/40×. Strongly birefringent particles with interference masked by the blue body colour in cross-polarised light. Same view as Fig. 3.

7. P1111: PPL/40×. Anhedral blue particle with irregular surface.

8. P1111: XPL/40×. Same grain as Fig. 7 in cross-polarised light.

9. P1111: PPL/CF. Same grain as in Fig. 7 seen through the Chelsea filter.

Crystallinity	Optic sign	n_ω	n_ε	δ
Tetragonal	[unknown]	1.633	1.593	0.040
				(data from Giester and Rieck, 1994)

The blue form of barium copper silicate ('Han blue') is a synthetic analogue of the rare tetragonal mineral effenbergerite; it is also structurally equivalent to the calcium copper silicate analogue of cuprorivaite ('Egyptian blue'). The pigment forms a range of crystal shapes from euhedral to subhedral plates to irregular anhedral masses. Such appearance of well-formed crystals is not to be expected in historical pigments but is typical of modern controlled methods of synthesis (see FitzHugh and Zycherman, 1992).

In plane-polarised light, the pigment is translucent to transparent, pale blue and strongly pleochroic from blue to almost colourless; particles appear blue or weakly blue in reflected light. Relief is low with the RI less than that of the medium; the RI has been reported variously as 1.620–1.630 (FitzHugh and Zycherman, 1983) and with $n_\omega = 1.633$ and $n_\varepsilon = 1.593$ (Giester and Rieck, 1994). Particle surfaces, like crystal shape, are very irregular and uneven and the particles may be inclusion-rich. Particle size is typically medium, but with a broad size distribution.

Under crossed polars this compound is strongly birefringent, high third and fourth order interference colours being typical, with some masking by the body colour. Particles may be finely polycrystalline, therefore appearing not to go into extinction; well-formed, individual crystals have straight extinction.

Particles appear weakly blue to blue in reflected light. The pigment transmits red when viewed through the Chelsea filter.

Colourless, purple barium copper silicate (Han purple') and opaque phases are typically present in samples. In historical material other impurities are also likely to be found including, according to FitzHugh and Zycherman (1983), yellow and brown particles that are related silicates.

Synthetic barium copper silicate has been described by FitzHugh and Zycherman (1983, 1992), while the optical properties of the mineral effenbergerite have been characterised by Giester and Rieck (1994).

Lit.: FitzHugh & Zycherman (1983); FitzHugh & Zycherman (1992); Giester & Rieck (1994)

The Pigment Compendium

BARIUM COPPER SILICATE, PURPLE TYPE

BaCuSi$_2$O$_6$

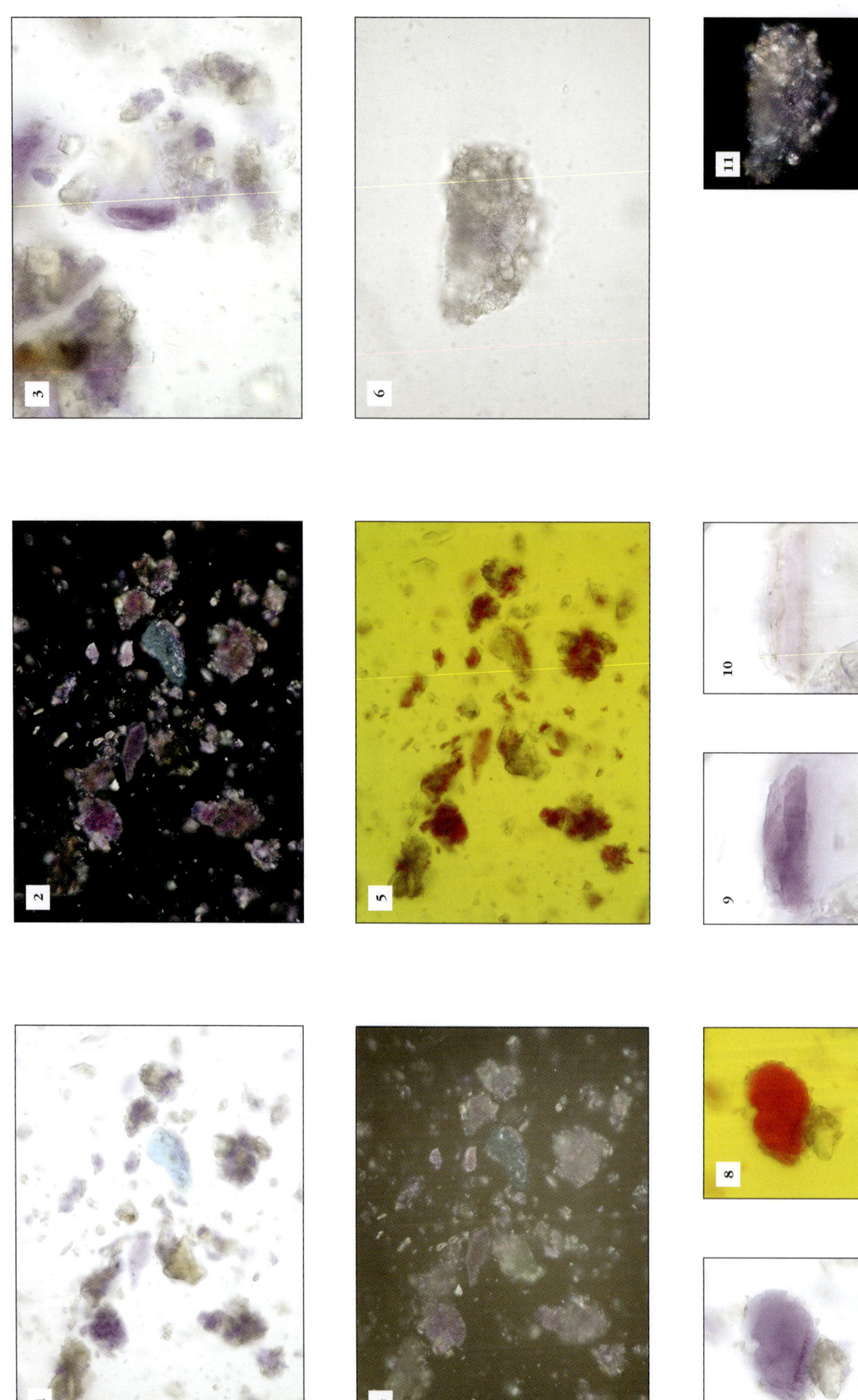

Key:

1. P1081: PPL/40×/H. Anhedral grains predominantly of purple colour, but blue and brown grains are also present.

2. P1081: XPL/40×/H. The same view as Fig. 1 under crossed polars with strong birefringence masked by body colour.

3. P1081: PPL/100×/H. Subhedral and anhedral purple grains plus colourless and yellow-brown particles.

4. P1081: RPL/40×/H. The same view as Fig. 1 in reflected light.

5. P1081: PPL/CF/40×/H. The same view as Fig. 1 through the Chelsea filter.

6. P1082: PPL/100×/H. Irregular spongy particle containing bubbles.

7. P1081: PPL/H. Shows purple particle.

8. P1081: PPL/CF/H. The particle of Fig. 7 showing very strong red transmission under the Chelsea filter.

9. P1081: PPL/H. A subhedral bladed particle.

10. P1081: PPL/H. Crystal in Fig. 9 with polariser rotated through 90° to show pleochroism.

11. P1082: XPL/H. The grain in Fig. 6 under crossed polars.

Crystallinity
Tetragonal

n_ω/n_ϵ
~1.72–1.74

δ
High

(data from FitzHugh and Zycherman, 1992; Janczak & Kubiak (1992))

The purple form of barium copper silicate (Han purple') is very closely related to the blue form (Han blue') and this pigment is generally found in trace amounts in preparations of that pigment and vice versa.

Under plane-polarised light, the pigment appears as distinctive translucent particles, generally purple but that may be from deep to light purple or colourless. Particles are strongly pleochroic from intense purple to colourless, though FitzHugh and Zycherman (1992) report that while they found pleochroism was common it was not present in all the samples that they studied. Particles have low relief, with RI just greater than that of the medium. Particle shape varies from anhedral and irregular, spongy appearing particles to euhedral rectangular crystals. Particle surfaces range from being smooth and clear, to heavily pitted and irregular; many of the latter particles are inclusion rich. Particle size distribution is broad, with particles ranging from fine to very coarse.

Under crossed polars this pigment is strongly birefringent; high order interference colours are typical, although these are substantially masked by the body colour. Many particles are polycrystalline and therefore do not go into complete extinction. Well-formed crystals show straight extinction.

Particles appear purple in reflected light, transmitting a strong red when viewed through the Chelsea filter. No UV fluorescence was observed in samples by the authors.

Other phases are normally present; typically these are trace amounts of the blue form along with colourless, pale brown and yellowish crystals, which are related silicates (FitzHugh and Zycherman, 1992).

The blue and purple forms of barium copper silicate have been described by FitzHugh and Zycherman (1983, 1992).

Lit.: FitzHugh & Zycherman (1983); FitzHugh & Zycherman (1992); Janczak & Kubiak (1992)

CALCIUM COPPER SILICATE, CUPRORIVAITE TYPE

CaCuSi$_4$O$_{10}$

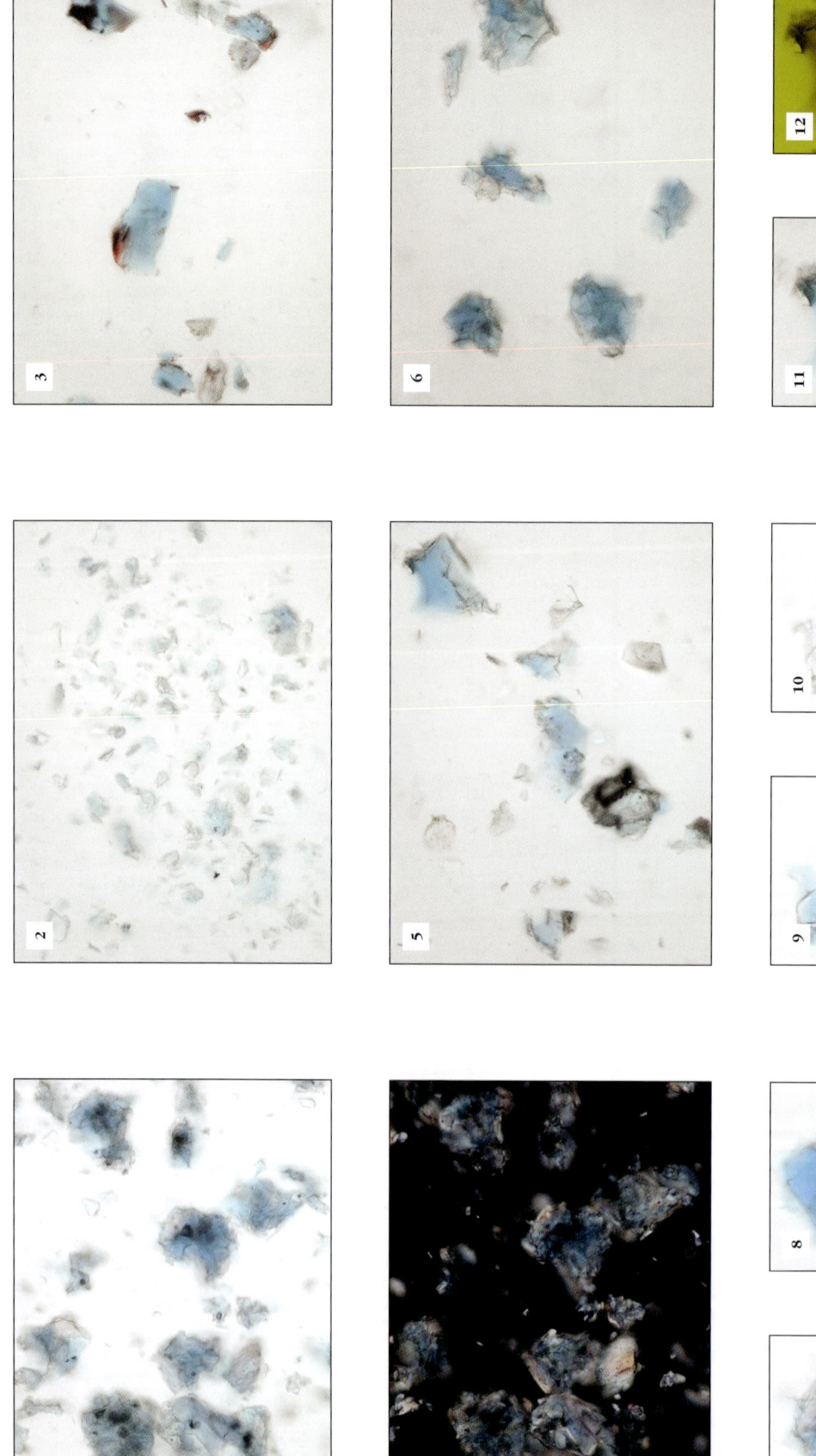

Key:

1. P1108: PPL/40×. Strongly irregular and inclusion-rich particles of 'Egyptian blue'.

2. P0180: PPL/40×. Weakly coloured to colourless grains of Egyptian blue.

3. P1374: PPL/40×/H. Angular particles of Egyptian blue showing patches of red coloration.

4. P1108: XPL/40×. The field of view in Fig. 1 under crossed polars.

5. P1379: PPL/40×/H. Angular particles of Egyptian blue.

6. P1498: PPL/40×. Irregular particles of Egyptian blue.

7. P1108: PPL. Detail of grain from Fig. 1.

8. P1379: PPL/ ○ /H. Tabular evenly coloured crystal of Egyptian blue.

9. P1379: PPL/ ○ /H. Unevenly coloured particle of Egyptian blue.

10. P1379: PPL/ ○ /H. The crystal in Fig. 9 rotated through 90° to show strong pleochroism.

11. P1379: PPL/H. Angular particle of Egyptian blue.

12. P1379: PPL/ CF/H. Same particle as Fig. 11 with Chelsea filter showing low red transmission.

Crystallinity	n_α	n_ε	δ
Tetragonal	1.636	1.591	0.045

(data from Pabst, 1959; cf. Riederer, 1997)

Calcium copper silicate ('Egyptian blue') is the synthetic analogue of the rare tetragonal mineral cuprorivaite. In plane-polarised light this compound occurs as pale blue particles, although the intensity of the colour is variable. For some examples the colour of the particles is intense and the particles are seen to have straight extinction. However, particles are not uncommonly polycrystalline and in such cases complete extinction does not occur.

Calcium copper silicate usually transmits a dull red colour when viewed through the Chelsea filter, though the intensity of the red transmission increases with the intensity of the body colour of the particles.

Various other phases are typically present along with the calcium copper silicate in examples of Egyptian blue. These are mostly colourless silicate phases, but may even include green or purple particles. Amorphous and opaque particles are not uncommon. Compounds known to occur during manufacture of the pigment include quartz and tridymite, wollastonite and the copper oxides cuprite and tenorite.

The optical properties of Egyptian blue have been widely discussed; see, for example, Pabst (1959), Riederer (1997) and Scott (2002) for further information.

Lit.: Pabst (1959); Riederer (1997); Scott (2002); Winchell (1931)

Under crossed polars, calcium copper silicate is strongly birefringent, with high third order and above interference colours, though these may be masked by the body colour. Where crystal faces are evident the particles are translucent while in other cases they may be weakly coloured and transparent. This phenomenon is unrelated to particle size. Crystals exhibit a pleochroism that is in some cases extreme, from blue to colourless. Relief is low, with RI less than that of the medium; various values have been given including $n_\omega = 1.6354$ and $n_\varepsilon = 1.6053$ with $\delta = 0.0301$ ('Vestorian Blue'; Winchell, 1931) and $n_\omega = 1.636$, $n_\varepsilon = 1.591$ and $\delta = 0.045$ (Pabst, 1959; cf. Riederer, 1997).

Particle surfaces range in texture from smooth and glassy to extremely rough with a pitted or spongy appearance. Particle shape is extremely variable. Perfect tabular crystals with platy, euhedral octagonal habits can be grown, although Riederer states that crystal faces are never visible in actual material (at least when observed using optical microscopy). Consequently samples typically show increasingly anhedral morphology, either broken angular shards or irregular, even spongy particles. Particle size is medium to coarse, with a broad particle size distribution.

The Pigment Compendium

CUPRO-WOLLASTONITE

β-CaSiO₃ + Cu

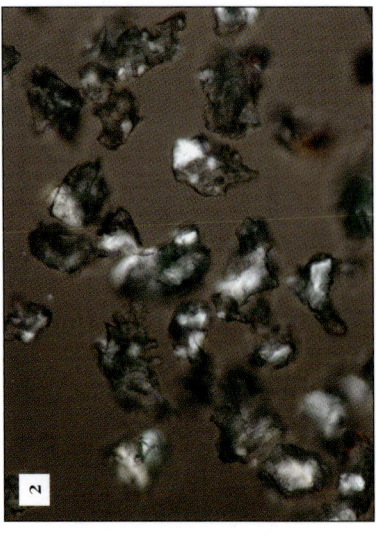

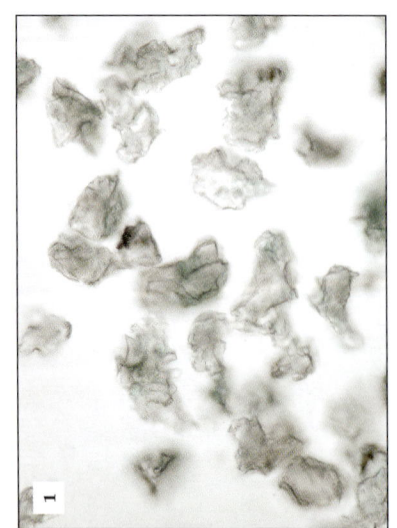

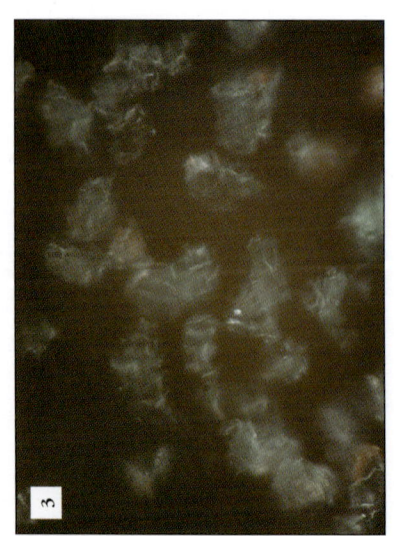

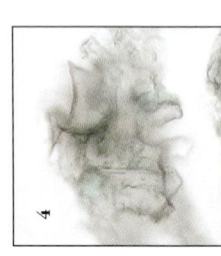

Key:

1. P1525: PPL/40×. Irregular particles of cupro-wollastonite showing variable coloration.

2. P1525: −XPL/40× ⦿ The field of view in Fig. 1 under partially crossed polars.

3. P1525: RPL/40×. The field of view in Fig. 1 in reflected light.

4. P1525: PPL. Enlargement of grain from Fig. 1 showing colour variation.

Crystallinity	Optic sign	2V	n_α	n_β	n_γ	δ
Triclinic		44–50	1.615–1.646	1.627–1.659	1.629–1.662	0.014–0.016
					(data for wollastonite-7A from Webmineral, 2003)	

So-called 'Egyptian green' consists of wollastonite-7A (also known as pseudowollastonite or 'β-wollastonite', variants of wollastonite, $CaSiO_3$ though neither term is a currently accepted mineral name) containing a few per cent of copper substituted into the structure – thus cupro-wollastonite; an amorphous phase is also present. The structure of pseudowollastonite has been most recently studied by Yang and Prewitt (1999). More information on the microscopy of this pigment can be found in Pagès-Camagna et al. (1999).

In plane-polarised light, cupro-wollastonite forms translucent, colourless to pale emerald green particles with moderate relief and RI less than that of the medium. Particle shapes are anhedral and irregular and are rich in inclusions, having a spongy appearance. Particle surfaces are rough and uneven. Particle size is medium to very coarse.

Under crossed polars cupro-wollastonite exhibits a low birefringence, with first order interference colours that are masked by the body colour (where present) such that particles appear grey-green. Some particles are isotropic, these being the amorphous portion of the pigment. Anisotropic particles show complete extinction.

With reflected light, particles of this pigment appear pale green with white raised edges. When viewed with the Chelsea filter cupro-wollastonite remains green.

Lit.: Pagès-Camagna et al. (1999); Yang & Prewitt (1999)

BARIUM MANGANATE (VI) SULFATE

xBaSO$_4$. yBaMnO$_4$

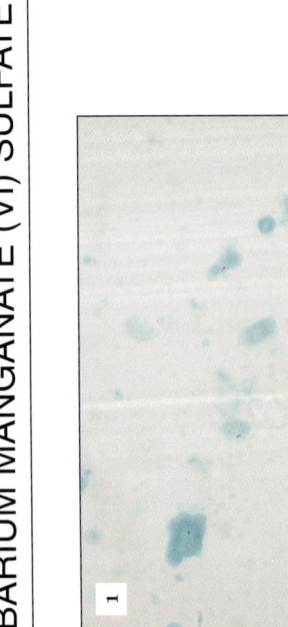

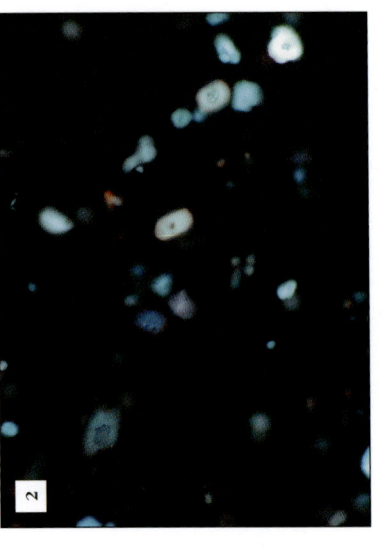

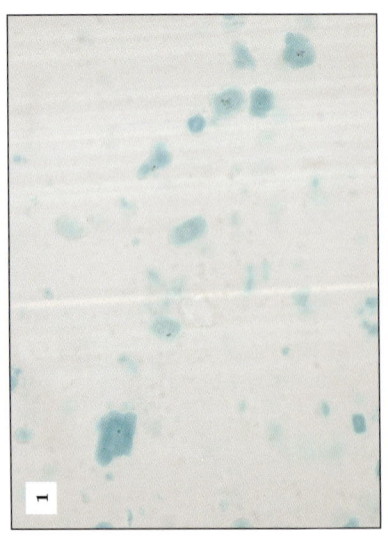

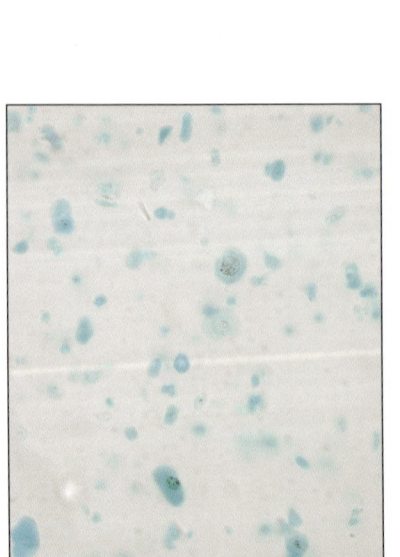

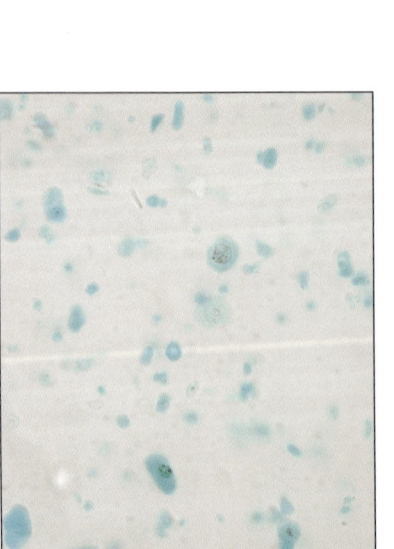

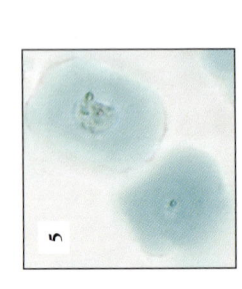

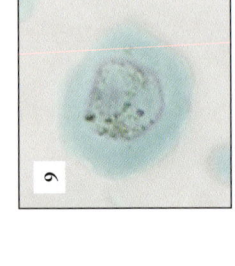

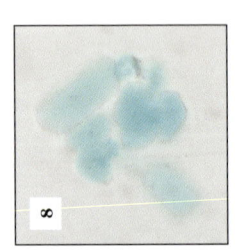

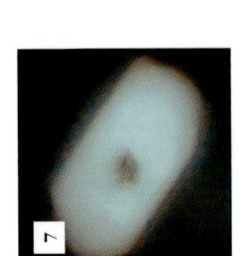

Key:

1. P0202: PPL/100×/ **O** Euhedral and subhedral plates of barium manganate (VI) sulfate.

2. P0202: XPL/100×/ **O**. The same field of view as in Fig. 1 under crossed polars.

3. P0202: PPL/100×. Euhedral and subhedral particles of barium manganate (VI) sulfate.

4. P0202: XPL/100×. The field of view as in Fig. 3 under crossed polars, note anomalous brown interference colours.

5. P0202: PPL/ **O**. Euhedral particles with rounded edges and inclusion-rich cores.

6. P0202: PPL/ **O**. Euhedral crystal of manganese blue.

7. P0202: XPL/ **O**. The crystal in Fig. 6 under crossed polars.

8. P0102: PPL. A group of euhedral barium manganate (VI) sulfate plates, the variation in colour intensity is due to pleochroism.

9. P0202: PPL. Large plate of barium manganate (VI) sulfate showing inclusion-rich core.

Crystallinity	**n**	**δ**
Anisotropic	~1.65	High
		(data from Mactaggart, 2002)

Barium manganate (VI) sulfate, commonly known as manganese blue is highly distinctive when viewed under the microscope. In plane-polarised light the particles show very low relief, having a refractive index similar to that of 1.662 RI mounting media. The particles are translucent, pale turquoise blue. The compound is strongly pleochroic, exhibiting colours from dark to very pale turquoise. Characteristically, particles have a dark spot in the centre of the particle and often show well-developed concentric growth zones. Most particles are euhedral, platy and hexagonal although faces do not have sharp angles, giving the particles a rounded appearance. Many particles show a high sphericity, while others are slightly elongated. Particle surfaces are smooth. In samples studied the particle size was fine to medium.

Under crossed polars the particles are strongly birefringent, but interference colours are masked by the body colour, producing anomalous bright green, bright pale blue and red-brown colours. The central dark spots are very clear under crossed polars, as is the concentric zoning. Barium manganate (VI) sulfate crystals have straight extinction.

Particles appear pale blue and grey when viewed through the Chelsea filter.

Lit.: Mactaggart (2002)

The Pigment Compendium

BARIUM MANGANATE (VI)

BaMnO$_4$

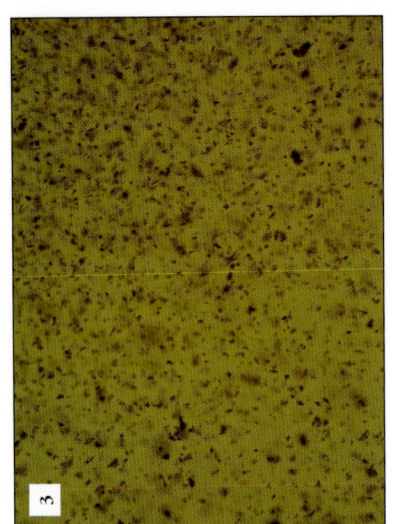

Key:
1. P0876: PPL/100×. Fine-grained rounded and acicular particles of barium manganate (VI).

2. P0876: ~XPL/100×. Same field of view as in Fig. 1 under crossed polars, note anomalous pink interference colours.

3. P0876: PPL/CF/100×. Barium manganate (VI) showing strong red transmission through the Chelsea filter.

Crystallinity
Anisotropic

n	δ
>1.662	Low
	(data from authors)

Although most modern examples of manganese blue are likely to be barium manganate (VI) sulfate (*q.v.*), there is some historical evidence in earlier documentary sources for the use of barium manganate (VI). These pigments are quite distinct both compositionally and microscopically.

In plane-polarised light, barium manganate (VI) forms weakly translucent to opaque, ink blue particles. Relief is moderate and RI greater than that of the medium. Particle size is very fine, and grains exist as individual grains and aggregates. Particle size is close to that resolvable with the optical microscope and grains appear rounded, bacterioid and acicular in shape. The aggregates may be coarse grained.

Under crossed polars barium manganate (VI) oxide is weakly birefringent but the interference colours are anomalous reds. Particles have complete and straight extinction.

Barium manganate (VI) appears a dark, dull red when viewed through the Chelsea filter.

The Pigment Compendium

COBALT ALUMINIUM OXIDE

CoAl$_2$O$_4$

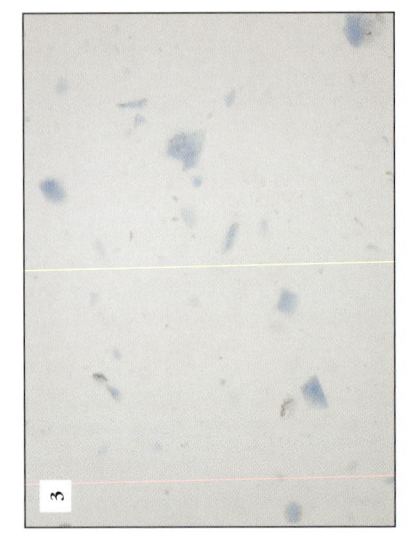

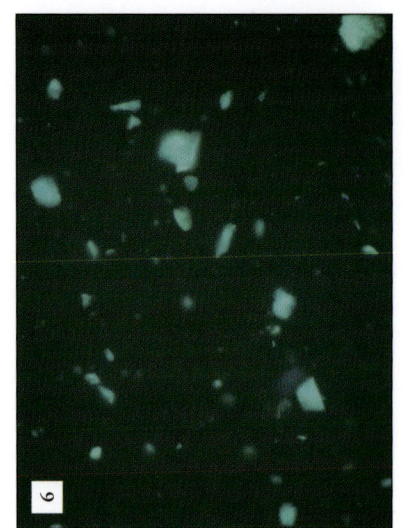

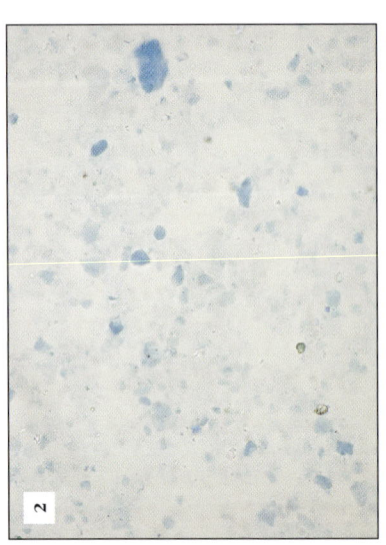

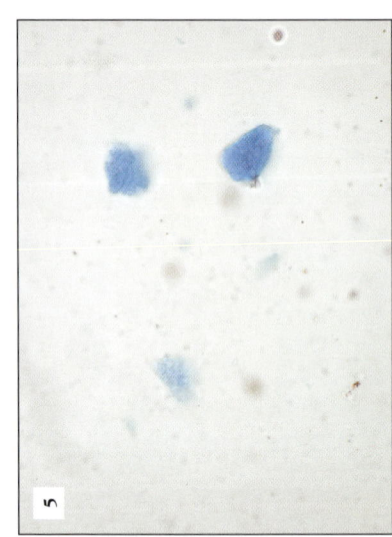

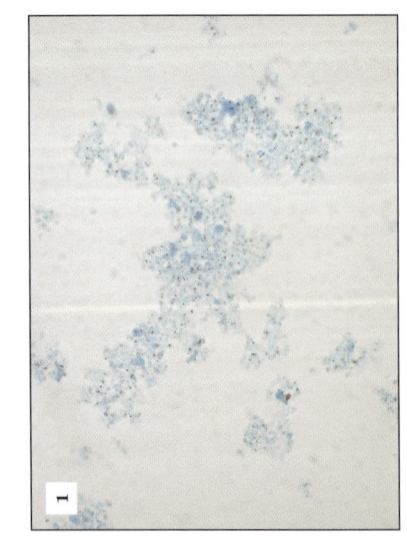

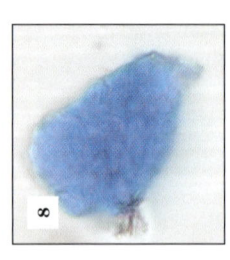

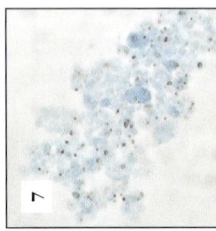

1. P0105: PPL/100×. Particles of cobalt aluminium oxide.

2. P0206: PPL/100×. Anhedral plates of cobalt aluminium oxide.

3. P0976: PPL/40×/H. Anhedral plates of cobalt aluminium oxide.

4. P0105: RPL/100×. The field of view in Fig. 1 under crossed polars.

5. P1003: PPL/100×/H. Anhedral particles of cobalt aluminium oxide.

6. P0976: UV/40×/H. The field of view in Fig. 3 showing UV fluorescence.

7. P0105: PPL. Detail of particle morphology from Fig. 1.

8. P1003: PPL/H. Detail of particle morphology from Fig. 5.

Crystallinity

Cubic	*n*	δ
	<1.662–1.72	Isotropic

(data from Mactaggart, 2002)

In plane-polarised light, cobalt aluminium oxide ('cobalt blue') has low relief and forms transparent, bright blue, angular to subangular particles. Relief varies from sample to sample from low to moderate; the RI is equal to or just greater than that of the medium. McCrone *et al.* (1973–80) report an RI of 1.69 for blue light, 1.70 for red, though Winchell (1931) gives the RI as around 1.74. Particle size is also variable, ranging from coarse to very fine. For the very fine-grained pigments most particles are at the resolution of the optical microscope and consequently appear to be of rounded to bacterioid form.

Cobalt aluminium oxide is cubic and therefore appears isotropic under crossed polars, although blue internal reflections are observable.

Cobalt aluminium oxide has high red transmission, thus appearing red when viewed using the Chelsea filter. However, violet-blue fringes are visible around the margins of the particles.

Lit.: Mactaggart (2002); McCrone *et al.* (1973–80); Winchell (1931)

The Pigment Compendium

COBALT TIN OXIDE

CoSnO$_3$

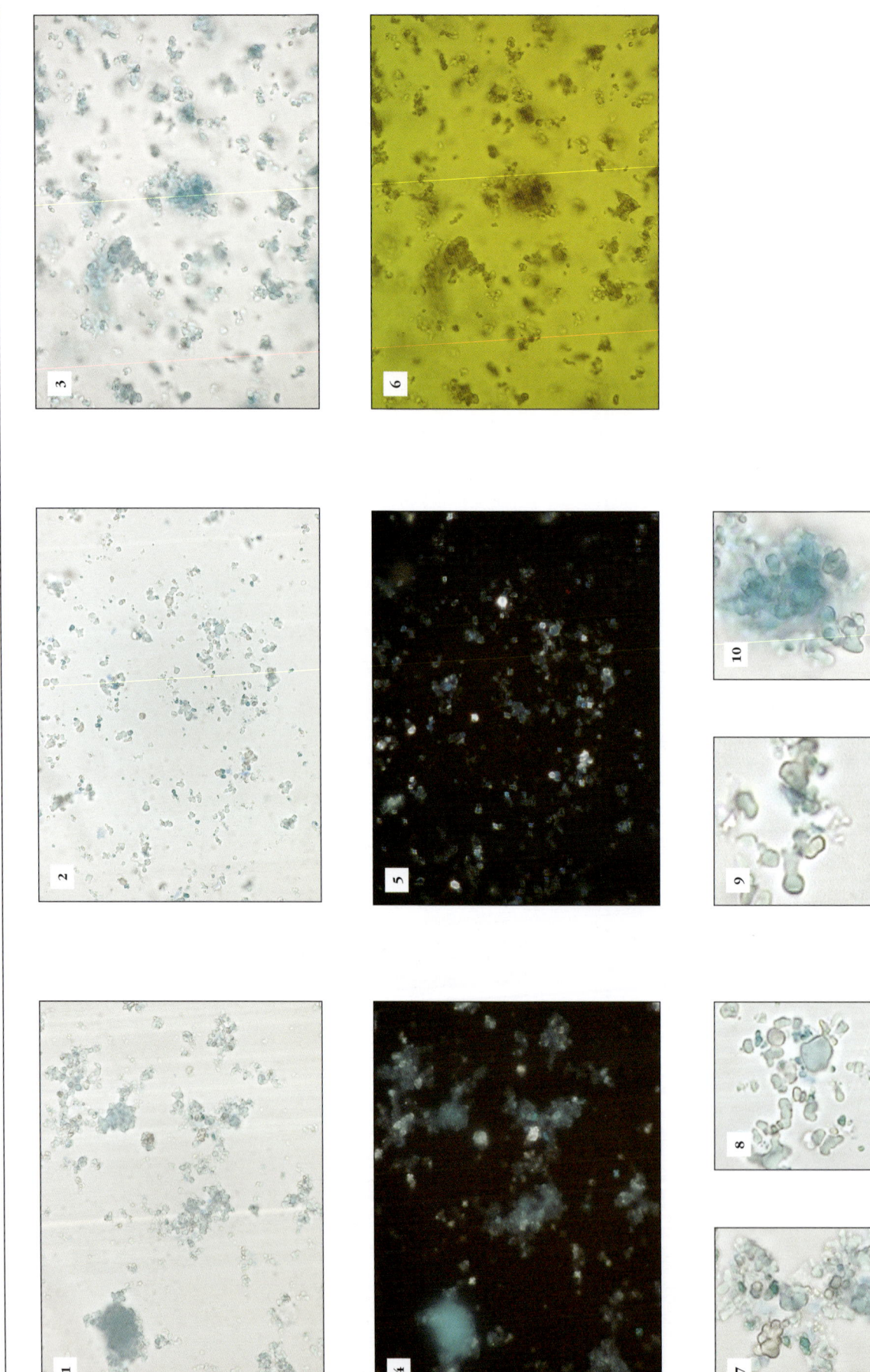

Key:

1. P0108: PPL/100×. Fine-grained platy particles of cobalt tin oxide.

2. P0205: PPL/100×. Platy crystals of cobalt tin oxide.

3. P1295: PPL/100×/H. Rounded crystals of cobalt tin oxide.

4. P0108: RPL/100×. The field of Fig. 1 under crossed polars.

5. P0205: RPL/100×. The field of view in Fig. 2 under crossed polars.

6. P1295: PPL/CF/100×/H. The field of view from Fig. 3 seen through the Chelsea filter showing red transmission.

7. P0108: PPL. Detail of particle morphology from Fig. 1.

8. P0205: PPL. Detail of particle morphology from Fig. 2.

9. P0205: PPL. Detail of particle morphology from Fig. 2.

10. P1295: PPL/H. Detail of particle morphology from Fig. 3.

Crystallinity

Isotropic/Anisotropic

n	δ
1.84	Low

(data from McCrone *et al.*, 1973–80)

In plane-polarised light cobalt tin oxide ('cerulean blue') forms translucent, turquoise blue grains with moderate relief and ar. RI greater than that of the medium. Particle shape is typically of subangular, anhedral to subhedral polyhedral plates along with clusters and aggregates of these. Particle surfaces are smooth, but appear slightly concave. Particle size is fine and the distribution narrow.

Under crossed polars cobalt tin oxide should be isotropic but commonly contains a proportion of birefringent particles that exhibit bright, high first order interference colours. This phenomenon has also been observed by

McCrone *et al.* (1973–80) and Mactaggart (2002). These birefringent particles appear finely polycrystalline.

Cobalt tin oxide has a moderate red transmission and thus appears a dull, deep red when observed through the Chelsea filter.

Cobalt tin oxide may contain trace amounts of free tin(IV) oxide (*q.v.*).

Lit.: Mactaggart (2002); McCrone *et al.* (1973–80)

COBALT ZINC OXIDE

CoZnO$_3$

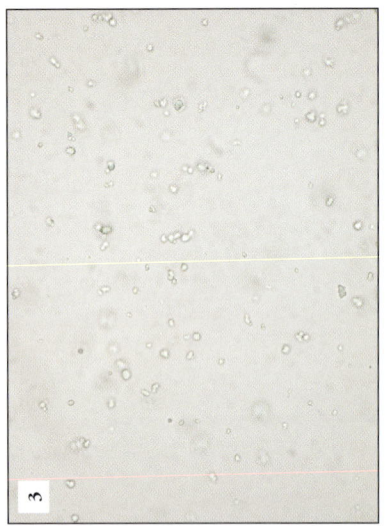

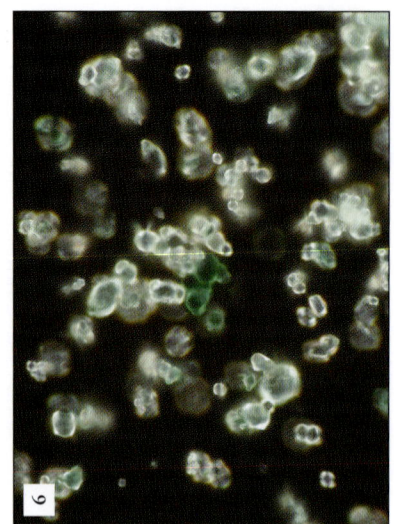

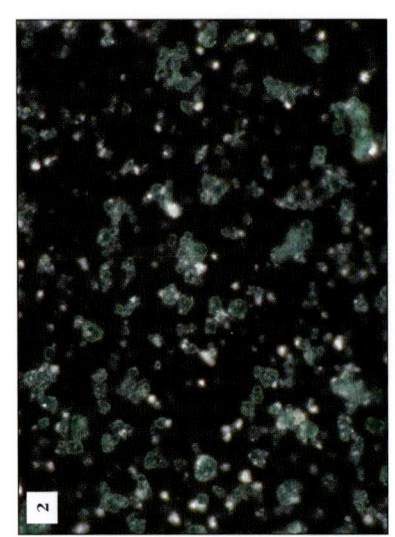

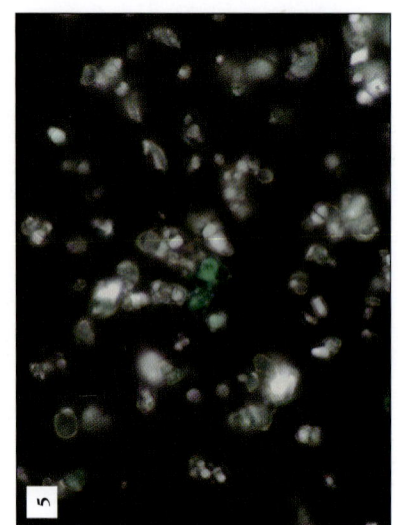

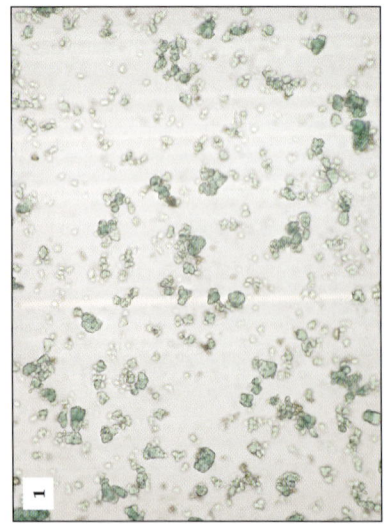

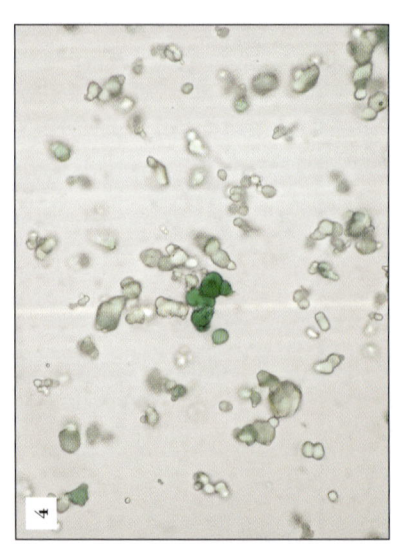

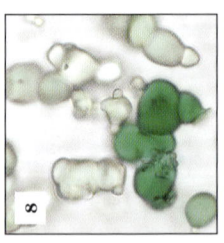

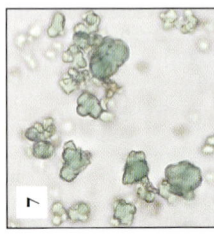

Key:

1. P0213: PPL/100×. Platy particles of green cobalt zinc oxide.

2. P0213: XPL/100×. The field of view in Fig. 1 in crossed polars.

3. P1300: PPL/100×/H. Pale green to colourless particles of cobalt zinc oxide.

4. P1299: PPL/100×/H. Variably coloured grains of cobalt zinc oxide.

5. P1299: XPL/100×/○/H. The field of view in Fig. 4 under crossed polars.

6. P1299: RPL/100×/H. The field of view in Fig. 4 in reflected light.

7. P0213: PPL. Subangular platy crystals of cobalt zinc oxide from Fig. 1.

8. P1299: PPL/H. Enlargement of particles from Fig. 4 showing rounded grain morphologies.

Crystallinity
Anisotropic

n	δ
1.94–2.0	High

(data from Gettens and Stout, 1966 and authors)

In plane-polarised light cobalt zinc oxide ('Rinmann's green') appears as crystals ranging in colour from very pale green and almost colourless to an intense, translucent blue-green. The particles are very weakly pleochroic. Relief is moderate to high, with the RI greater than that of the medium. The particles form angular to subrounded anhedral grains; many surfaces show a poorly defined conchoidal fracture; others are pitted and inclusion rich. Particle size may vary between very fine to coarse and size distribution within a sample may be uneven.

Under crossed polars, cobalt zinc oxide is strongly birefringent, although interference colours are masked by the body colour. Some particles are shown under these conditions to be polycrystalline.

Cobalt zinc oxide transmits a weak, dull pink colour when viewed through the Chelsea filter. It also has dull green fluorescence with UV excitation.

SMALT

$SiO_2(vit)Co_x$

Key:

1. P0613: PPL/100×/H. Pale blue and colourless elongate shards of smalt with curved fractured edges visible.

2. P1055: PPL/10×. Anhedral grains of intense blue smalt with many inclusions and conchoidal fracture.

3. P0621: PPL/100×/H. Colourless angular fragments containing many inclusions.

4. P0613: PPL/CF/100×/H. Weak red transmission of the blue smalt grains using the Chelsea filter. Same field of view as Fig. 1.

5. P1055: PPL/CF/10×. Red transmission of grains using the Chelsea filter, with grain edges a darker red colour. Same field of view as Fig. 2.

6. P0620: ~XPL/40×/H. Weakly anisotropic smalt grains under partially crossed polars.

7. P1055: PPL ●. Anhedral inclusions within a larger anhedral smalt grain.

8. P1055: ~XPL ●. Same particle as in Fig. 7, isotropic in crossed polars with strongly birefringent inclusions.

9. P1055: PPL/CF. Red transmission of grain in Fig. 7.

10. P1052: PPL. Unevenly coloured smalt grain with conchoidal fracture.

11. P1052: PPL/CF. Weakened transmission of unevenly coloured blue smalt grains.

12. P0001: PPL. Angular fragment showing uneven colour distribution and conchoidal fracture.

Crystallinity
Amorphous

δ
Isotropic

(data from Mühlethaler and Thissen, 1993)

n
1.46–1.55

Under plane-polarised light the cobalt-doped glass smalt appears as strongly transparent, blue coloured particles. However, it is notable that colour intensity increases with size, the effect of which is to produce particles that have an uneven colour because of variable thickness; particle margins are for this reason generally only weakly coloured. In addition, smalt that either has low intrinsic colour intensity, or has faded or discoloured, occurs as transparent, colourless or very pale blue particles. Consequently, particles may be encountered as faded or mottled between deep blue, pale blue to colourless.

Relief is moderate and the RI is less than that of the medium; RI values given in the literature are 1.49–1.52 (Gettens and Stout, 1966) and 1.46–1.55 (Mühlethaler and Thissen, 1993), the latter probably being more representative of the true range. Smalt is a glass and therefore breaks on crushing with a pronounced conchoidal fracture. This phenomenon is clearly visible on the particle surfaces where the characteristic curved fracture surfaces often exhibit concentric pressure ridges. Additionally, the particle habit is one of angular shards with curved particle boundaries. The presence of inclusions is typical of many examples of smalt. These may be simply gas bubbles (which are in turn characteristic of any glass), which have a spherical form; these may also be observed as 'craters' on particle surfaces. Crystalline inclusions may also be present, often forming radiating aggregates of tiny crystals – crystallites. Particle size is usually medium to very coarse, smalt substantially losing its colour when finely ground. Especially coarse grades were also used historically for 'strewing', the sprinkling of smalt onto a surface to achieve an intense blue.

Under crossed polars, as an amorphous substance, smalt is isotropic. However, deformed particles may show a weak anisotropism, typically low first order interference colours being observed with extinction that is undulose. Inclusions of crystallites, if present, will often be strongly anisotropic and have high birefringence, appearing as bright spots.

When viewed through the Chelsea filter smalt can be seen to transmit red light; however, particle margins especially may appear purplish blue. The colour is less strongly red than that for ultramarine and lazurite (*q.q.v.*). With reflected light the particles appear transparent blue to black, perhaps with white edges.

The most likely associated material found with smalt is quartz, residual material from the formation process. In some instances this appears to form a birefringent core, the smalt glass coating it.

Differentiating smalt from lazurite (natural ultramarine) may be difficult as both phases have similar optical properties. The presence of circular bubbles and other inclusions is usually diagnostic of smalt. Conchoidal fracture is also far more pronounced in smalt than in lazurite, which demonstrates only a weak conchoidal fracture. Smalt is easily distinguished from cobalt aluminium oxide (*q.v.*) by its lower refractive index.

The optical characteristics and analysis of smalt have been described by Mühlethaler and Thissen (1993).

Lit.: Gettens & Stout (1966); Mühlethaler & Thissen (1993)

LAZURITE

$(Na,Ca)_8[(Al,Si)_{12}O_{24}](S,SO_4)$

Key:

1. P1291: PPL/40×/H. Shows angular blue lazurite crystals some with uneven colour; also includes many opaque (pyrite) and colourless impurities.

2. P0321: PPL/40×/H. Note unevenly coloured particles with conchoidal fracture; colourless impurities of similar habit.

3. P1219: PPL/40×. Note strong coloured particles in this sample as well as uneven colour; conchoidal fracture is evident here.

4. P1291: ~XPL/40×/H. Under partially crossed polars this shows isotropic to weakly anisotropic particles of lazurite. Strongly birefringent particles are impurities.

5. P0321: ~XPL/40×/H. Shows particles of lazurite as isotropic; impurities show low first order interference colours.

6. P0009: XPL/40×/ ❍. Birefringent impurities in natural ultramarine sample in crossed polars: calcite (centre) and grains of diopside (clinopyroxene; left and right).

7. P1219: PPL. Uneven coloured grains of lazurite, note conchoidal fracture.

8. P1291: PPL. An intensely coloured lazurite particle.

9. P0321: PPL/H. Uneven distribution of colour in natural lazurite.

10. P1291: PPL/H. Strong variation in colour from colourless to blue in a lazurite crystal. Note the crystal also contains fluid inclusions.

11. P0009: PPL. The diopside crystal from the right of Fig. 6 enlarged, containing inclusions of lazurite.

12. P0009: XPL/ ❍. The diopside crystal from Fig. 11 in crossed polars, note the prism parallel cleavage.

Crystallinity
Cubic/Pseudo-cubic

n
1.50

δ
[Isotropic to weakly anisotropic]
(data from Tröger, 1952)

In plane-polarised light, lazurite has an intense blue, sometimes almost violet-blue, body colour, appearing deep opaque blue in reflected light. It is not pleochroic. Typically, the colour may be patchy across crystals and is often observed to fade dramatically across single particles through pale blue to colourless. Poorer grades, such as the so-called 'ultramarine ash', will have a predominance of the paler particles present. Lazurite as moderate relief. Particle shape is generally representative of the crushing used to reduce the rock (*lapis lazuli*) to a pigment and therefore angular shards are normal. Surfaces often show conchoidal fractures.

Particle size distribution of lazurite is likely to be highly variable with broad size ranges. However, this aspect has been studied in more detail by Asperen de Boer (1974) for some early Netherlandish paintings, while an unusually pure example of lazurite with a narrow particle size distribution has been noted in paintings of Perugino by Bomford *et al.* (1980).

In cross-polarised light lazurite should be isotropic since it has a cubic crystal structure. However, in naturally occurring minerals of the sodalite group, of which lazurite is a member, substitutions and distortions in the crystal structure cause phase changes that often make them 'pseudo-cubic', belonging to the

orthorhombic or tetragonal systems. Therefore lazurite may show low first order birefringence in these pseudo-cubic crystals, though the interference colours will be masked by the strong body colour.

Lazurite has a high transmission at the red end of the visible spectrum such that it exhibits a distinctive strong red colour when viewed with a Chelsea filter.

Lazurite is derived from the rock lapis lazuli, and is therefore never a pure material; the commonly occurring associated minerals are calcite and pyrite, both of which are normally observed in samples of natural material. Lazurite itself belongs to the sodalite group of feldspathoid framework silicates and is a sulfur-bearing variety of the closely related mineral haüyne. Sodalite and haüyne may consequently also be present in the sample; for a discussion of these, see the section on Feldspathoid minerals. Plesters (1993) further reports from X-ray diffraction studies that diopside, forsterite, muscovite and wollastonite may be present, also deriving from the lapis lazuli matrix.

Lit.: Asperen de Boer (1974); Bomford *et al.* (1980); Plesters (1993); Tröger (1952)

ULTRAMARINE

$Na_7Al_6Si_6O_{24}S_3$

1. P0750: PPL/100×/H. Angular particles of synthetic ultramarine.

2. P0750: ~XPL/100×/H. The same field of view as Fig. 1 in partially crossed polars.

3. P0750: PPL/CF/100×/H. Same field of view as Fig. 1 observed through the Chelsea filter.

4. P1288: PPL/100×/H. Angular particles and aggregates of fine particles.

5. P1289: PPL/100×/H. Aggregates of very fine particles.

6. P0584: PPL/10×/H. Uneven distribution of colour in early ultramarine.

7. P0750: PPL/H. Enlarged view of particle morphologies in Fig. 1.

8. P1289: PPL/H. Enlarged particle aggregate from Fig. 5.

9. P0584: PPL/H. Enlarged crystals from Fig. 6.

10. P0584: PPL/ CF/H. Crystals from Fig. 9 viewed through the Chelsea filter.

Crystallinity

	n	δ
Cubic	1.50	0.0

(data for lazurite from Tröger, 1952)

Ultramarine is the synthetic analogue of the mineral lazurite (*q.v.*); it belongs to the cubic crystal system. Synthetic ultramarine is generally more finely divided and lower in impurities than the naturally occurring equivalent.

In plane-polarised light, the particles are even-coloured, of an intense violet-blue hue; early samples may show uneven coloration. The pigment is non-pleochroic. Particle morphology is angular, with some particles present in crumb-like aggregates. However, it is typically a fine-grained pigment of narrow particle size range. This pigment may be distinguished from the natural analogue, lazurite, by the finer grain size and purity. McCrone *et al.* (1973–80) also notes that the uneven particle surfaces of ultramarine differentiate it from smooth-surfaced lazurite particles. Plesters (1993) reports that one manufacturer (Reckitt's) documented the particle size of their product in the 1950s as being in the range 0.5–5.0 µm, but also states that she had observed particles as large as 30 µm in older material. It should be noted though that larger particles are usually agglomerates.

In cross-polarised light, as ultramarine has a cubic crystal structure, it appears isotropic.

Ultramarine has high red transmission and therefore produces a distinctive and strong red when viewed using the Chelsea filter.

Synthetic ultramarine is typically low in impurities, though X-ray diffraction studies of samples by the present authors identified calcite and sodalite in several commercial products. Use of so-called 'laundry blue' as a pigment in ethnographic artefacts may also introduce sodium hydrogen carbonate, a common functional admixture to that product.

Differentiation of the more common blue pigments lazurite, ultramarine, smalt and cobalt aluminium oxide (*q.q.v.*) may sometimes present difficulties if particle sizes are small and few in number. The last of these is differentiable on the basis of an RI greater than the medium; the others must normally be separated on the basis of particle morphology and associated minerals.

Lit.: McCrone *et al.* (1973–80); Plesters (1993); Tröger (1952)

ULTRAMARINE: GREEN, VIOLET AND RED FORMS

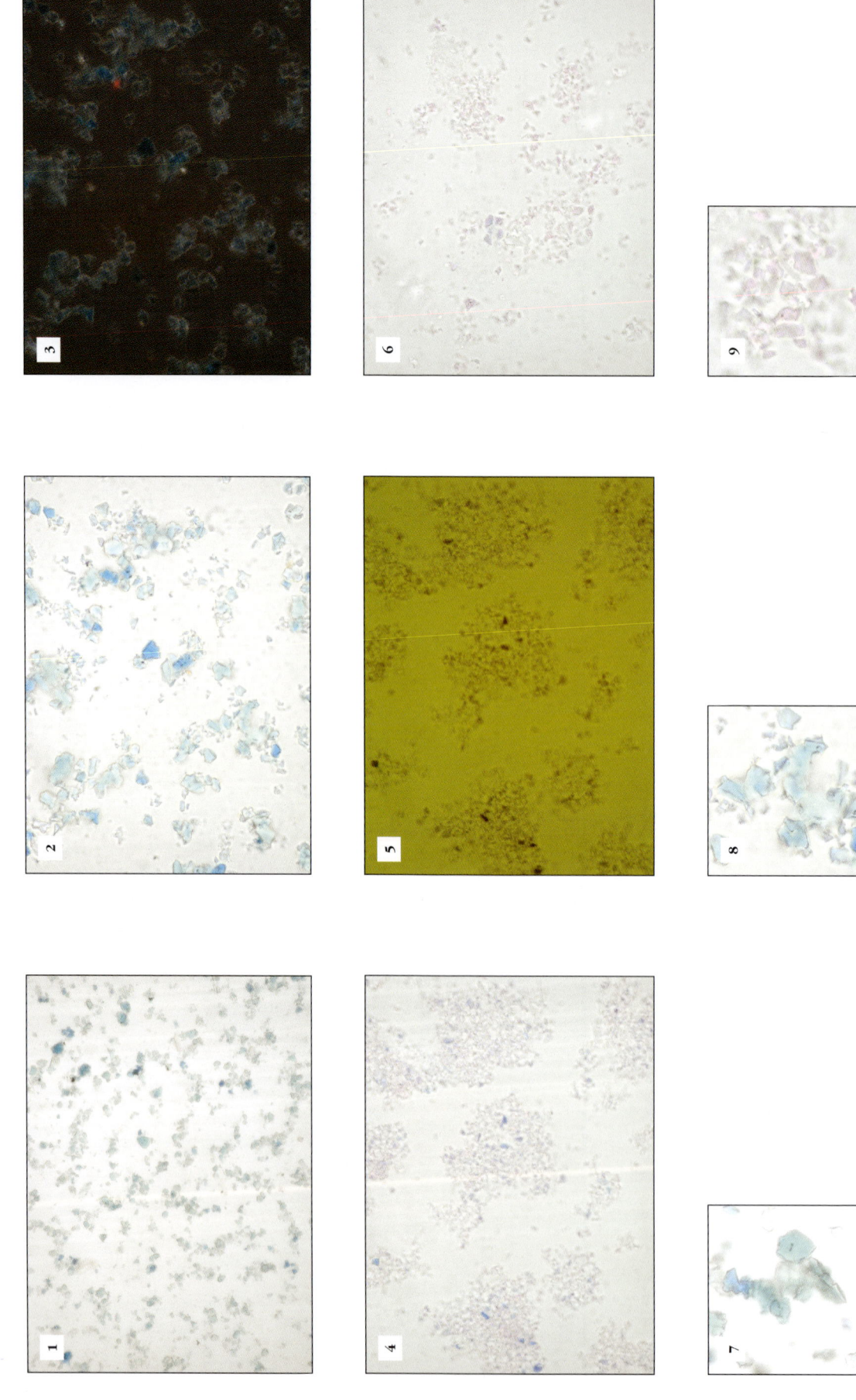

Key:

Key:

1. P0201: PPL/40×. Ultramarine green: a mixture of blue-green and deep blue anhedral particles, with angular colourless particles also present.

2. P0201: PPL/100×. Ultramarine green: mixture of pale blue-green and blue anhedral particles, with colourless grains also present as a third phase.

3. P0201: XPL/100×. Isotropic particles of ultramarine green.

4. P0100: PPL/100×. Ultramarine violet: a mixture of finer-grained anhedral pale violet grains and larger anhedral blue grains.

5. P0100: PPL/CF/100×. Shows the transmission of red light by the blue particles in ultramarine violet. Same field of view as in Fig. 4.

6. P0072: PPL/100×. Shows anhedral angular pale pink to pale purple grains of ultramarine red.

7. P1093: PPL. Ultramarine green: a mixture of pale blue-green, blue and colourless anhedral particles.

8. P0201: PPL. Anhedral blue-green, blue and colourless grains of ultramarine green.

9. P0072: PPL. Anhedral angular pale pink crystals.

Crystallinity
Cubic

n	δ
<1.662	Isotropic (data from authors)

Ultramarine red is a synthetic pigment with a chemistry and crystal structure related to ultramarine (*q.v.*) and the sodalite group of the feldspathoid framework silicates. Under plane-polarised light, the pigment forms fine to medium grained particles with low relief, the RI slightly less than that of the medium. The particle habits are angular to subangular and irregular shards. In colour, they are pale pink and translucent. Particle surfaces appear slightly rough and pitted. Under crossed polars, ultramarine red is isotropic. The colour of ultramarine red is unaffected when viewed through the Chelsea filter.

In plane-polarised light, the ultramarine violet illustrated here appears to be formed from two similar phases. The first has a fine to very fine grain size and forms transparent, pale violet grains while the second has medium grained particles and is coloured an intense, deep blue. Both phases have irregular, angular grain shapes and low relief, with an RI less than that of the medium. Under crossed polars, ultramarine violet is isotropic. When viewed through the Chelsea filter, the blue grains present in ultramarine violet transmit red light.

Under plane-polarised light three phases can be observed as present in the samples of ultramarine green illustrated here. The majority is composed of transparent, pale blue-green, weakly coloured grains. These form subangular anhedral to subhedral polyhedral crystals, with apparently undulating grain surfaces. The second phase is most distinctive and forms deep blue translucent particles with anhedral morphologies. Finally, the third phase is composed of angular shards of colourless crystals. Grain size of all phases ranges from fine to medium. All have low to moderate relief, with RI less than that of the medium. Under crossed polars, ultramarine green is isotropic. However, a few of the colourless grains are weakly anisotropic and show low first order grey interference colours. This is presumably the effect of distortions of the crystal structure in this phase. When viewed through the Chelsea filter, the green phase transmits red light.

The Pigment Compendium

COPPER CARBONATE HYDROXIDE, AZURITE TYPE

$Cu_2CO_3(OH)_2$

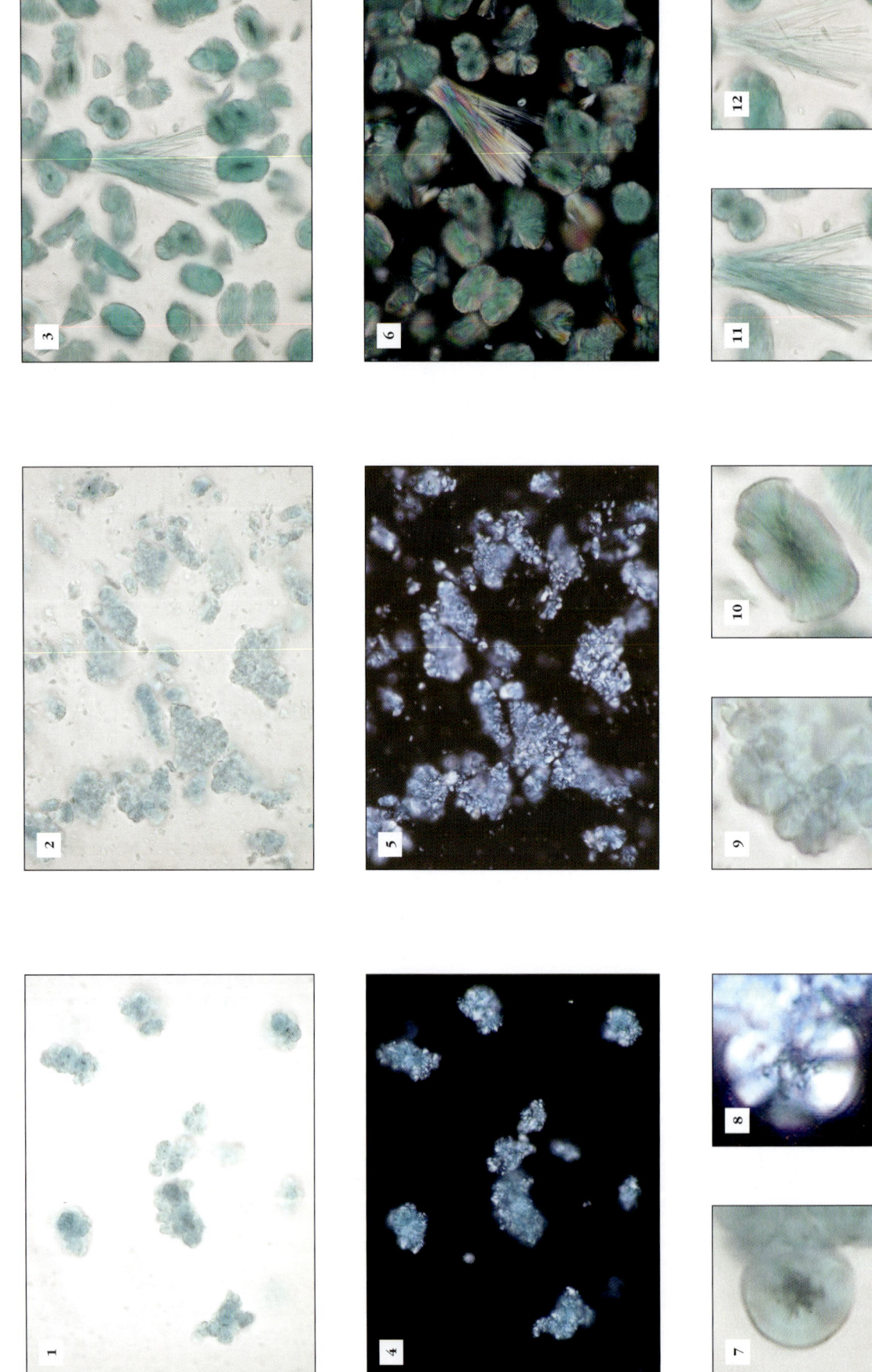

Key:

1. P1406: PPL/100×. Spherulitic particles of copper carbonate hydroxide, azurite type.

2. P0207: PPL/100×. Spherulitic particles of copper carbonate hydroxide, azurite type.

3. P1103: PPL/100×. An unusually well-formed example of copper carbonate hydroxide, azurite type.

4. P1406: XPL/100×/●. The field of view in Fig. 1 under crossed polars.

5. P0207: XPL/100×/●. The field of view in Fig. 2 under crossed polars.

6. P1103: XPL/100×/●. The field of view in Fig. 3 in crossed polars, rotated through 45°.

7. P1406: PPL. Detail of spherulitic particle.

8. P1406: XPL/●. The particle in Fig. 7 under crossed polars, note poorly defined extinction cross.

9. P0207: PPL. Detail of spherulite aggregate from Fig. 2.

10. P1103: PPL. Detail of spherulitic particle from the sample in Fig. 3.

11. P1103: PPL. Variolitic aggregate in Fig. 3.

12. P1103: PPL. The crystals in Fig. 11 with the polariser rotated through 90° to show change in body colour due to pleochroism.

Crystallinity	Optic sign	2V	n_α	n_β	n_γ	δ
Monoclinic	+ve	64°–68°	1.730	1.758	1.838	0.1080

(data for mineral from Mason, 1968; cf. Webmineral, 2003)

The synthetic blue copper carbonate hydroxide analogue ('blue verditer') of the mineral azurite appears under plane-polarised light as distinct from the latter in that it forms finer grained particles, typically 5–10 μm diameter, and has a characteristic habit. This compound forms spherulitic particles, aggregates of these and broken spherules, producing variolitic sheaves. Spherules may also show concentric rings of growth zones and characteristically have a dark central spot. Like azurite, the body colour is blue, though paler, and the pleochroism is less obvious due to the very fine, fibrous crystal habit. This copper carbonate hydroxide has moderate relief, with a RI greater than that of the medium. As a result of the orientation of crystallites within the spherules and the variation of RI exhibited by carbonates, a single RI is generally observed for the spherules. Mactaggart and Mactaggart (1980) note this to be 1.72, which is notably lower than that quoted above for azurite mineral.

Under crossed polars the compound is seen as strongly birefringent, though due to the fine particle size of the crystals it is second order interference colours that are typical. These are often masked by the body colour. The polycrystalline nature of the spherules is evident such that they appear

mottled at low magnification; a standing extinction cross is often clear at higher magnifications orientated such that the straight extinction of the crystals is clear. Individual fibres are length fast.

This pigment has low transmission at the red end of the spectrum such that it is unaffected when viewed through the Chelsea filter.

The most likely compound to be found associated with this pigment, and the one with which it is most likely to be confused, is the green malachite-structured analogue. A related copper zinc carbonate hydroxide, analogue of the mineral rosasite, is also known to have been used as a pigment (Dunkerton and Roy, 1996).

The optical properties of blue verditer in the context of pigments have been described by Gettens and FitzHugh (1993a) and Scott (2002).

Lit.: Dunkerton & Roy (1996); Gettens & FitzHugh (1993a); Mactaggart & Mactaggart (1980); Scott (2002)

COPPER CARBONATE HYDROXIDE, AZURITE TYPE

The Pigment Compendium

AZURITE

$Cu_2CO_3(OH)_2$

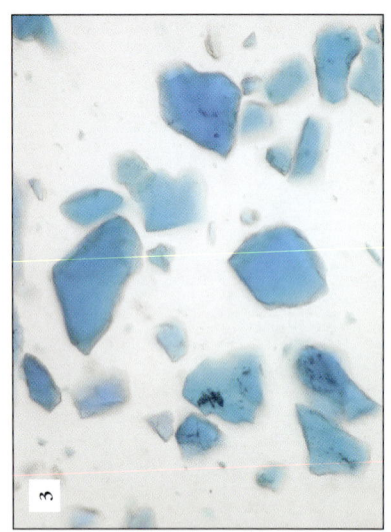

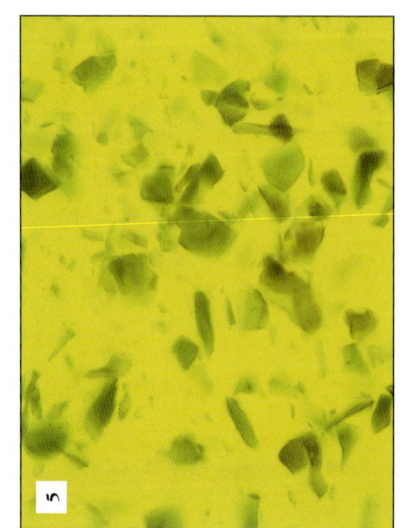

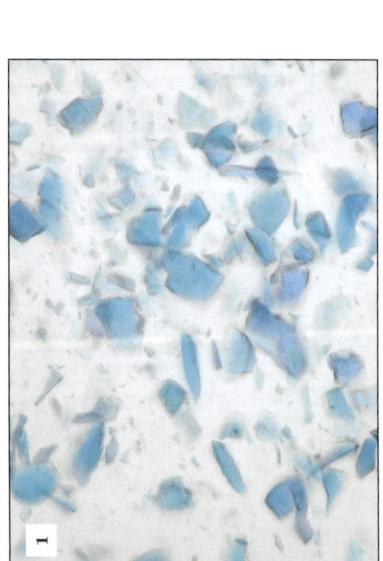

Key:

1. P0817: PPL/40×. Angular particles of azurite.

2. P0817: XPL/40× / ◒. The field of view in Fig. 1 under crossed polars.

3. P0818: PPL/40×. Coarse-grained particles of azurite, note the opaque inclusions of copper oxides to left of centre.

4. P0817: RPL/40×. The field of view in Fig. 1 in reflected light.

5. P0817: PPL/CF/40×. The field of view in Fig. 1 through the Chelsea filter.

6. P0818: XPL/40× / ◒. The field of view in Fig. 3 under crossed polars.

7. P0818: PPL. Coarse-grained particles of azurite.

8. P0818: PPL. The crystals in Fig. 7 with polariser rotated through 90° to show change in body colour due to pleochroism.

9. P1149: PPL. Crystal of azurite showing conchoidal fracture surfaces.

10. P1149: XPL/◒. The crystal in Fig. 9 under crossed polars.

Crystallinity	Optic sign	2V	n_α	n_β	n_γ	δ
Monoclinic	+ve	64°–68°	1.730	1.758	1.838	0.1080

(data from Mason, 1968; cf. Webmineral, 2003)

In plane-polarised light, azurite appears as translucent blue crystals, often with slight greenish overtones; larger particles will be deep blue, finer particles paler. Larger, strongly coloured particles exhibit pleochroism from blue to green-blue. It has moderate relief that varies slightly as the stage is rotated; RI is greater than the medium and dispersion is relatively weak. Crystals are typically angular shards produced by crushing. However, azurite may also occur as encrusting and botryoidal masses; therefore fibrous and earthy, granular masses may equally occur. Euhedral crystals have bladed to columnar habits. Particle size typically ranges from 5 to 40 μm, the coarser the particle size, the stronger the colour. Particle size has been studied in more detail by Asperen de Boer (1974) for some early Netherlandish paintings.

Under crossed polars, the high birefringence is masked by the body colour. However, characteristically, the particles have second to third order colours. Azurite has straight extinction.

Azurite may be found in association with malachite, cuprite and chalcopyrite. Samples that have been heated to temperatures exceeding 300°C revert to cuprite. Additionally, azurite has been known to alter to paratacamite, clinoatacamite and tenorite. Organo-copper complexes may form through reactions with organic media.

The appearance of azurite is unaffected when viewed through the Chelsea filter.

The optical properties of azurite in the contexts of a pigment have been described by Gettens and FitzHugh (1993a) and Scott (2002).

Lit.: Asperen de Boer (1974); Gettens & FitzHugh (1993a); Scott (2002)

The Pigment Compendium

COPPER CARBONATE HYDROXIDE, MALACHITE TYPE

$Cu_2CO_3(OH)_2$

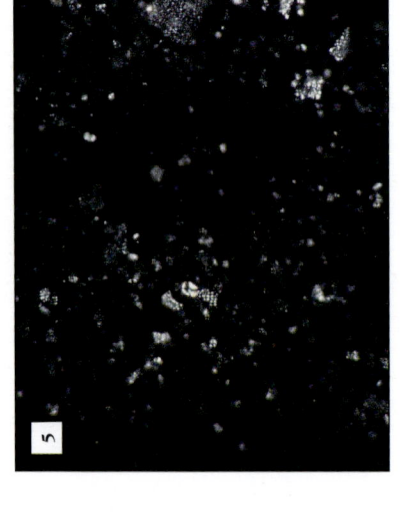

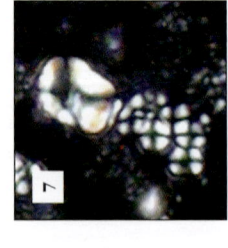

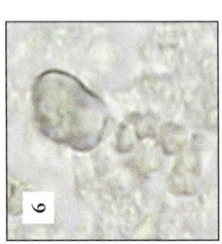

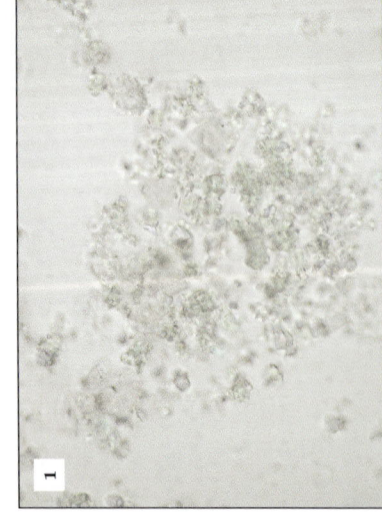

Key:

1. P0176: PPL/100×. Pale green spherulitic particles of copper carbonate hydroxide, malachite type.

2. P0176: XPL/100×/ ⊙ The field of view in Fig. 1 under crossed polars, note standing extinction crosses; the highly birefringent material is calcite.

3. P0176: XPL/100×/ ⊙ Copper carbonate hydroxide, malachite type under crossed polars with well-defined spherulitic particles; the highly birefringent material is calcite.

4. P0215: PPL/100×. Very pale green low relief particles of copper carbonate hydroxide, malachite type.

5. P0215: XPL/100×/ ⊙ The field of view in Fig. 4 under crossed polars.

6. P0215: PPL. Detail of spherulitic particles from Fig. 4.

7. P0215: XPL/ ⊙ The particles in Fig. 6 under crossed polars.

8. P0215: XPL ⊙ Aggregate of very fine spherulites.

Crystallinity	Optic sign	2V	n_α	n_β	n_γ	δ
Monoclinic	−ve	38°–43°	1.655	1.875	1.909	0.2540

(data for mineral from Mason, 1968; cf. Webmineral, 2003)

Under plane-polarised light, the copper carbonate hydroxide analogue of the mineral malachite (q.v.; 'green verditer' or 'spherical malachite') is distinct from the mineral itself in that it forms finer grained particles (typically 2–10 μm diameter) and it has a characteristic habit where it forms spherulitic particles, aggregates of these and broken spherules, producing variolitic sheaves. These spherules may also show concentric rings of growth zones and characteristically have a dark central spot. Mactaggart (2002) has observed in samples he examined that the general morphology was similar to the blue form (copper carbonate hydroxide, azurite type) but that the particle size was usually smaller and more misshapen. Like malachite the body colour is pale green, but the pleochroism is less obvious due to the very fine, fibrous crystal habit. The pigment has moderate to low relief, with RI greater than the medium.

Under crossed polars this pigment is strongly birefringent, though due to the fine particle size of the crystals, second and third order interference colours are typical, often masked by the body colour. The polycrystalline nature of the spherules is evident; they appear mottled at low magnification, with a standing extinction cross often clear at higher magnifications, orientated such that the straight extinction of the crystals is clear. The degree of crystallisation affects the sharpness of the extinction cross.

There is some confusion in the literature over the nomenclature of so-called 'green verditer' and 'spherical malachite'. Although Mactaggart has proposed that green verditer is distinct from spherical malachite on the basis of crystallite morphology (perhaps resulting from different manufacturing processes) it is otherwise difficult to differentiate these on the basis of chemistry and crystallography. Both are synthetic spherulitic preparations of copper carbonate hydroxide, malachite type. It is probable that Mactaggart's green verditer is cryptocrystalline. These morphologies are not differentiated by Gettens and FitzHugh (1993b).

The most likely compound to be found associated with this pigment, and the one with which it is most likely to be confused, is the blue azurite-structured analogue. A related copper zinc carbonate hydroxide, analogue of the mineral rosasite, is also known to have been used as a pigment (Dunkerton and Roy, 1996).

The optical properties of green verditer in the context of a pigment have been described by Gettens and FitzHugh (1993b) and Scott (2002).

Lit.: Dunkerton & Roy (1996); Gettens & FitzHugh (1993b); Mactaggart (2002); Scott (2002)

The Pigment Compendium

MALACHITE

$Cu_2CO_3(OH)_2$

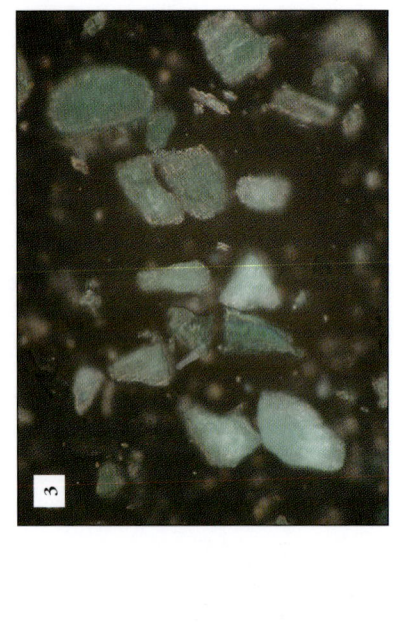

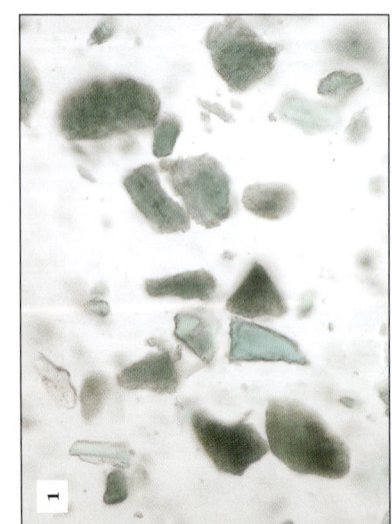

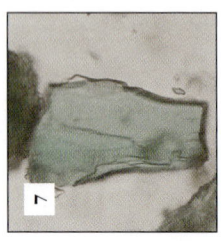

Key:

1. P0005: PPL/40×. Large clear crystals and fibrous aggregates of malachite.

2. P0005: UV/40×. The field of view in Fig. 1 under crossed polars.

3. P1226: PPL/40×. The field of view in Fig. 1 in reflected light.

4. P1226: PPL/40×. Fibrous aggregates of malachite.

5. P1226: XPL/40×/●. The field of view in Fig. 4 under crossed polars.

6. P0537: PPL/10×/H. Fine-grained malachite sample rich in impurities.

7. P0005: PPL/●. A translucent pale green single crystal of malachite.

8. P0005: PPL/●. The crystal in Fig. 7 rotated through 90° showing change in body colour due to pleochroism.

9. P0005: XPL/●. The crystal in Fig. 7 with polariser under crossed polars.

Crystallinity	Optic sign	2V	n_α	n_β	n_γ	δ
Monoclinic	–ve	38°–43°	1.655	1.875	1.909	0.2540

(data for mineral from Mason, 1968; cf. Webmineral, 2003)

In plane-polarised light, malachite is a very distinctive mineral. In colour it is green, with some particles having blue-green overtones; it also exhibits pleochroic colours ranging from pale to mid-green. Colour intensity decreases with particle size and therefore crushed malachite used as a pigment is typically coarse, although within a sample particle size distribution may be uneven. Crystal habit is typically encrusting, botryoidal masses, which appear as finely fibrous, variolitic aggregates when viewed under the microscope. These aggregates may also show banding. Growth zones may be apparent, forming normal to the long axes of the fibres. Malachite also has perfect cleavage. Euhedral single crystals of malachite are less common, but these may occur as platy to tabular forms. Additionally, earthy, granular masses occur. Relief is moderate and slightly variable as the stage is rotated. The RI is greater than that of the medium. Dispersion is considered to be weak.

Under crossed polars, malachite has very high birefringence, and shows high third order, fourth order and higher interference colours. This is only weakly masked by the body colour; anomalous blue colours may also be produced. Malachite has straight extinction; however, McCrone et al. (1973–80) note inclined extinction in some particles. Malachite is also known to develop twins (Gettens and FitzHugh, 1993b) although the typical fine acicular crystals observed are too small for this phenomenon to be commonly observed.

Malachite is often found in association with azurite (q.v.); Gettens and FitzHugh (1993b) also remark that chrysocolla and cuprite are commonly found with it.

The optical properties of malachite in the contexts of a pigment have been described by Gettens and FitzHugh (1993b) and Scott (2002).

Lit.: Gettens & FitzHugh (1993b); McCrone et al. (1973–80); Scott (2002)

The Pigment Compendium

CHALCONATRONITE

$Na_2Cu(CO_3)_2 \cdot 3H_2O$

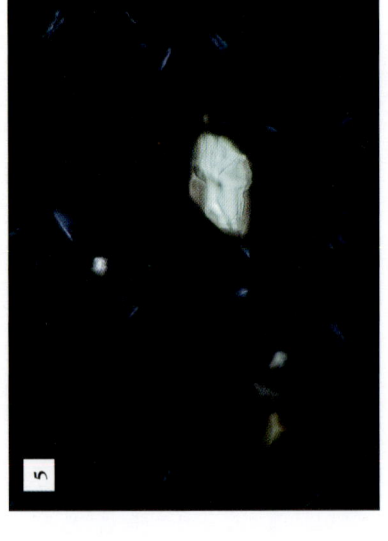

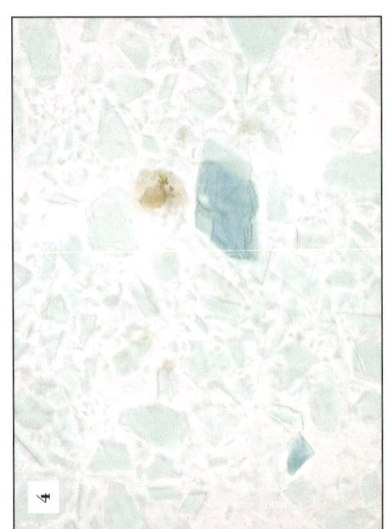

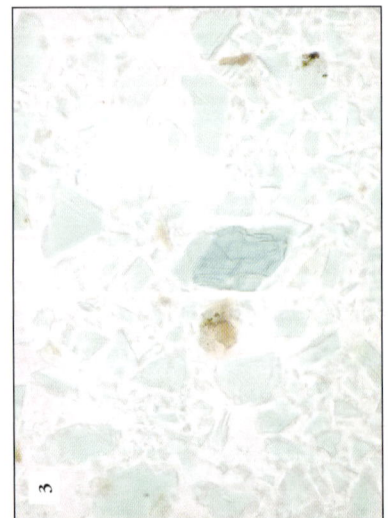

Key:

1. P1168: PPL/10×. Translucent sky blue particles of chalconatronite.

2. P1168: XPL/10×. The field of view in Fig. 1 under crossed polars.

3. P0303: PPL/100×. Anhedral fragments of chalconatronite, with a strongly coloured particle at the centre.

4. P0303: PPL/100×. The field of view in Fig. 3 rotated through 90° to show the intense pleochroism.

5. P0303: XPL/100×. ● The field of view in Fig. 4 under crossed polars.

Crystallinity	Optic sign	2V	n_α	n_β	n_γ	δ
Monoclinic	+ve	70°–86°	1.483	1.530	1.576	0.0930
					(data for mineral from Dana, 1997; cf. Webmineral, 2003)	

In plane-polarised light chalconatronite can be seen as forming transparent, pale blue crystals. Crystals are strongly pleochroic in some sections from pale blue to colourless. Relief is moderate and RI is less than that of the medium. Particle shape is of anhedral angular shards and plates, grain surfaces are smooth and glassy and show a weak conchoidal fracture; many grains have curved boundaries. Particle size ranges from fine to very coarse.

Under crossed polars, chalconatronite has moderate birefringence, with particles showing up to third order interference colours. Particle have complete and straight extinction.

The appearance of chalconatronite is unaffected when viewed through the Chelsea filter.

The Pigment Compendium

COPPER(II) HYDROXIDE

Cu(OH)$_2$

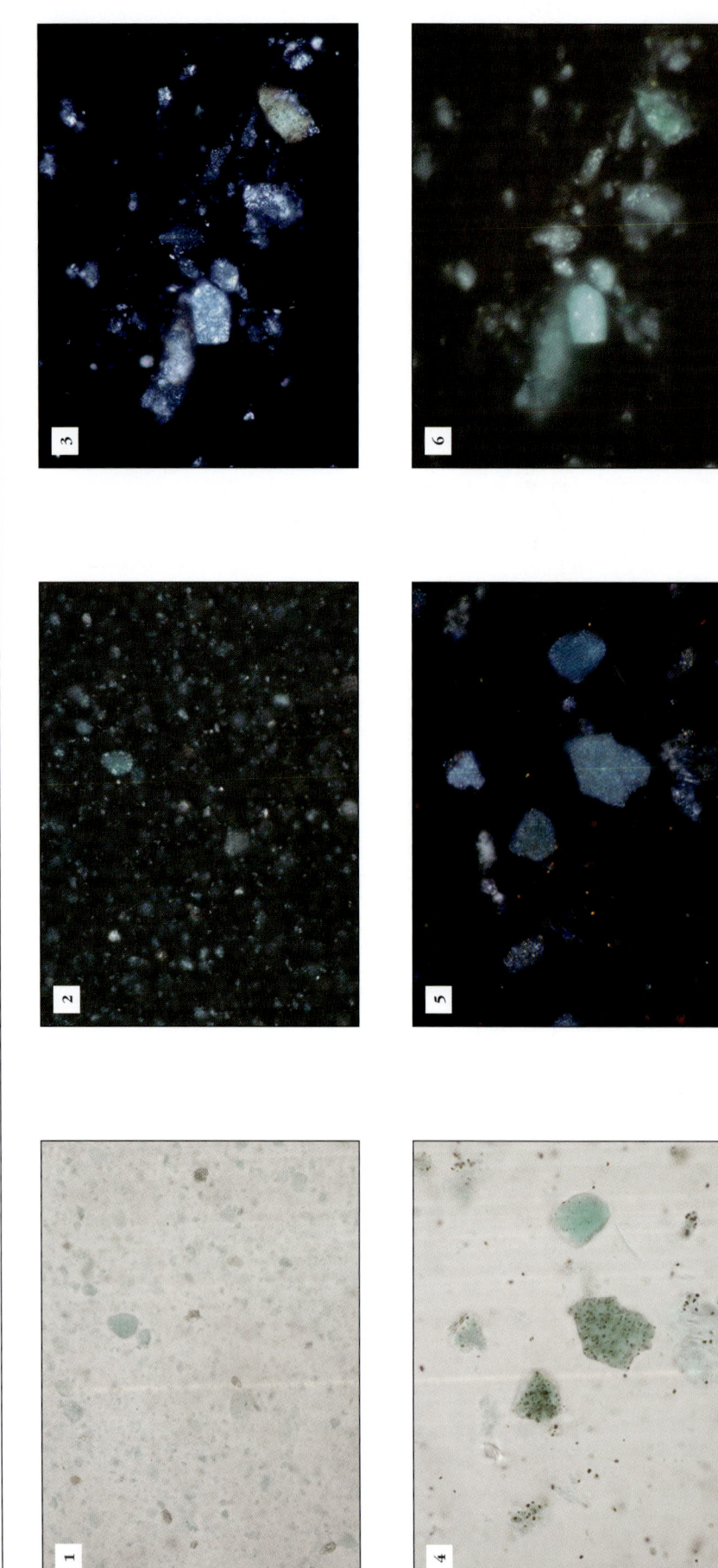

Key:

1. P0890: PPL/40× ⊙. Pale blue-green rounded particles with low relief.

2. P0890: XPL/40× ⊙. The field of view under Fig. 1 in crossed polars showing rounded particles with low birefringence masked by body colour.

3. P0311: XPL/100× ⊙. Interference colours masked by body colour and strong blue internal reflections.

4. P0299: PPL/100×. Particles with multiple inclusions.

5. P0299: XPL/100× ⊙. The field of view in Fig. 4 under crossed polars showing blue-green internal reflections and reddish inclusions in crossed polars.

6. P0311: RPL/100×. In reflected light the particles appear a strong blue-green colour. Same field of view as Fig. 3.

Crystallinity	Optic sign	2V	n_α	n_β	n_γ	δ
Orthorhombic [unknown]	[unknown]	[unknown]	1.72	[–]	[–]	[Low–moderate] (data from Scott, 2002)

Under plane-polarised light, copper(II) hydroxide, the synthetic analogue of the mineral spertiniite, appears as pale green-blue (the colour also observed in reflected light) particles with no pleochroism. The particles are rounded and have low relief, with the RI being just greater than that of the medium. The surface texture is pitted and rough.

Under crossed polars, the particles are shown to be finely polycrystalline, so that complete extinction does not occur, but the particles twinkle as the stage is rotated. Birefringence is moderate to low and masked by the body colour. Copper(II) hydroxide shows strong blue green internal reflections under cross-polarised light.

The compound appears a blue-green in reflected light.

Copper(II) hydroxide may form in some preparations of copper acetate hydroxide ('basic verdigris'). Scott (2002) also comments that it is generally a transitory phase and its long-term stability in a pigment context is therefore unknown.

The chemical properties of this compound have been discussed by Scott (2002).

Lit.: Scott (2002)

The Pigment Compendium

ATACAMITE AND COPPER CHLORIDE HYDROXIDE, ATACAMITE TYPE

$Cu_2Cl(OH)_3$

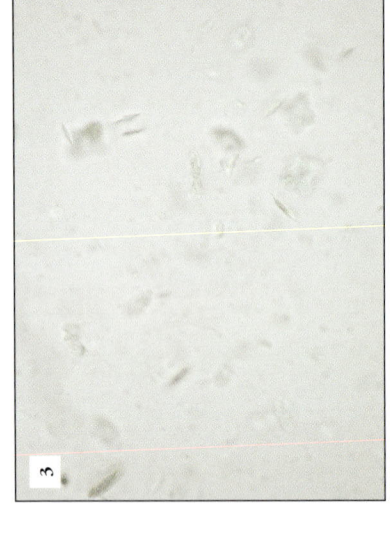

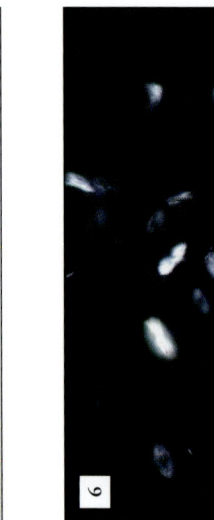

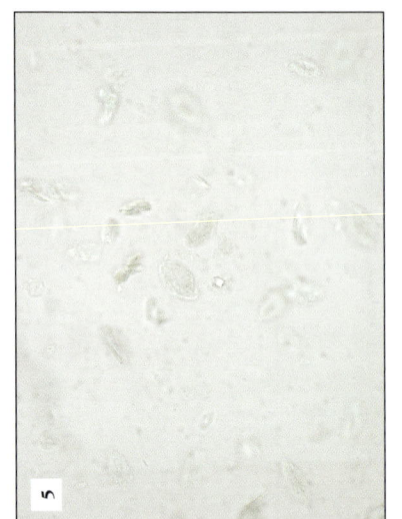

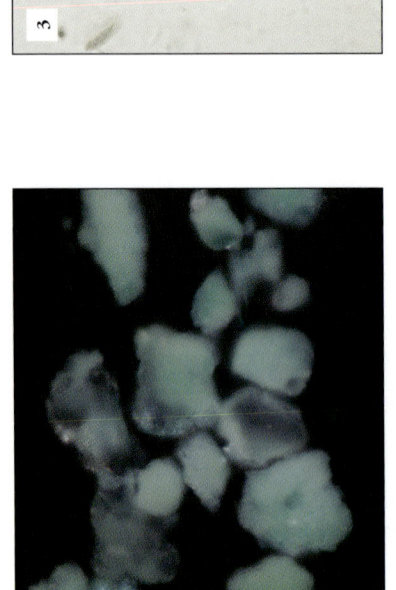

Key:
1. P0306: PPL/40×. Natural atacamite crystals and impurities.
2. P0306: XPL/40×. The field of view in Fig. 1 under crossed polars.
3. P0292: PPL/100×. Variolitic aggregates of fibrous crystals of copper chloride hydroxide.
4. P0300: PPL/100×/ **O**. Plates composed of fibrous mats of copper chloride hydroxide.
5. P0292: PPL/100×. Ovoid plates of copper chloride hydroxide, the fibrous aggregates are basic verdigris (A).
6. P0292: XPL/100×. The field of view in Fig. 5 under crossed polars.

Crystallinity	Optic sign	2V	n_α	n_β	n_γ	δ
Orthorhombic	−ve	74°	1.831	1.861	1.880	0.0490
					(data for mineral from Ford, 1932; cf. Webmineral, 2003)	

In plane-polarised light, atacamite forms translucent green crystals with high relief and RI greater than that of the medium. Dispersion is strong. For natural samples, particle shapes are typically fibrous, and occur in aggregates with radiating variolitic and botryoidal habits. Particle surfaces are striated and rough; as a crushed mineral, particle size varies from fine to very coarse. Naumova and Pisareva (1994) describe a number of examples that they believe to be synthetic in origin, noting in particular a 'circular' morphology that occurred in material prepared according to a recipe of Theophilus.

Under crossed polars, atacamite has moderate birefringence and second order interference colours are typical. Interference colours are masked by the body colour and therefore some particles appear bright green and occasionally anomalous blues are produced. Extinction is sweeping or twinkling. Fibrous particles can be seen to be length slow.

Synthetic atacamite may occur with copper acetate compounds (Naumova and Pisareva). Differentiation of atacamite from other copper chlorides may be difficult on optical grounds alone although the birefringence of this group varies significantly.

Lit.: Naumova & Pisareva (1994); Scott (2002)

The Pigment Compendium

BOTALLACKITE

$Cu_2Cl(OH)_3$

Key:

1. P1516: PPL/40×. Anhedral shards of botallackite showing range of colours due to pleochroism.

2. P1516: XPL/40×/ ○ . The view in Fig. 1 under crossed polars.

3. P1516: PPL/40×. Rhombic crystal of botallackite.

4. P1516: PPL/40×. The crystal in Fig. 3 rotated through 90° to show pleochroism.

5. P1516: XPL/40×/ ○ . The crystal in Fig. 3 under crossed polars rotated slightly for maximum illumination.

Crystallinity	Optic sign	2V	n_α	n_β	n_γ	δ
Monoclinic	+ve	[unknown]	1.775	1.800	1.846	0.0710
					(data for mineral from Webmineral, 2003)	

In plane-polarised light the mineral botallackite can be seen as translucent, pale blue-green crystals with low to moderate relief and an RI greater than that of the medium. Particles are weakly pleochroic. Particle morphology is exhibited as angular, bladed to prismatic grains. However, crystals are generally anhedral, with grain shape strongly influenced by the one perfect cleavage. Cleavage planes are very clear as striations on the crystal surfaces. Particle size distribution may be broad, varying from fine to very coarse.

Under crossed polars, botallackite has high birefringence with particles showing up to third order interference colours. Finer plates show first order, slightly anomalous, blue-greys.

The appearance of botallackite is unaffected when viewed through the Chelsea filter.

The Pigment Compendium

PARATACAMITE

$Cu_2Cl(OH)_3$

1

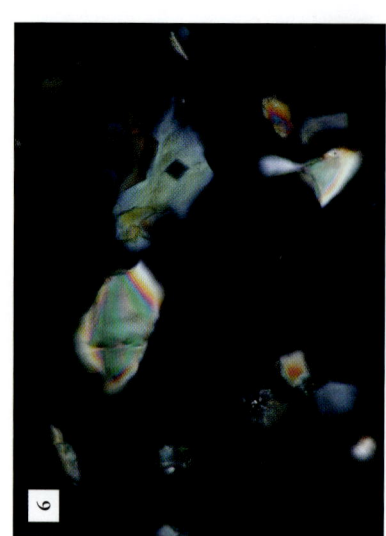

2

3

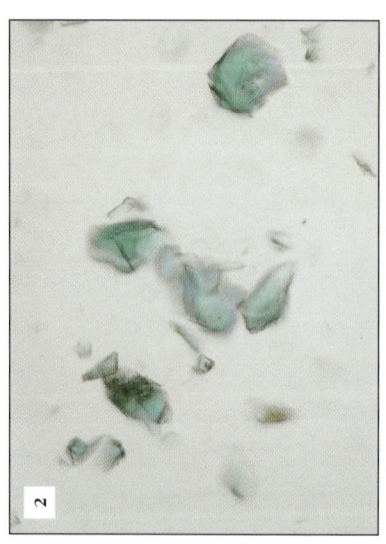

4

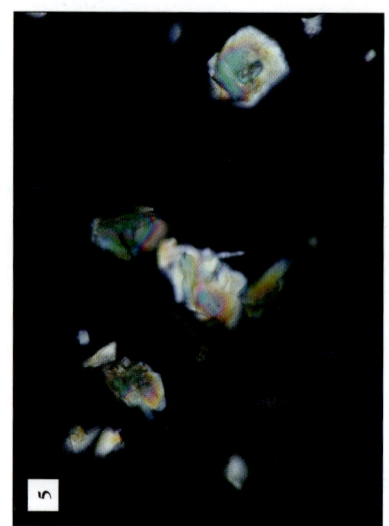

5

6

Key:

1. P1515: PPL/40×. Anhedral green crystals of paratacamite.

2. P1515: PPL/40×. Anhedral green crystals of paratacamite.

3. P1515: PPL/40×. Anhedral green crystals of paratacamite.

4. P1515: XPL/40×/ **o**. The field of view in Fig. 1 under crossed polars.

5. P1515: XPL/40×/ **o**. The field of view in Fig. 2 under crossed polars.

6. P1515: XPL/40×/ **o**. The field of view in Fig. 3 under crossed polars.

Crystallinity	Optic sign	n_ε	n_ω	δ
Trigonal	−ve	1.848	1.843	0.005

(Jambor *et al.*, 1996; cf. Webmineral 2003)

In plane-polarised light, paratacamite can be seen as translucent bright emerald green crystals. Relief is high and RI is greater than that of the medium. Particle shape when crushed is of regular, subangular particles, with uneven surface textures. Grain size distribution is broad from fine to very coarse, with colour intensity increasing with grain size.

Under crossed polars paratacamite has high birefringence; however, interference colours are anomalous bright greens and blues. Extinction is strongly sweeping to undulose.

The appearance of paratacamite is unaffected when viewed through the Chelsea filter.

PSEUDOMALACHITE $Cu_5(PO_4)_2(OH)_4$

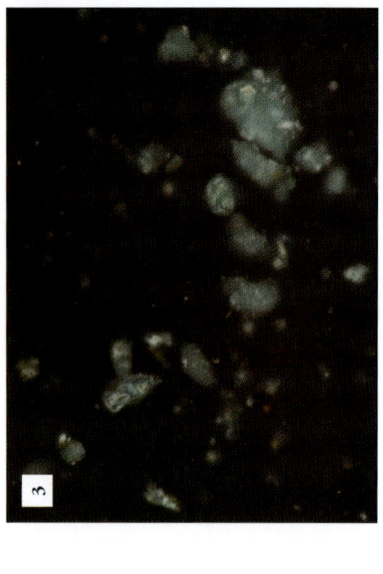

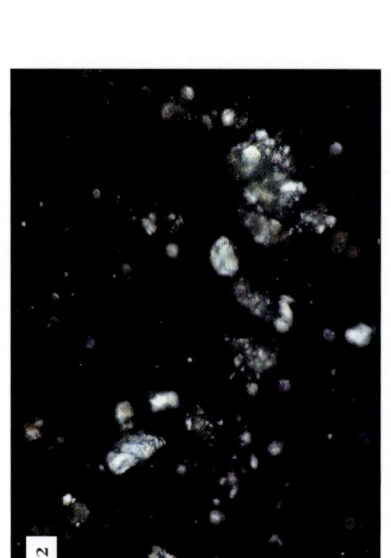

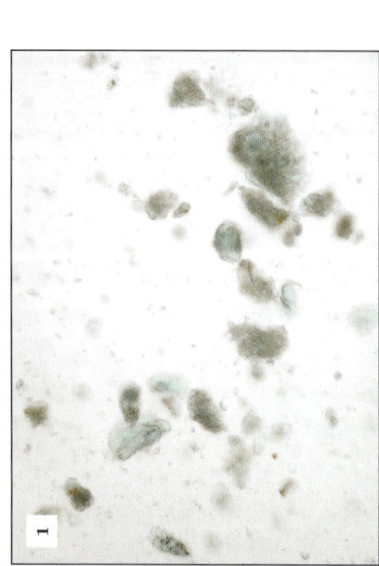

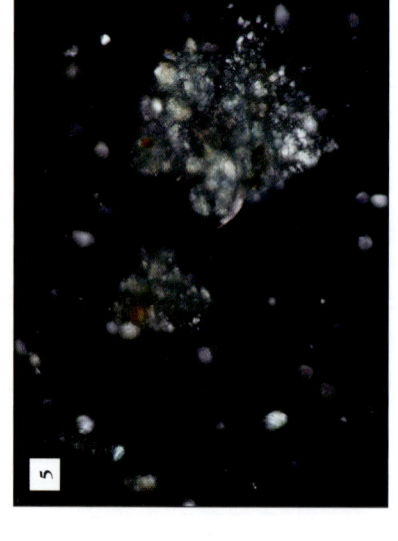

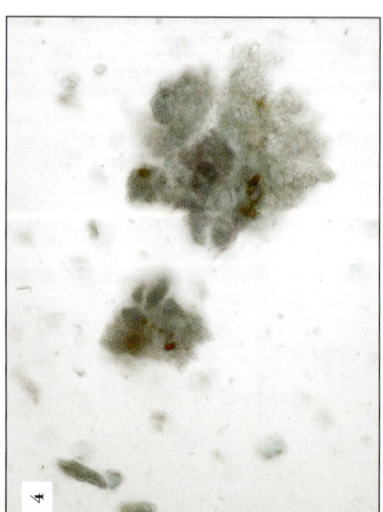

Key:

1. P1244: PPL/40×. Translucent green and cloudy particles of pseudomalachite.

2. P1244: XPL/40×/ ⊙. The field of view in Fig. 1 under crossed polars.

3. P1244: RPL/40×. The field of view in Fig. 1 in reflected light.

4. P1244: PPL/40×. Irregular polycrystalline but strongly coloured particles of pseudomalachite.

5. P1244: XPL/40×/ ⊙. The field of view in Fig. 4 under crossed polars.

Crystallinity	Optic sign	2V	n_α	n_β	n_γ	δ
Monoclinic	−ve	42°–48°	1.791	1.856	1.867	0.0760

(data from Webmineral, 2003)

In plane-polarised light, pseudomalachite forms translucent, cloudy green crystals, with moderate relief and RI greater than that of the medium. Crystals form fibrous and granular masses; grain surfaces are rough and uneven. Particle size varies from fine to coarse, with a broad size distribution.

Under crossed polars, pseudomalachite has moderate birefringence and high first order interference colours are typical. These may be masked by the body colour. The grains are clearly polycrystalline in crossed polars and extinction is sweeping and mottled.

The Pigment Compendium

ANTLERITE

$Cu_3(SO_4)(OH)_4$

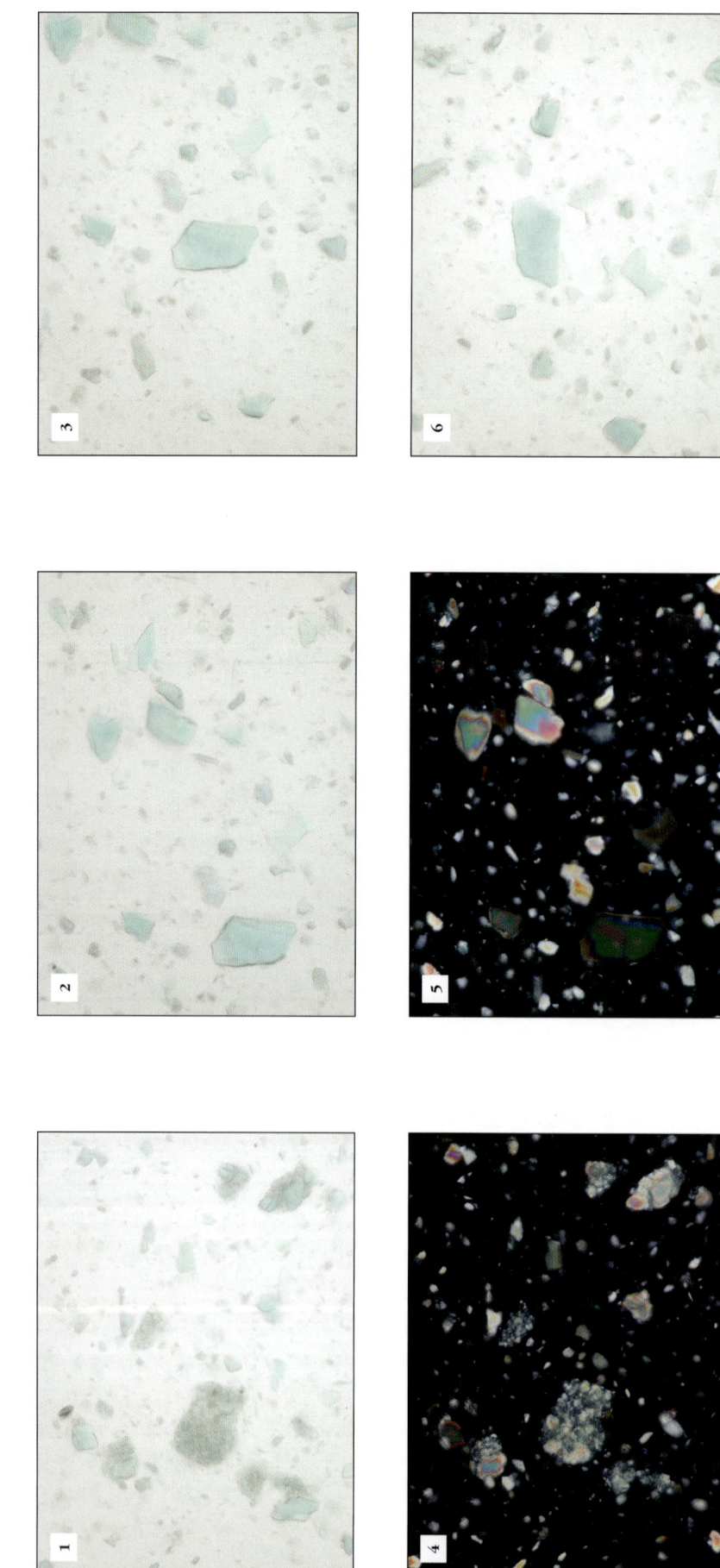

Key:
1. P1145: PPL/40×. Translucent clear and cloudy particles of antlerite.

2. P1145: PPL/40×. Translucent sky blue crystals of antlerite.

3. P1145: PPL/40×. Centre, large crystal of antlerite.

4. P1145: XPL/40×/ ○. The field of view in Fig. 1 under crossed polars.

5. P1145: XPL/40×/ ○. The field of view in Fig. 2 under crossed polars.

6. P1145: PPL/40×. The crystal in Fig. 3 rotated through 90° through show pleochroism.

Crystallinity	Optic sign	2V	n_α	n_β	n_γ	δ
Orthorhombic	+ve	53°–54°	1.726	1.738	1.789	0.0630
					(data from Ford, 1932; cf. Webmineral, 2003)	

In plane-polarised light, the copper sulfate hydroxide antlerite forms translucent to transparent pale blue-green crystals, which are weakly pleochroic from blue-green to yellow-green. The dispersion is relatively strong. Particle shapes are of angular, anhedral shards, while particle surfaces are glassy. Particle size ranges from fine to very coarse.

Under crossed polars, antlerite has high birefringence, with particles showing up to high third order interference colours. Finer grained particles show first order interference colours. Extinction is complete and straight.

Copper sulfates and their natural analogues have been described by Scott (2002).

Lit.: Scott (2002)

COPPER SULFATE HYDRATE, BROCHANTITE TYPE

$Cu_4(SO_4)(OH)_6$

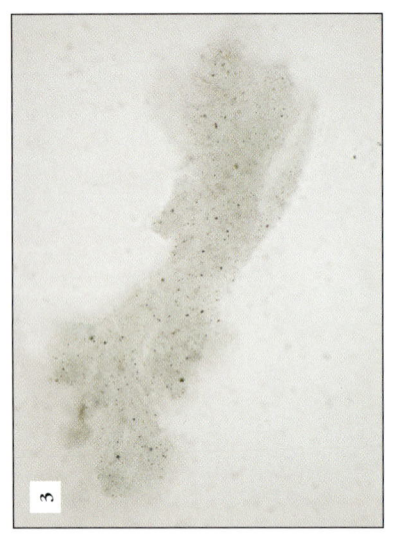

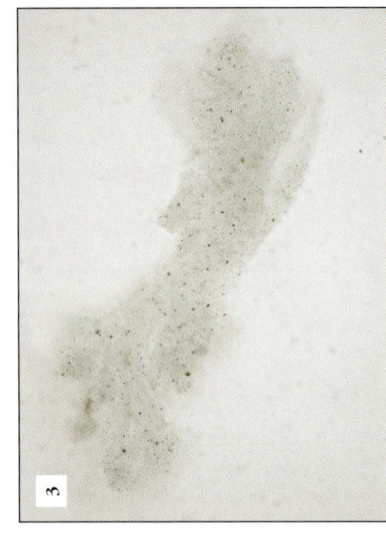

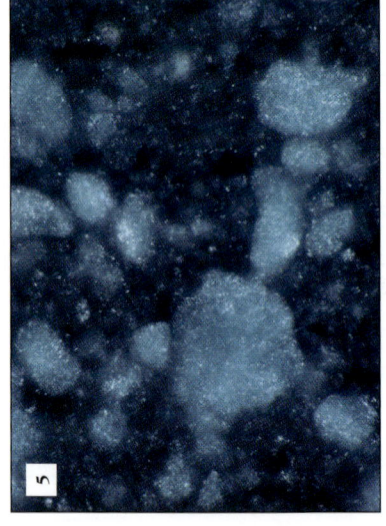

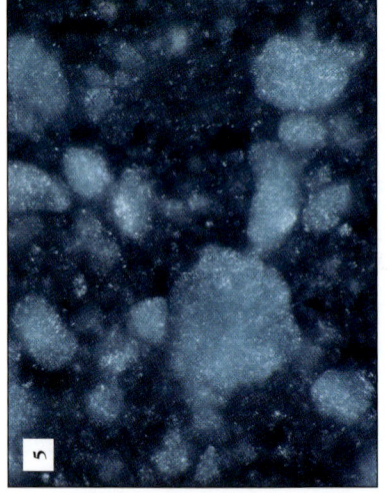

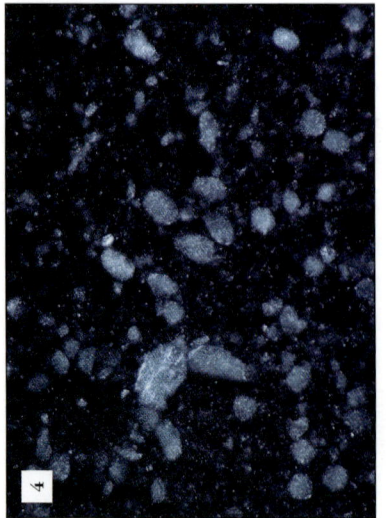

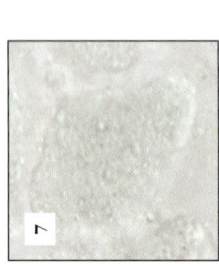

Key:

1. P0315: PPL/40×. Polycrystalline particles of copper sulfate hydrate.

2. P0315: PPL/100×. Polycrystalline particles of copper sulfate hydrate.

3. P0281: PPL/100×. Irregular particle of copper sulfate hydrate with opaque inclusions.

4. P0315: XPL/40×/ ○. The field of view in Fig. 1 under crossed polars.

5. P0315: XPL/100×. The field of view in Fig. 2 under crossed polars.

6. P0281: XPL/100×. The field of view in Fig. 3 under crossed polars.

7. P0315: PPL. Detail of particle in Fig. 2.

Crystallinity	Optic sign	2V	n_α	n_β	n_γ	δ
Monoclinic	−ve	72°–76°	1.728	1.771	1.800	0.0720
					(data for mineral from Mason, 1968; cf. Webmineral, 2003)	

Under plane-polarised light the synthetic copper sulfate hydrate analogue of the mineral brochantite (*q.v.*) ranges in shape from anhedral, subrounded to rounded particles, to almost spherical or elongated ovals. Relief is low, with the RI slightly greater than that of the medium; dispersion is considered weak. The surfaces of particles are rough and pitted, giving a finely speckled overall texture. Copper sulfate hydrate is weakly coloured, particles appearing as a very pale green. Particle size in samples examined was strongly uneven, with medium to coarse particles as well as very fine particles present.

Under crossed polars, the particles are demonstrated to be finely polycrystalline. The pigment is also birefringent, with moderate interference colours. Total extinction cannot be achieved and the particles appear to twinkle as the stage is rotated.

Copper sulfates and their natural analogues have been described by Scott (2002).

Lit.: Scott (2002)

BROCHANTITE

$Cu_4(SO_4)(OH)_6$

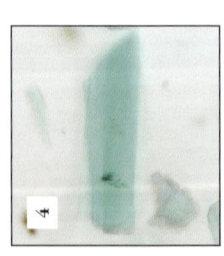

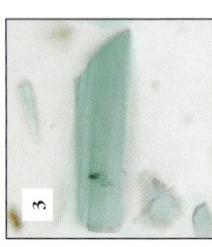

Key:

1. P1513: PPL/10×. Bladed and anhedral crystals of brochantite. Colour intensity increases with crystal size.

2. P1513: XPL/10×/ ⊙. The field of view in Fig. 1 under crossed polars.

3. P1513: PPL. Bladed crystal of brochantite showing prism-parallel cleavage.

4. P1513: PPL. The crystal in Fig. 3 with polariser rotated through 90° showing pleochroism.

Crystallinity	**Optic sign**	**2V**	n_α	n_β	n_γ	δ
Monoclinic	−ve	72°–76°	1.728	1.771	1.800	0.0720
					(data from Mason, 1968; cf. Webmineral, 2003)	

In plane-polarised light, brochantite is visible as translucent, blue-green crystals with low to moderate relief and an RI greater than that of the medium. Particles are weakly pleochroic and dispersion is relatively weak. Crystal shape is strongly influenced by the perfect cleavage, leading to bladed to tabular and platy crystal forms. Particle surfaces are smooth, but cleavage is visible as weak striations. Particle size distribution is broad and particle size ranges from fine to coarse.

Under crossed polars, brochantite has high birefringence with crystals showing up to third or fourth order interference colours. Fine or thin grains show high first order interference colours. Strongly coloured particles have masked interference colours. Extinction is straight and complete.

The appearance of brochantite is unaffected when viewed through the Chelsea filter.

The Pigment Compendium

COPPER ARSENITE GROUP

[Various]

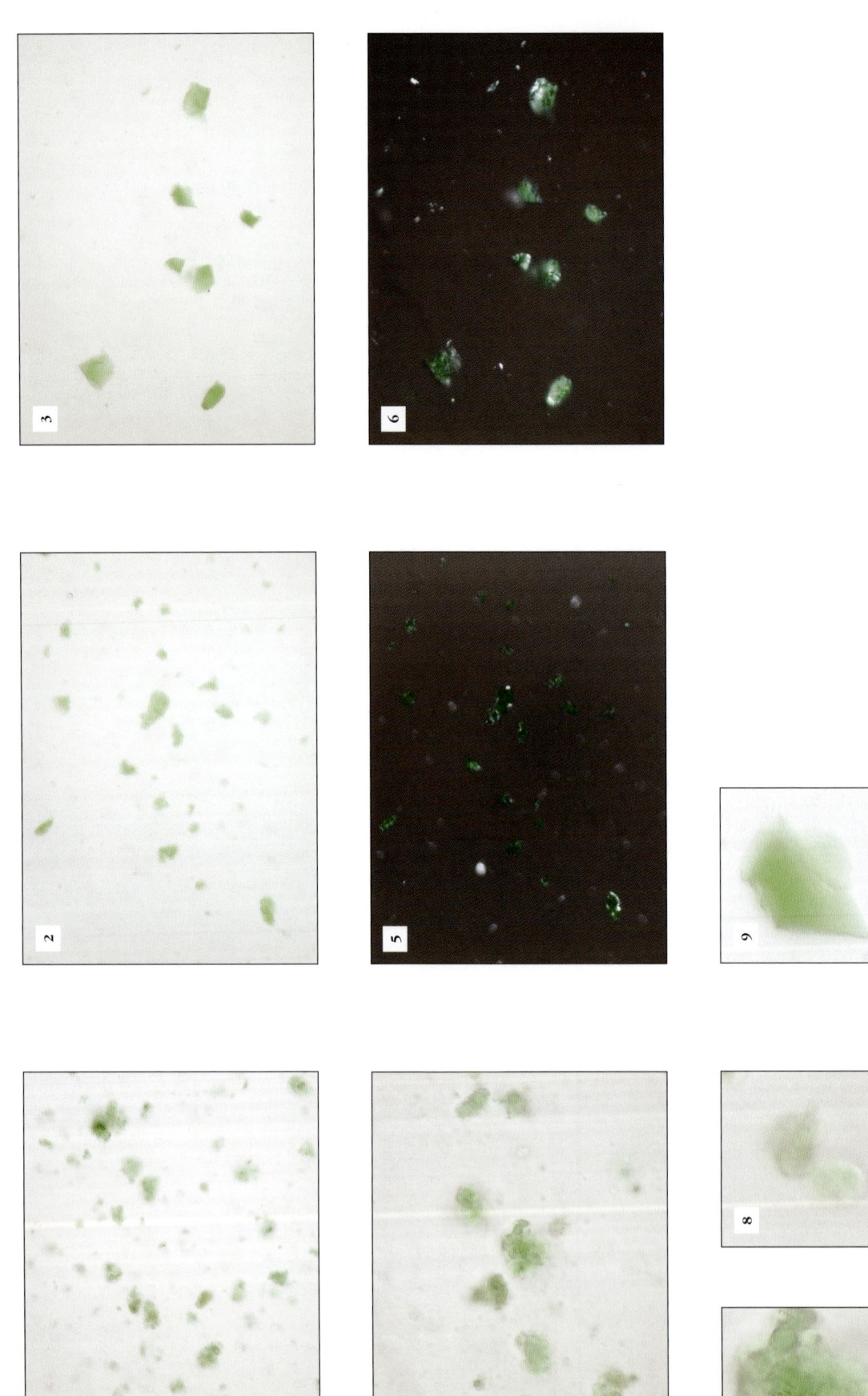

1. P0534: PPL/40×/H. Spherulitic and platy particles of copper arsenite.

2. P0532: PPL/40×/H. Spherulitic and platy particles of copper arsenite.

3. P0535: PPL/40×/H. Platy crystals of copper arsenite.

4. P0534: PPL/100×/H. Spherulitic particles of copper arsenite.

5. P0532: ~XPL/40×/H. The field of view in Fig. 2 under crossed polars.

6. P0535: XPL/40×/H. The field of view in Fig. 3 under crossed polars.

7. P0534: PPL/H. Enlargement of spherulitic aggregate from Fig. 4.

8. P0532: PPL/H. Detail of particle morphology of copper arsenite plates.

9. P0535: PPL/H. Detail of particle morphology from Fig. 3.

Crystallinity
Amorphous/Crystalline

n
1.550–1.749

δ
Isotropic/Anisotropic
(data from Fiedler and Bayard, 1997)

The term 'copper arsenite' covers a group of some ten compounds generally associated with the historical pigment term *Scheele's green* and it is thought that that pigment is a mixture of several of these compounds (Fiedler and Bayard, 1997). A number of examples were examined by the authors; these were established to be largely amorphous in nature by X-ray diffraction. However, Fiedler and Bayard have noted other varieties in which the material is more explicitly birefringent.

Under plane-polarised light the examples of copper arsenite examined by the authors appeared as pale translucent yellow-green particles; this, though, can be a deeper green in other material. The pigment has moderate to low relief and an RI that may be either less than or greater than that of the mounting medium. Particles have anhedral, subangular flake-like shapes with surfaces that are rough and pitted, while particle size is variable but usually medium to coarse. Fiedler and Bayard differentiated two phases within their samples: first, larger flakes with irregular outlines; second, very fine equant to needle-shaped crystals and rounded polycrystalline nodules.

Under crossed polars, the flakes appear isotropic (amorphous phase) or weakly birefringent (crystalline phase). In the latter case, first order grey interference colours are typical. Particles are observed to be polycrystalline and therefore do not undergo complete extinction. Birefringent samples studied by Fiedler and Bayard contained 'very small, equant to needle-shaped [crystals and] rounded polycrystalline nodules'; they further note that the birefringence was moderate and extinction of the needles parallel.

Samples may well be found with copper acetate arsenite ('emerald green'). There is also a significant resemblance to the organo-copper complexes (copper 'resinates', 'oleates' and 'proteinates') from which the arsenite may prove difficult to differentiate conclusively without other analyses.

Copper arsenites and copper acetate arsenite, known generically as Scheele's green and emerald green respectively, have been discussed by Fiedler and Bayard as well as Scott (2002).

Lit.: Fiedler & Bayard (1997); Scott (2002) 307–310

COPPER ACETATE ARSENITE

$Cu_4(OAc)_2(AsO_2)_2)_6$

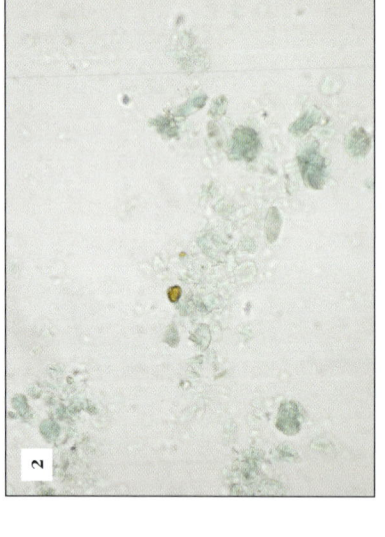

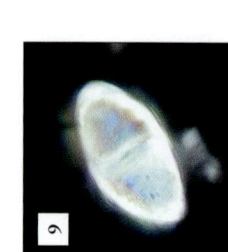

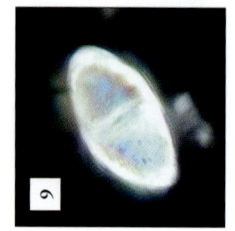

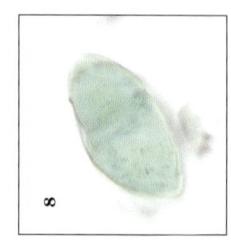

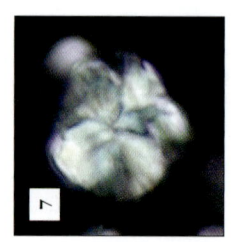

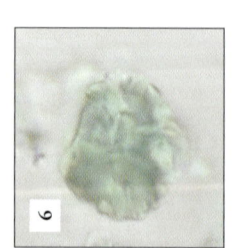

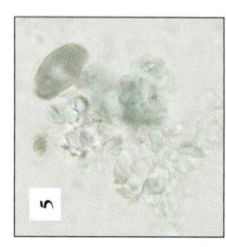

Key:

1. P0217: PPL/100×. Spherulitic and platy particles of copper acetate arsenite.

2. P0214: PPL/100×. Spherulitic and platy particles of copper acetate arsenite with yellow impurities.

3. P0217: XPL/100×. The field of view in Fig. 1 under crossed polars.

4. P0214: XPL/100×/ ⊙. The field of view in Fig. 2 under crossed polars.

5. P0214: XPL/ ⊙. Platy crystals of copper acetate arsenite.

6. P0214: PPL. Spherulite of copper acetate arsenite.

7. P0214: XPL/ ⊙. The spherulite in Fig. 6 under crossed polars.

8. P1302: PPL/H. Plate of copper acetate arsenite.

9. P1302: XPL/ ⊙/H. The crystal in Fig. 8 under crossed polars.

10. P1302: PPL/H. Barrel-shaped crystal of copper acetate arsenite.

Crystallinity Monoclinic	Optic sign [unknown]	2V [unknown]	n_α 1.71–1.72	n_β 1.77–1.78	n_γ	δ 0.06–0.07

(data from Fiedler and Bayard, 1997)

In plane-polarised light, copper acetate arsenite ('emerald green') forms pale green particles with low relief and an RI greater than that of the medium. Weak pleochroism is also observable, from bright blue-green to pale light green. Particle size was fine to very fine in samples examined by the authors, although Fiedler and Bayard (1997) note that individual crystal plates are from 2 to 20 μm wide and of submicron thickness while single spherulites have diameters of 8 to 40 μm. Particle habit is variable, and has been discussed in some detail by Sattler (1888) and Fiedler and Bayard (1997). These may include (rare) individual crystals with euhedral acicular or platy habits, the latter exhibiting rectangular, barrel-shaped (in outline) or oval forms; such crystals may also show twinning. Otherwise, aggregates of plates or needles are more common and these appear as spherulites and spherulite aggregates. Broken spherules are also common. More complex forms are also recognised, but require electron microscopy for identification. All crystal forms may be found in the same sample. These morphologies may permit differentiation of manufacturing process; it is thought that the presence of well-formed plates or prisms indicates crystallisation in undisturbed conditions, the presence of arsenic(III) oxide suggests copper acetate as a precursor and the size of the spherulites is related to the rate of crystal growth, larger spherulites having had a slower rate (Fiedler and Bayard).

Under crossed polars, copper acetate arsenite has low birefringence, with first order greys to second order reds and yellow interference colours, the latter only in the largest particles. Some particles show anomalous blues. The crystal form of the spherulites is not of regular needles and so standing extinction crosses are not typically observed. However, particles do appear finely polycrystalline, with complete extinction not achieved. Individual platy and acicular crystals normally have straight, complete extinction although Fiedler and Bayard note that undulose, irregular or inclined extinction can also be observed. Acicular particles are length slow.

Samples of the pigment may be found in conjunction with copper arsenites (q.v.; 'Scheele's green'). Fiedler and Bayard add that the pigment was commonly also extended with barium sulfate, calcium sulfate and chromium pigments. Excess of raw materials during the manufacturing process could also lead to the presence of arsenic(III) oxide, clays, lead sulfate, calcium carbonate, copper carbonate, magnesium carbonate and aluminium oxide.

Copper arsenites and copper acetate arsenite, known generically as Scheele's green and emerald green respectively, have been discussed by Fiedler and Bayard as well as Scott (2002).

Lit.: Fiedler & Bayard (1997); Sattler (1888); Scott (2002) 310–314

The Pigment Compendium

COPPER(II) ACETATE A

[Cu(CH$_3$COO)$_2$]$_2$·Cu(OH)$_2$·5H$_2$O

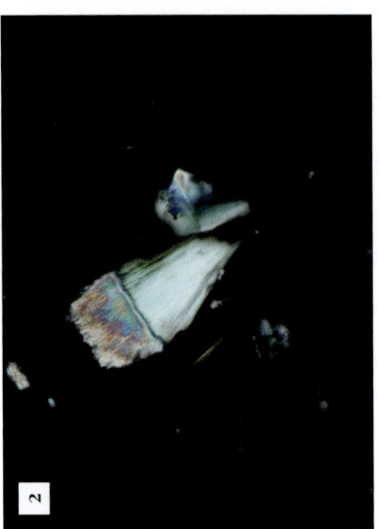

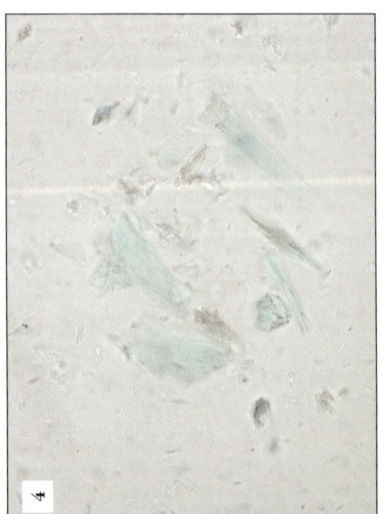

Key:

1. P0301: PPL/40×. Classic 'paintbrush' morphology of copper(II) acetate A fibres.

2. P0301: XPL/40×. The same view as in Fig. 1 under crossed polars.

3. P0301: RPL/40×. The same view as in Fig. 1 in reflected light.

4. P0301: PPL/40×. Pale blue fibrous aggregates of copper(II) acetate A.

5. P0301: XPL/40×. The same field of view as in Fig. 4 under crossed polars.

6. P0515: PPL/40×/H. Fibres of copper(II) acetate A.

Crystallinity

	n	δ
Anisotropic	<1.662	High
		(data from authors)

Under plane-polarised light, copper(II) acetate hydroxide hydrate ('copper(II) acetate A') appears as finely fibrous crystals. Typically these form lamellar aggregates of fibrous crystals, sometimes with a radiating, variolitic habit that often resemble the form of a paintbrush. In colour they are transparent and pale blue-green, or even colourless. The crystals are weakly pleochroic, from almost colourless to blue green. Relief is moderate to low with RI less than that of the medium. Particle size may be coarse where the brush-like aggregates are prevalent, or very fine in the case of individual fibrous or acicular crystals.

Under crossed polars, the compound is strongly birefringent. It is noticeable that in the variolitic aggregates the 'brush tips' are far more intensely birefringent, with high third order colours, than the main part of the brush, which shows slightly anomalous blue-grey colours. Individual fine crystals show high birefringence. The fibres have inclined extinction with a low

extinction angle. For fibrous aggregates extinction is sweeping; fibres are length fast.

With reflected light, particles of this compound are seen to be white to blue in colour.

Copper(II) acetate ('basic verdigris') may be any one, or a mixture, of several compounds. Copper(II) acetate A is also often found in association with copper(II) acetate hydrate F ('neutral verdigris').

The chemistry and optical properties of the verdigris compounds have been reviewed by Kühn (1993b) and Scott (2002).

Lit.: Kühn (1993b); Scott (2002)

The Pigment Compendium

COPPER(II) ACETATE B

Cu(CH₃COO)₂·Cu(OH)₂·5H₂O

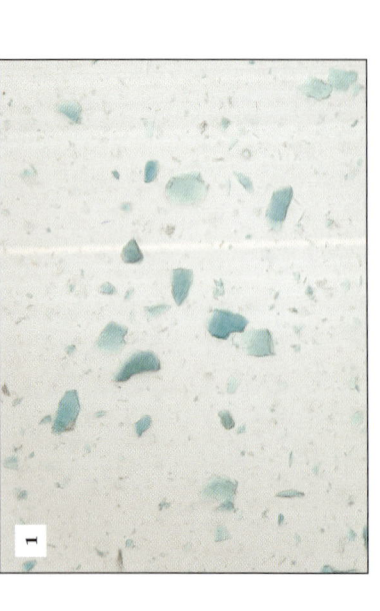

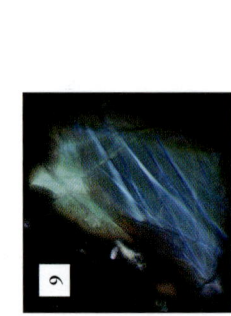

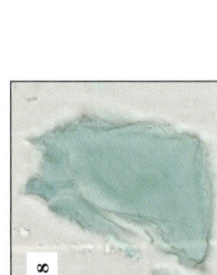

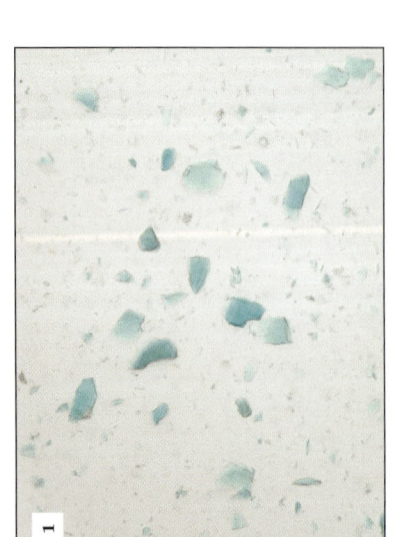

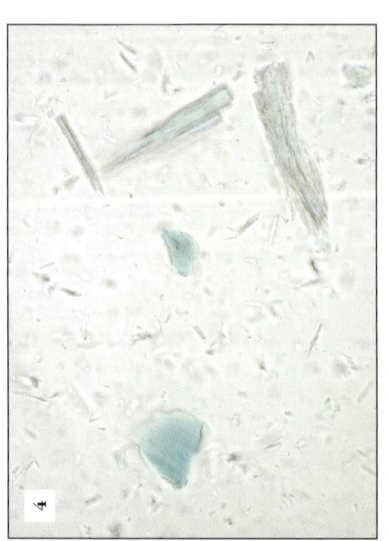

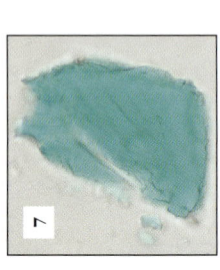

Key:

1. P0278: PPL/40×. Mixture of copper(II) acetate B with copper(II) acetate A and F.

2. P0278: PPL/100×. Crystals and fibrous aggregates of copper(II) acetate B, showing range of body colour due to pleochroism.

3. P0278: XPL/100×/ O. The same field of view as in Fig. 2 under crossed polars.

4. P0280: PPL/100×. Fibrous aggregates of copper(II) acetate B.

5. P0280: XPL/100×. The same field of view as in Fig. 4 under crossed polars.

6. P0280: RPL/100×. The same field of view as in Fig. 4 in reflected light.

7. P0278: PPL. Particle of copper(II) acetate B.

8. P0278: PPL. The same particle as in Fig. 7 with polariser rotated through 90° to show pleochroism.

9. P0278: XPL. The crystal in Fig. 7 under crossed polars.

Crystallinity
Anisotropic

n	δ
1.548 and 1.53	High
	(data from Scott, 2002)

Under plane-polarised light, copper(II) acetate B ('basic verdigris') may be seen as forming two distinct habits, either aggregates of fibrous crystals or medium- to coarse-grained fragments of single crystals. It is strongly coloured, with translucent, pleochroic, intense blue-green to pale green crystals. It has moderate relief, with the RI less than that of the medium.

Under crossed polars the fibrous nature of the bulk of the crystals is evident. It has high birefringence, though typically anomalous green and straw yellow interference colours are produced. Extinction is inclined and sweeping across the particles. Fibrous particles are length slow.

Particles appear a blue-green colour in reflected light.

The chemistry and optical properties of the verdigris compounds have been reviewed by Scott (2002) and Kühn (1993b).

Lit.: Kühn (1993b); Scott (2002)

The Pigment Compendium

COPPER(II) ACETATE D

$Cu(CH_3COO)_2 \cdot [Cu(OH)_2]_3 \cdot 2H_2O$

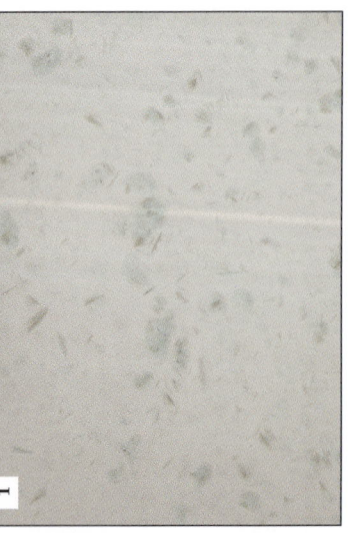

1. P0283: PPL/40×. Needles and radiating aggregates of copper(II) acetate D.

3. P0283: PPL/100×. Needles and radiating aggregates of copper(II) acetate D.

2. P0283: XPL/40×./ ○. The same view as in Fig. 1 under crossed polars.

4. P0283: PPL/100×. The same field of view as in Fig. 3 rotated through 90° to show weak pleochroism.

5. P0283: XPL/100×./ ○. The same field of view as in Fig. 4 under crossed polars.

Crystallinity
Anisotropic

n
<1.662

δ
Moderate to High
(data from authors)

In plane-polarised light, copper(II) acetate D ('basic verdigris') forms crystals with low relief and an RI slightly less than that of the medium. The crystals are transparent and very pale green in colour and are weakly pleochroic to almost colourless. The crystals are acicular and form aggregates of radiating particles and lamellar sheaves. Particle size among samples examined was medium to very fine.

Under crossed polars, the crystals have moderate to high birefringence, although the interference colours produced are anomalous, with pale and dark blues and browns typical. Extinction is sweeping across the particle aggregates and weakly inclined. Spherulitic forms may demonstrate poorly defined standing extinction crosses. Crystals are length slow.

Particles are a blue-white colour in reflected light.

The chemistry and optical properties of the verdigris compounds have been reviewed by Scott (2002) and Kühn (1993b).

Lit.: Kühn (1993b); Scott (2002)

The Pigment Compendium

COPPER(II) ACETATE F

$Cu(CH_3COO)_2.H_2O$

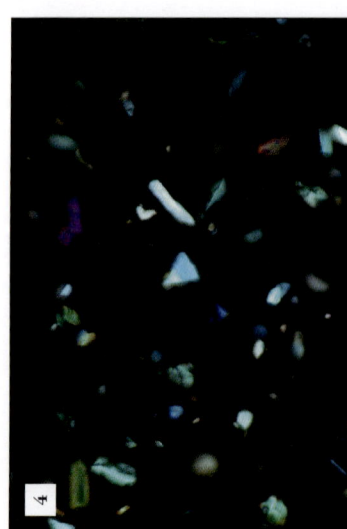

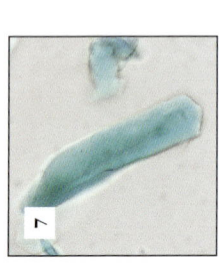

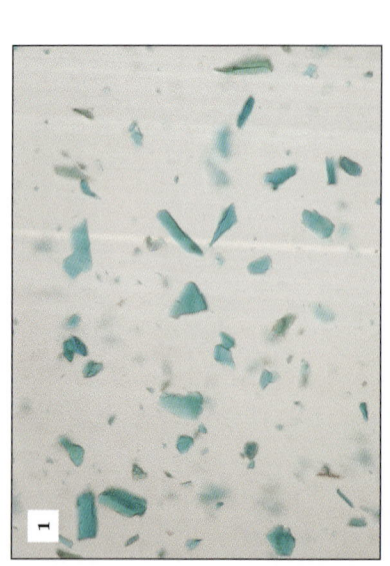

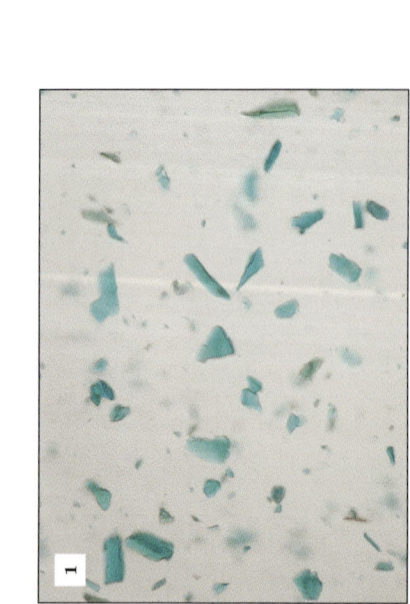

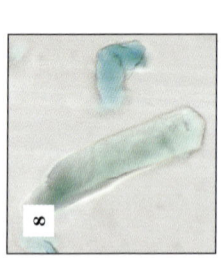

Key:

1. P0096: PPL/40×. General view showing platy to tabular, angular subhedral to anhedral shards of copper(II) acetate F.

2. P0307: PPL/100×. Similar morphology to Fig. 1.

3. P0312: PPL/100×. Similar morphology to Fig. 1.

4. P0096: XPL/40×. Same field of view as Fig. 1 showing anomalous interference colours.

5. P0307: XPL/100×. Anomalous brown interference colours in crossed polars. Same field of view as Fig. 2.

6. P0312: XPL/100×. Anomalous brown interference colours in crossed polars. Same field of view as Fig. 3.

7. P0096: PPL. Prismatic particle.

8. P0096: PPL. Same particle as in Fig. 7 with polariser rotated through 90° to show pleochroism.

9. P0312: PPL. Particles of copper(II) acetate F.

10. P0312: PPL. Same particles as in Fig. 9 with polariser rotated through 90° to show pleochroism.

Crystallinity	Optic sign	2V	n_α	n_β	n_γ	δ
Monoclinic	[unknown]	[unknown]	1.53	[–]	1.56	0.030

(data from Kühn, 1993b)

Copper(II) acetate hydrate ('neutral verdigris') is the commonest of the copper acetates known generically as verdigris. Under plane-polarised light copper(II) acetate hydrate appears as stubby, prismatic crystals with pyramidal terminations. The euhedral form is, however, generally lost on crushing; more typically, platy to tabular, angular subhedral to anhedral shards are seen. Copper(II) acetate hydrate has a weak conchoidal fracture and curved particle boundaries are common. On the whole, particle surfaces are smooth and clear. Particle size among samples examined had a broad distribution, from fine grained to very coarse grained.

In colour it is a translucent blue green. Crystals are strongly pleochroic from pale green to turquoise blue. Colour intensity increases with increasing particle size. The compound has moderate to low relief, with an RI less than that of the medium.

Under crossed polars, copper(II) acetate hydrate has moderate to high birefringence, with the second to third order colours strongly masked by the body colour. Anomalous colours are normally produced; these include blue-greens and, distinctively, browns. Extinction is straight and may be complete; however, undulose extinction is not uncommon.

With reflected light, particles of this compound vary in colour from pale to dark blue.

Neutral copper(II) acetate hydrate may be found alone or in association with basic copper acetate varieties of verdigris, especially the basic copper(II) acetate A (q.v.).

The chemistry and optical properties of the verdigris compounds have been reviewed by Kühn (1993b) and Scott (2002).

Lit.: Kühn (1993b); Scott (2002)

The Pigment Compendium

COPPER(II) ACETATE H

$Cu(CH_3COO)_2 \cdot [Cu(OH)_2]_4 \cdot 3H_2O$

Key:

1. P0294: PPL/100×. Pale green aggregates of fibrous and acicular crystals.

2. P0294: XPL/100×. Same field of view as in Fig. 1 under crossed polars.

3. P0298: PPL/40×. Pale green aggregates of fibrous and acicular crystals.

4. P0298: XPL/40× / ⊙. The same field of view as in Fig. 3 under crossed polars.

Crystallinity	n	δ
Anisotropic	≤1.662	Low
		(data from authors)

Under plane-polarised light copper(II) acetate H appears very similar to copper(II) acetate D (*q.v.*). However, rather than a very pale green, it is very pale blue in colour. The crystals are fibrous to acicular and form aggregates of lamellar sheaves. Particle size is medium to very fine among examples examined. Relief is very low and variable, from very slightly less than that of the medium to equal to that of the medium.

Under crossed polars birefringence is seen as low, so that the finest particles appear isotropic. Larger particles exhibit low first order interference colours and anomalous browns and blues. Extinction is straight, undulose in some

particles or, more frequently, sweeping across fibrous aggregates. Crystals are length slow.

With reflected light, particles of this compound are seen to be a pale blue colour. The appearance of copper(II) acetate H is only slightly modified when viewed through the Chelsea filter.

The chemistry and optical properties of the verdigris compounds have been reviewed by Scott (2002) and Kühn (1993b).

Lit.: Kühn (1993b); Scott (2002)

The Pigment Compendium

COPPER FORMATES (COPPER FORMATE 1) [Various]

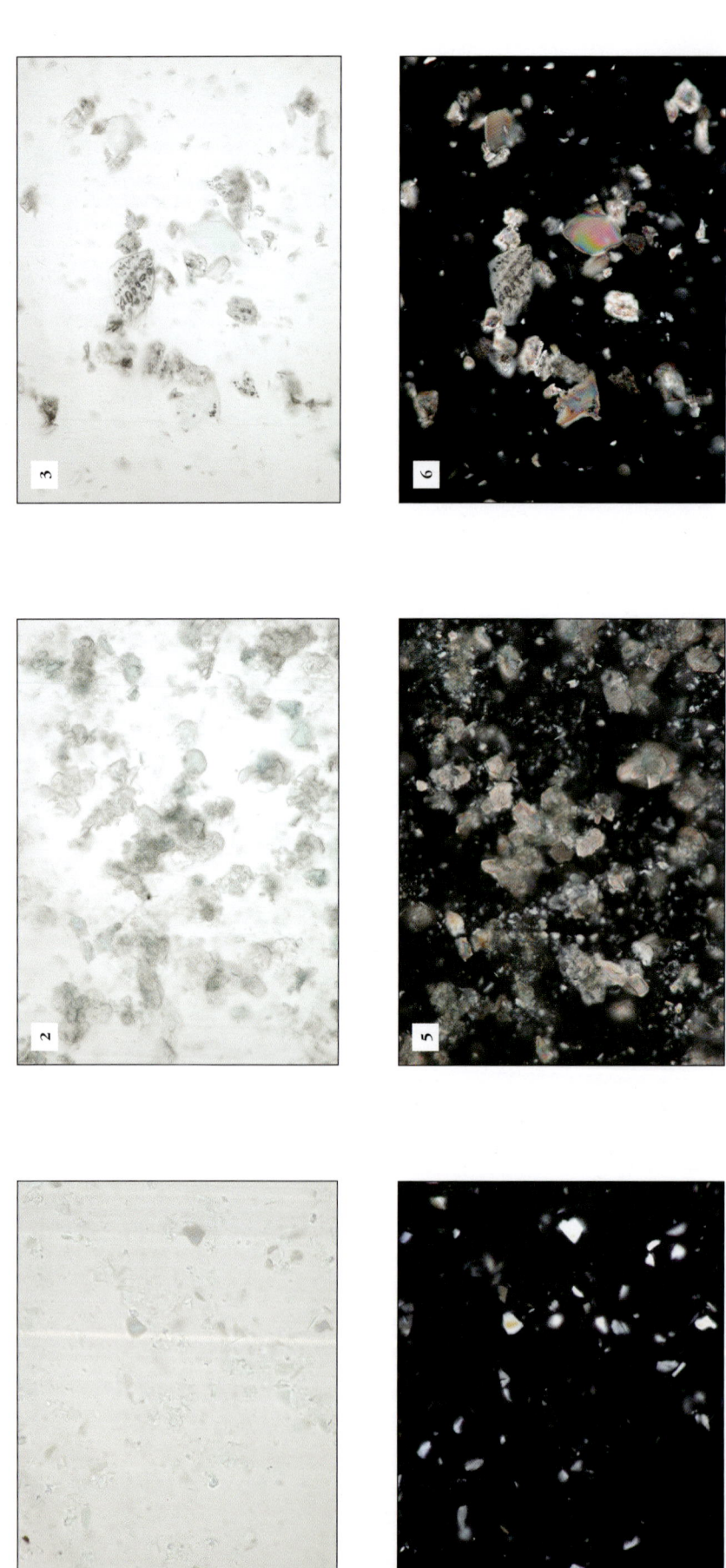

Key:

1. P0285: PPL/100×. Pale blue angular shards of neutral copper formate I.

2. P0286: PPL/40×. Colourless and pale blue particles of neutral copper formate II.

3. P0287: PPL/40×. Fluid inclusion-rich angular particles of colourless neutral copper formate III.

4. P0285: ~XPL/100×/ ◘ . The field of view in Fig. 1 under crossed polars.

5. P0286: XPL/40×/ ◘ . The field of view in Fig. 2 under crossed polars.

6. P0287: XPL/40×/ ◘ . The field of view in Fig. 3 under crossed polars.

7. P0286: PPL. Detail of skeletal crystal in neutral copper formate II.

Similar to the copper acetates, a number of copper formates exist. Their occurrence as pigments is relatively unstudied (although see Scott, 2002), though their optical appearance is sufficiently distinctive within the group to warrant illustrating a series of examples. The principal confusion is likely to be with the acetates.

Under plane-polarised light, neutral copper formate I forms transparent, colourless to very pale blue crystals. The pale blue crystals are strongly pleochroic from blue to colourless. Crystal habit ranges from plate- and flake-like particles to inclusion-rich, spongy appearing grains. Relief is moderate and RI is less than that of the medium. Grain size is medium to coarse. Under crossed polars, neutral copper formate I is moderately birefringent, with interference colours ranging from high first order to third order colours. Crystals have inclined and complete extinction.

Under plane-polarised light, it may be seen that three phases are present in the sample of neutral copper formate II examined. All have coarse grain size and form platy crystals. Phase A forms transparent, colourless, euhedral to subhedral hexagonal plates. Phase B forms subhexagonal, subhedral and anhedral plates which are a very pale blue colour. Finally, and most distinctively, but of lowest abundance is Phase C, bright, turquoise blue forming elongated hexagonal and lath-shaped particles and also spongy, inclusion-rich particles. Phases A and B have moderate relief with RI less than that of the medium. Phase C has variable relief with RI ranging from equal to that of the medium to less than that of the medium. Under crossed polars, the differences between phases A and B are more distinctive. Phase A has low birefringence and shows low first order grey colours. In contrast,

Phase B shows fourth order interference colours. Phase C also shows high interference colours, but strongly masked by the body colour. Phase C has sweeping extinction. Phases A and B have complete extinction, but this is disrupted by the overlapping particles at different crystallographic orientations.

Under plane-polarised light neutral copper formate III forms translucent, turquoise blue crystals. The crystals are only weakly pleochroic and this phenomenon is visible only in the most strongly coloured grains. Relief is variable and RI ranges from equal to that of the medium to less than that of the medium. Particle shapes are distinctively euhedral, with habits ranging from acicular with pyramidal terminations to hexagonal prisms. Particle surfaces are smooth and many grains contain fluid inclusions. Particle size is coarse to very coarse. Under crossed polars the particles have high birefringence, but interference colours are masked by the strong body colour such that anomalous blues and pale brown are typically seen. Crystals have complete and straight extinction.

Under plane-polarised light, neutral copper formate IV forms colourless particles, which are rendered almost opaque by dense clusters of inclusions. Particle margins have lowest inclusion densities. Relief is very low and RI is extremely close to that of the medium. Particle shapes are angular and anhedral and grain surfaces appear extremely rough and pitted, despite the low relief. Particle size is medium to coarse. Under crossed polars, the grains have high birefringence, with third order and greater interference colours visible. The presence of inclusions destroys the birefringence of the central portions of the grains, and these regions show brown reflections.

The Pigment Compendium

COPPER FORMATES (COPPER FORMATE 2)

[Various]

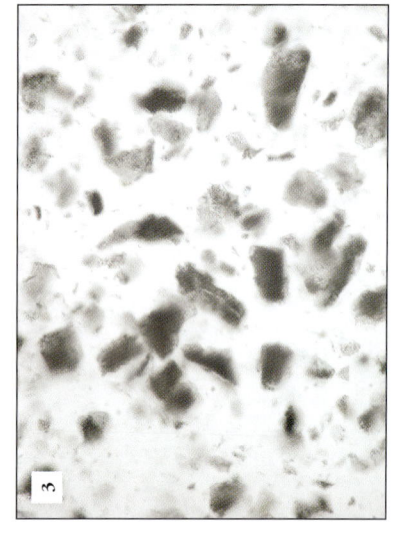

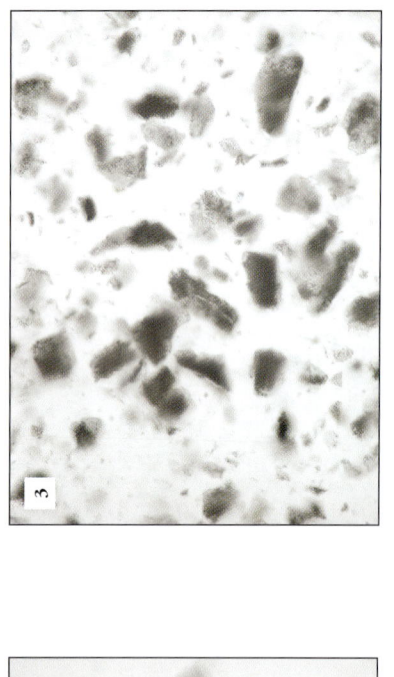

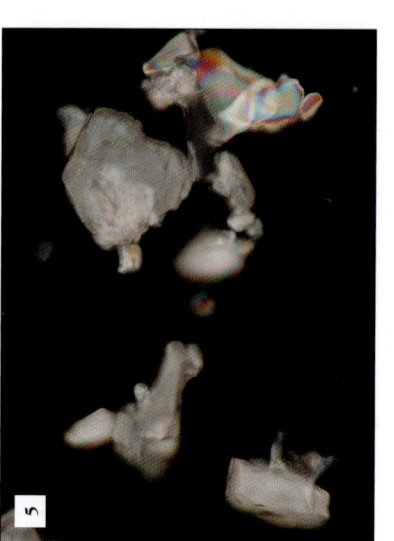

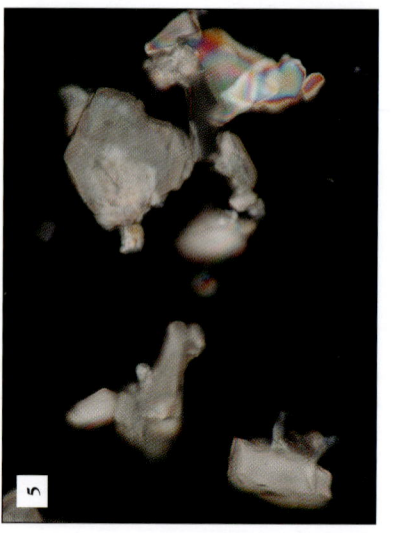

Key:

1. P0288: PPL/40×. Blue, subhedral crystals of neutral copper formate IV.

2. P0289: PPL/40×. Pale blue and colourless particles of basic copper formate I.

3. P0290: PPL/40×. Clouded, inclusion-rich particles of basic copper formate II.

4. P0288: XPL/40×/ ● . The field of view in Fig. 1 under crossed polars.

5. P0289: XPL/40×/ ● . The field of view in Fig. 2 under crossed polars.

6. P0290: XPL/40×/ ● . The field of view in Fig. 3 under crossed polars.

7. P0290: PPL. Detail of crystal of basic copper formate II, showing blue coloration.

Under plane-polarised light, basic copper formate I forms transparent, very pale blue and weakly pleochroic particles. Relief is low and slightly variable, with RI just greater than that of the medium. Particle shapes are predominantly euhedral to subhedral although anhedral angular particles are also present. Euhedral particles have acicular and rectangular prismatic habits. Particle surfaces are smooth and grain size is medium to fine. Under crossed polars, basic copper formate I has low birefringence, and shows first order interference colours. Particles have complete and straight extinction.

Under plane-polarised light, basic copper formate II forms colourless, transparent crystals, with moderate to high relief and RI less than that of the medium. Particle shapes are distinctive, with large skeletal and hopper crystals formed of 'building blocks' of square and rectangular forms. Additionally, a groundmass of finely particulate grains occurs. Particle surfaces are smooth and grain size ranges from coarse to very fine. Under crossed polars, the particles are strongly birefringent and have fourth order interference colours. Clear, rectangular grains have straight and complete extinction.

Lit.: Scott (2002)

COPPER CITRATE

$Cu_4[HOC(CH_2COO)_2(COO)]_2$

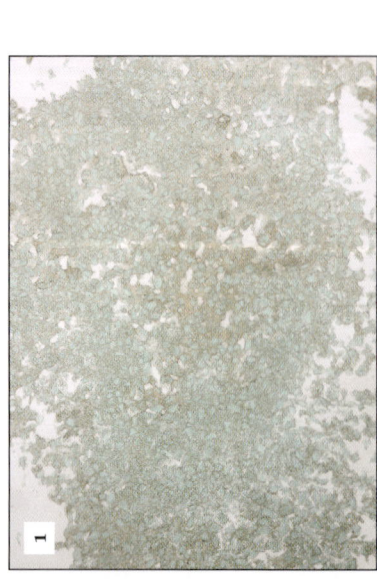

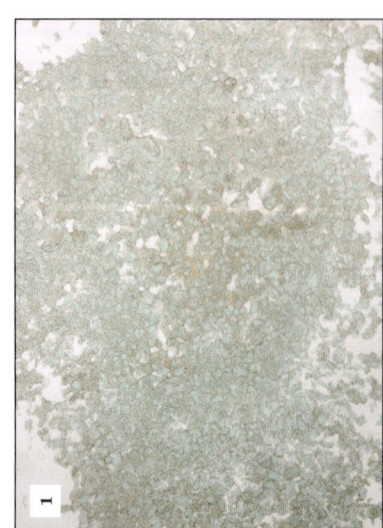

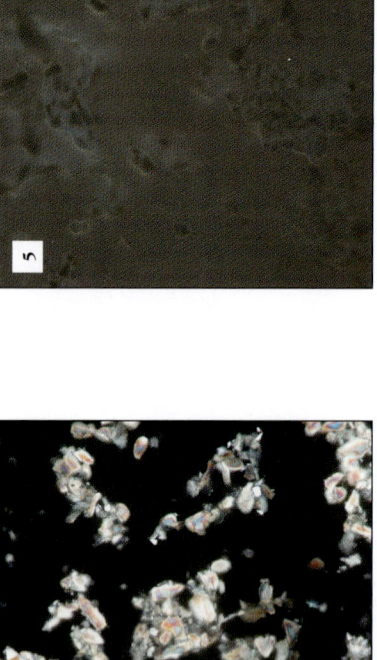

Key:

1. P0316: PPL/10×/●. Angular anhedral green grains set in a finer-grained matrix.

2. P0316: XPL/10×/●. The same field of view as Fig. 1 in cross-polarised light.

3. P0316: PPL/40×/●. Anhedral angular green grains often bound in a paler coloured matrix.

4. P0316: XPL/40×/●. Strongly birefringent grains with the binding matrix showing lower order interference colours. Same field of view as in Fig. 3.

5. P0316: UV/40×. Shows under UV illumination that the matrix is weakly fluorescent. Same field of view as in Fig. 3.

Crystallinity
Anisotropic

n
<1.662

δ
High
(data from authors)

Copper citrate was prepared by dissolving a copper acetate compound in lemon juice; a recipe for its preparation is given, for example, in Scott (2002).

Under plane-polarised light this compound looks as Scott (2002) describes it – a 'sticky green mass'. It appears as angular anhedral shards set in a mass of similar relief. The entire mass has low to moderate relief with the RI less than that of the medium. The angular particles are pale green in colour and non-pleochroic. They are translucent to transparent and have smooth surfaces. The surrounding groundmass is also pale green and translucent. It has an uneven, frothy, surface appearance.

Under crossed polars, the angular particles are strongly birefringent, showing second to third order colours, oranges, yellows, blues and pinks. These particles have complete and inclined extinction. In contrast, the groundmass shows lower interference colours, from amorphous isotropic to low first order greys. Extinction of this substance is mottled.

The compound appears bright green when viewed through the Chelsea filter. The matrix was also noted to fluoresce a weak blue-white under UV excitation.

Lit.: Scott (2002)

The Pigment Compendium

ORGANO-COPPER COMPLEXES

[Various]

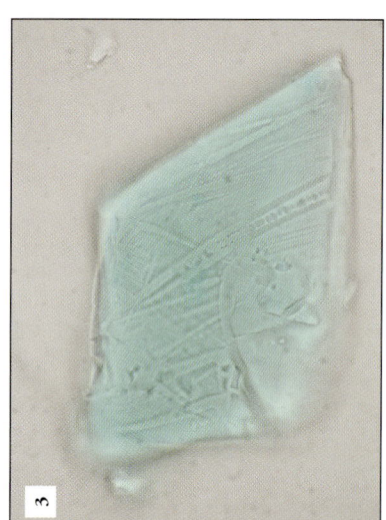

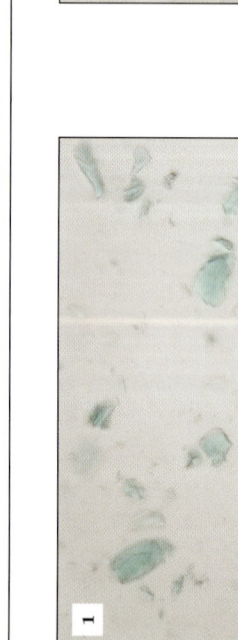

Key:
1. P0871: PPL/40×. Particles of blue-green copper resinate.
2. P0871: PPL/100×. Particle of copper resinate, note striated surface.
3. P0871: PPL/100×/ ⦿ Particle of copper resinate, note striated surface.
4. P0871: PPL/100×. Flake of copper resinate, note saw-tooth grain margins.

Crystallinity	n	δ
Amorphous	~1.52	Isotropic
		(data from Kühn, 1993b)

In plane-polarised light and when fresh the organo-copper complexes form transparent pale blue-green particles. However, noted for their discoloration, in aged samples these pigments are more often seen as a brown matrix with embedded regions of residual green. The compounds have moderate relief and the RI is less than that of the mounting medium. Particle shape is exhibited as an irregular, anhedral habit. For some particles surfaces are smooth and glassy, and particles can have a well-developed conchoidal fracture. Other particles, however, are inclusion rich and have a spongy, pitted texture in this context. Particle size is likely to be coarse. Kühn (1993b) has discussed the appearance of recently prepared 'copper resinate' under the microscope, where he notes the formation of glassy, irregular green fragments. Discoloration of this type of pigment in paintings is discussed by

Kockaert (1979), who also presents thin sections of paint samples showing that the phenomenon is a surface one.

The organo-copper complexes are amorphous and therefore isotropic under crossed polars. It is not uncommon, though, to observe anisotropic inclusions that represent a residual phase of the copper salt (such as a copper acetate) used to prepare the pigment.

The organo-copper complexes are notable under UV in that they do not fluoresce; in the case where these compounds have discoloured, this property might be used to suggest that the material is not a simple resin or oil.

Lit.: Kockaert (1979); Kühn (1993b)

The Pigment Compendium

AERINITE

$Ca_4(Al,Fe^{3+},Mg,Fe^{2+})_{10}Si_{12}O_{36}(CO_3).12H_2O$

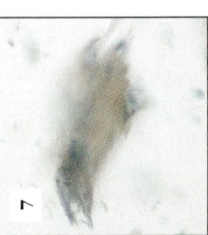

Key:

1. P0867: PPL/40×. Fibrous crystals of blue aerinite.

4. P0867: XPL/40× ⬤ The field of view in Fig. 1 under crossed polars.

7. P0867: PPL. The fibrous sheaf of crystals from Fig. 1 rotated through 90° to show change in colour due to pleochroism.

2. P0867: PPL/40×. Fibrous and polycrystalline aggregates of aerinite.

5. P0867: PPL/40×. The field of view in Fig. 2 rotated through 90°, note the dramatic pleochroism of the fibrous aggregate of crystals now lying at bottom centre of the field of view.

3. P1132: PPL/40×. Polycrystalline aggregates of aerinite.

6. P1132: ~XPL/40×. The field of view in Fig. 3 under crossed polars.

Crystallinity	Optic sign	2V	n_α	n_β	n_γ	δ
Monoclinic	−ve	62°–63°	1.51	1.56	1.58	0.0700

(data from Webmineral, 2003)

Under plane-polarised light, the rare silicate mineral aerinite can be seen as intense sky blue crystals, which are strongly pleochroic from blue to almost colourless. Crystals are fibrous and arranged in irregular aggregates with a subparallel fibre orientation; consequently, particle surfaces have a rough, striated appearance. Relief is low to moderate and RI is less than that of the medium. Particle size is variable and ranges from very fine to coarse.

Under crossed polars, aerinite is distinctive. Although it has high birefringence, it emits bright anomalous yellow-orange-brown interference colours. Some regions of the crystal are masked by the body colour and appear

blue-grey to turquoise. Aerinite has straight extinction although this is sweeping across the crystal aggregates. Fibres are length fast.

The appearance of aerinite is unaffected when viewed through the Chelsea filter.

Microscopic appearance of aerinite from paintings has been discussed by Casas and Llopis (1992).

Lit.: Casas & Llopis (1992)

The Pigment Compendium

CHRYSOCOLLA

$(Cu,Al)_2H_2Si_2O_5(OH)_4 \cdot xH_2O$

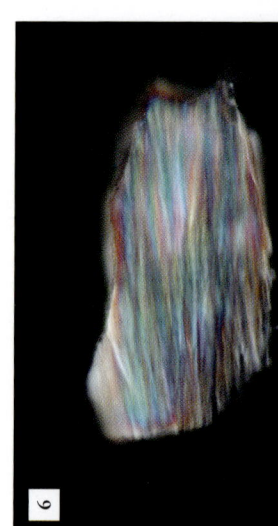

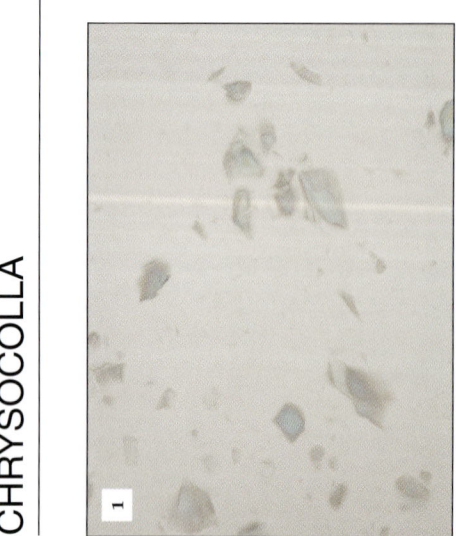

Key:

1. P1171: PPL/40×. Angular particles of sky blue chrysocolla.

2. P1171: XPL/40× ● The field of view in Fig. 1 under crossed polars.

3. P1475: XPL/10×. Typical appearance of chrysocolla in crossed polars.

4. P1475: XPL/100×. Angular crystal of chrysocolla.

5. P1475: PPL/100×. The crystal in Fig. 4 with the polariser rotated through 90° to show change in body colour due to pleochroism.

6. P1475: PPL/100×. The crystal in Fig. 4 under crossed polars where it is shown to be a finely fibrous aggregate.

Crystallinity
Orthorhombic to Amorphous

n_ε	n_ω	δ
1.57	1.46	0.1380–0.1750

(data from Ford, 1932; cf. Webmineral, 2003)

Chrysocolla is a copper silicate mineral with a striking turquoise blue colour when observed as a massive specimen, but which grinds down to a pale blue powder. It is extremely variable, which probably accounts for the confusion that exists in the literature dealing with its crystal structure and optical properties.

In plane-polarised light, examples of chrysocolla examined by the authors had low to moderate relief, with an RI less than that of the medium. It may occur in various habits, usually as fibrous botryoidal or stalactitic masses that under the microscope appear as fibrous sheaves, sometimes showing banding. It may also occur as earthy masses showing poor crystalline structure. It

is pleochroic from colourless to pale green-blue. However, colour intensity is variable, with some examples very pale.

Under crossed polars chrysocolla shows strikingly high birefringence with bright third order colours. It has straight extinction, but the common fibrous habit results in the extinction being sweeping or, more commonly, mottled. Crystal habit is very clear under cross-polarised light. The presence of fibres, or mottled cryptocrystalline structure, never going wholly into extinction, is clear. Fibrous particles are length slow.

Chrysocolla appears green when viewed with the Chelsea filter.

The Pigment Compendium

CELADONITE AND GLAUCONITE $K(Mg,Fe^{2+})(Fe^{3+},Al)Si_4O_{10}(OH)$ and $(K,Na)(Fe^{3+},Al,Mg)_2(Si,Al)_4O_{10}(OH)_2$

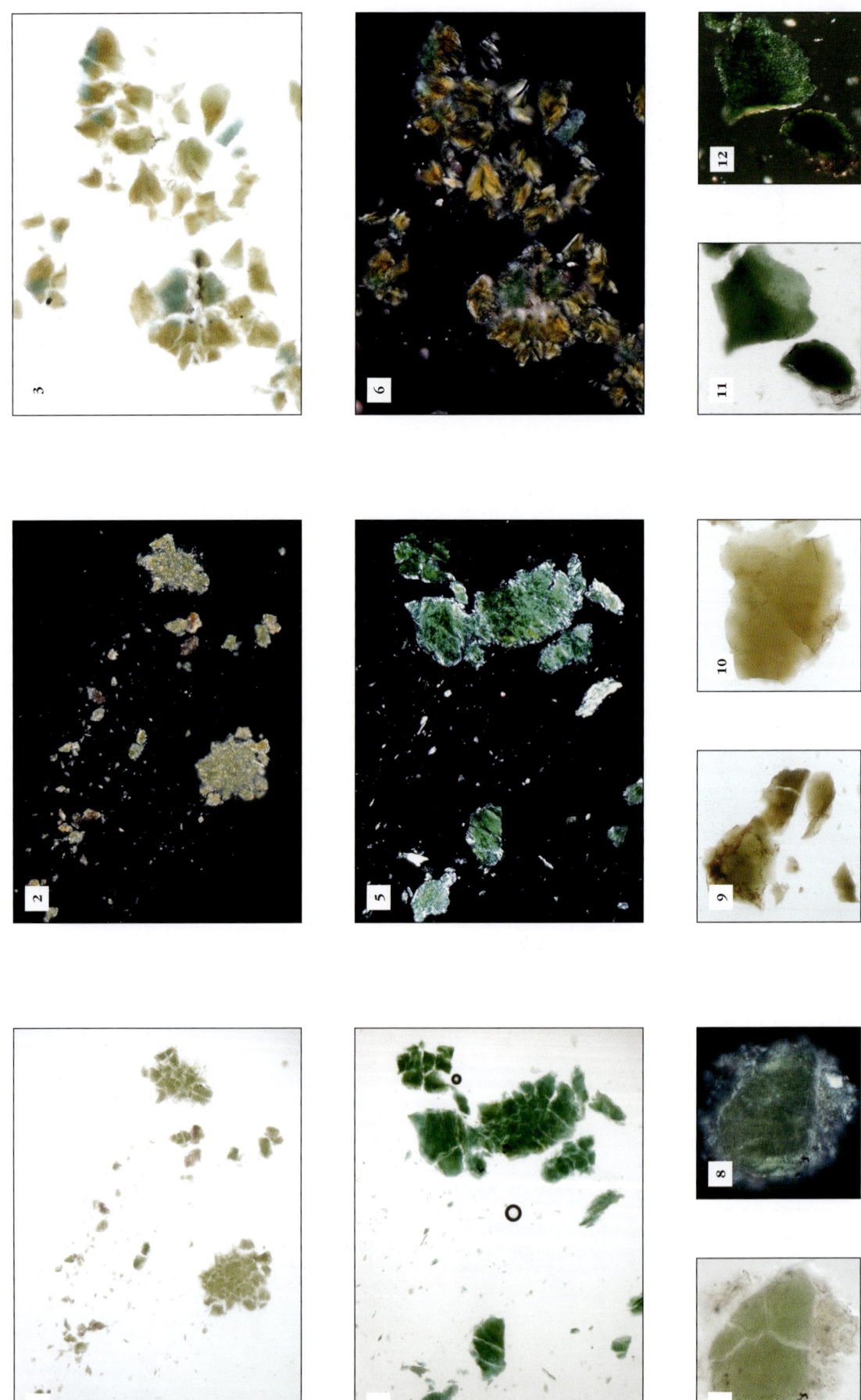

Key:

1. P1443: PPL/10×. Polycrystalline olive green platy aggregates of glauconite.

2. P1443: XPL/10×. Same view as in Fig.1 under crossed polars.

3. P1440: PPL/40×. Brown and green variolitic aggregates of celadonite.

4. P1441: PPL/10×. Strong green polycrystalline particles of celadonite.

5. P1441: XPL/10×. The same field of view as in Fig. 4 under crossed polars.

6. P1440: XPL/40×. The same view as in Fig. 3 under crossed polars showing anomalous brown, purple and green interference colours.

7. P1452: PPL. Rounded plate of glauconite.

8. P1452: XPL. Same grain as in Fig. 7 under crossed polars.

9. P1453: PPL. Finely crystalline brown-green aggregates of glauconite.

10. P1453: PPL. Finely crystalline pale brown platy aggregate of glauconite.

11. P1458: PPL. Dark green polycrystalline aggregates of glauconite.

12. P1458: RPL. Dark green reflections of glauconite aggregate (as in Fig. 11) in reflected light.

Crystallinity	Optic sign	2V	n_α	n_β	n_γ	δ
Celadonite: Monoclinic	−ve	0°–26°	1.606–1.644	1.63–1.662	1.63–1.663	0.0190–0.0240
Glauconite: Monoclinic	−ve	20°–24°	1.59–1.612	1.609–1.643	1.61–1.644	0.0200–0.0320

(data from Webmineral, 2003)

Glauconite and celadonite in pigments are normally indistinguishable by optical microscopy (Grissom, 1986). However, it may be possible to differentiate them based on their associated minerals when they occur in green earths.

In plane-polarised light, celadonite and glauconite both form translucent to weakly translucent particles with colour varying from bright emerald green, through green, olive green and brown. Pleochroism is reported in glauconite as ranging from lemon yellow, pale yellowish green or bright green for one axis to green, dark green, yellowish green, bluish green or olive green for the others; blue–green to green colours were noted by the authors. Particles have low relief and the RI is just less than that of the medium. Particle shape is variable. Individual crystals are very fine and therefore most observed particles are polycrystalline. Aggregates have a platy habit, occasionally fibrous and lamellar forms are observed and also variolitic masses. As particles tend to fall with plates flat to the microscope slide surface the lamellar habit is not commonly observed. Typically, particles are rounded, and this is particularly the case in glauconite-bearing greensands. The variety of habits observable in glauconite and celadonite has been described in detail in Grissom (1986) and Grissom (1970). Particle surfaces are rough. Particle size varies in natural specimens from fine to very coarse grained; however, Grissom (1986) remarks that in green earths used in pigments, celadonite and glauconite particles seemingly rarely exceed 10 μm, but Mactaggart (2002) more realistically gives this figure as up to 60 μm. McCrone et al. (1973–80) state that particles less than about 2 μm are generally colourless.

Under crossed polars, both glauconite and celadonite have moderate birefringence and up to second or third order interference colours are typically seen; these may be heavily masked by the high body colour. The polycrystalline nature of the particles is clear, and particles are observed to have mottled extinction, or sweeping extinction where radiating fibres are observed. Both compounds are length slow.

In reflected light these minerals appear weakly white or yellow to green. Celadonite and glauconite also appear dark blue-green when viewed using the Chelsea filter; Mactaggart has commented that blue-green particles may appear reddish. No fluorescence was observed by the authors among numerous samples of these two minerals under UV excitation.

Celadonite and glauconite may be found in association with iron oxides such as goethite, quartz and various clay minerals.

Properties and occurrences of glauconite and celadonite in green earth pigments have been reviewed by Grissom (1986).

Lit.: Grissom (1970); Grissom (1986); Mactaggart (2002); McCrone et al. (1973–80)

The Pigment Compendium

CHLORITE GROUP MINERALS

[e.g.: Clinochlore, $(Mg,Fe^{2+})_5Al(Si_3Al)O_{10}(OH)_8$]

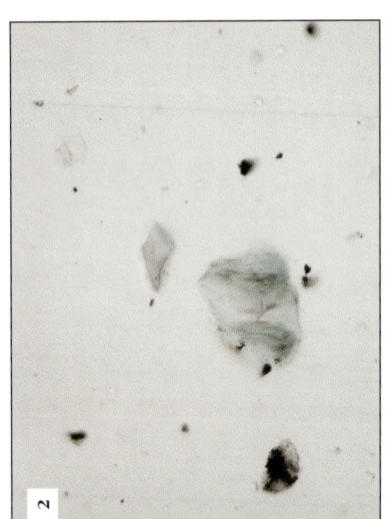

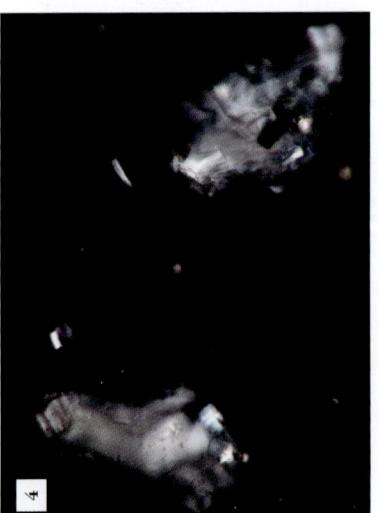

Key:

1. P1175: PPL/40×. Weakly and variably coloured subhedral plates of clinochlore.

2. 1175: PPL/40×. Polycrystalline fibrous plate of clinochlore.

3. P1175: PPL/40×. Plates of clinochlore containing opaque inclusions.

4. P1175: XPL/40×/ ○ . The field of view in Fig. 1 under crossed polars.

5. 1175: XPL/40×/ ○ . The field of view in Fig. 2 under crossed polars.

6. P1175: XPL/40×/ ○ . The field of view in Fig. 3 under crossed polars.

Crystallinity	Optic sign	2V	n_α	n_β	n_γ	δ
Monoclinic	±ve	20°–60°	1.57–1.67	1.57–1.69	1.57–1.69	0.00–0.02

(data from Deer et al., 1992)

In plane-polarised light the chlorite group minerals, such as chamosite, clinochlore and pennanite, appear translucent and range in colour from colourless to greens and blue-greens. They are pleochroic through this range. Chlorites have strong dispersion and the refractive index varies with the light source used. Relief varies from very low to low, with RI varying from less than that of the medium to just greater than that of the medium. Particles are anhedral plates. Edge-on views show particles to be lamellar and they appear fibrous under these conditions. Particle size may range from fine to very coarse.

Under crossed polars, chlorites have low birefringence, and low first order greys and whites are typical. However, many samples show anomalous colours. Iron-rich chlorites often exhibit anomalous blues whereas magnesium-rich chlorites exhibit anomalous browns. Chlorites have sweeping extinction.

Chlorites are common components of green earths and occur commonly in a wide variety of rock types. Discussion of the optical properties and occurrences of the chlorite minerals may be found in Deer et al. (1992).

Lit.: Deer et al. (1992)

GLAUCOPHANE

$Na_2(Mg,Fe^{2+})_3Al_2Si_8O_{22}(OH)_2$

1. P1500: PPL/10×. Bladed crystals of glaucophane.

2. P1500: PPL/10×. A similar field of view to Fig. 1, note the change in body colour of the two prominent crystals due to pleochroism after rotation of the stage.

3. P1500: XPL/10×/ ○. The same field of view as Fig. 2 under crossed polars, note that glaucophane has inclined extinction.

4. P1500: PPL/10×. Two bladed crystals of glaucophane in the centre of the field of view show extremes of the pleochroic scheme from lilac to blue.

5. P1500: PPL/10×. The field of view as in Fig. 4 rotated through 90°; the two bladed crystals now show different body colours due to pleochroism.

6. P1500: PPL/CF/10×. Glaucophane as seen through the Chelsea filter. Same field of view as in Fig. 5.

Crystallinity Monoclinic	Optic sign −ve	2V 0°–50°	n_α 1.594–1.630	n_β 1.612–1.650	n_γ 1.618–1.652	δ 0.023–0.020
						(data from Deer et al., 1992)

In plane-polarised light, glaucophane is one of the most distinctive of the amphibole minerals. It forms translucent and strongly pleochroic blue to lavender crystals. Relief is moderate to low and RI is less than that of the medium. Glaucophane has relatively strong dispersion, and therefore relief varies with the light source used. Crystal shapes are typically bladed prisms (with rhombic basal sections); crushed samples may be angular shards. Glaucophane has two perfect cleavages parallel to the prismatic section that strongly influences grain shape. Particle surfaces are smooth, though striations formed by the cleavage are often present. Glaucophane is frequently rich in fluid and mineral inclusions.

Under crossed polars, glaucophane has moderate birefringence, but the interference colours are masked by the strong body colour. Therefore grains often appear bright but blue, lavender, brownish shades. Glaucophane has straight and complete extinction.

Glaucophane appears a dull red when viewed through the Chelsea filter.

Glaucophane typically occurs in relatively rare (although locally abundant) metamorphic rocks in association with epidote, garnet, chorite, calcite, lawsonite and kyanite. The related mineral riebeckite is less common; however, the fibrous variety, crocidolite, is blue asbestos. It appears very similar to glaucophane except that it has blue-yellow pleochroism. The optical properties and occurrence of glaucophane and other sodic amphiboles have been discussed by Deer et al. (1992).

Lit.: Deer et al. (1992)

The Pigment Compendium

CHROMIUM(III) OXIDE

Cr_2O_3

1. P0209: PPL/100×. Particles of chromium(III) oxide.

2. P0091: PPL/40×. Dark olive green particles of chromium(III) oxide.

3. P0091: PPL/100×. Enlarged view of sample as in Fig. 2; note bimodal grain size.

4. P0209: XPL/100×. The field of view in Fig. 1 under crossed polars.

5. P0091: XPL/40×. The field of view as in Fig. 2 under crossed polars.

6. P0091: XPL/100×. The field of view as in Fig. 3 under crossed polars.

Crystallinity	Optic sign	n_ω/n_ε	δ
Trigonal-Hexagonal	+ve	Average value = 2.5	High

(data for eskolaite from Merwin, 1917; cf. Newman, 1997)

In plane-polarised light chromium(III) oxide ('opaque chrome oxide'), the synthetic analogue of the rare mineral eskolaite, forms very fine to fine green particles that are close to the resolution of the optical microscope; these therefore appear as rounded or bacterioid forms. Newman (1997) notes that particle sizes he observed in samples were in the range 0.1–1.0 μm although others report larger values (for example, Robinson, 1973, who gives an average particle size of 3.8 μm). According to McCrone *et al.* (1973–80) twinning is common, also noting hexagonal basal tablets or equant prisms with distinct rhombohedral cleavage. Newman has published electron micrographs showing particles that tend to be equant and often appear as hexagonal plates, similar to the hexagonal prisms and plates reported for the mineral analogue eskolaite. In colour chromium(III) oxide particles are translucent to opaque and a dull, almost olive green; Mactaggart (2002) notes that the smallest particles may be reddish or brownish in colour. Relief is high and RI is greater than that of the medium.

Under crossed polars, most distinctively chromium oxide green has moderate, emerald green internal reflections. The samples illustrated here have high birefringence, typically third order colours, but this is strongly masked by the body colour. Anomalous browns may be observed in a few particles.

When viewed through the Chelsea filter chromium oxide transmits a dull red in larger particles. This effect may not be apparent in smaller particles.

The optical and physical properties of chromium oxide and hydrated chromium oxide greens have been reviewed by Newman (1997).

Lit.: Mactaggart (2002); McCrone *et al.* (1973–80); Newman (1997); Robinson (1973)

CHROMIUM OXIDE HYDRATE

$Cr_2O_3 \cdot 2H_2O$

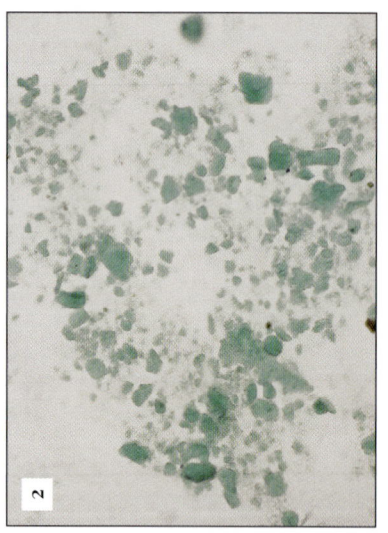

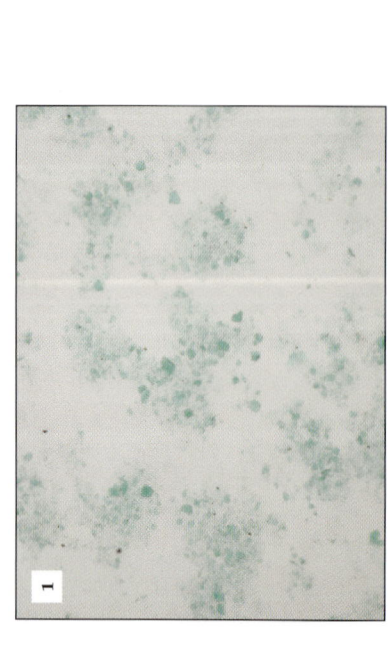

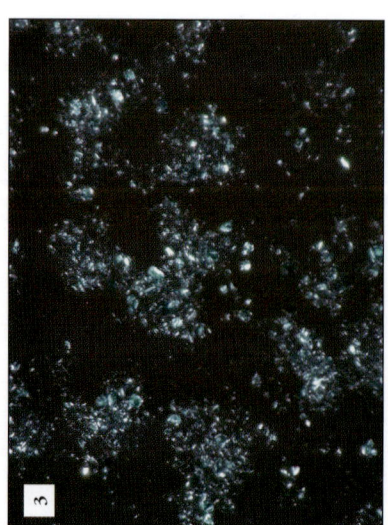

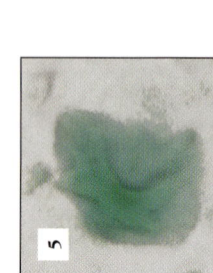

Key:

1. P0210: PPL/40×. Plates and fine-grained particles of chromium oxide hydrate.

2. P0210: PPL/100×. Polycrystalline plates of chromium oxide hydrate.

3. P0210: XPL/40×. The field of view as in Fig. 1 under crossed polars.

4. P0210: XPL/100×. The field of view as in Fig. 2 under crossed polars.

5. P0210: PPL. An enlargement of a particle from Fig. 2 showing the uneven surface texture.

Crystallinity	Optic sign	2V	n_α	n_β	n_γ	δ
Anisotropic (biaxial)	[unknown]	[unknown]	1.62	1.62	2.12	0.50

(data from Merwin, 1917; cf. Newman, 1997)

Under plane-polarised light, chromium oxide hydrate ('viridian') appears as anhedral, subangular to rounded plates and platelets. In colour they are typically blue-green and translucent. Particle surfaces appear bumpy and irregular. Relief is low to moderate and slightly variable and RI is greater than that of the medium. McCrone *et al.* (1973–80) state that the RI is variable from sample to sample, giving values of $n_{\alpha/\beta} = 1.76$ to slightly greater than 1.80 and $n_\gamma = 1.80$ to much greater than 1.90; Mactaggart (2002) has not noted the RI as less than 1.82. Particle size can show a broad distribution within a sample from very fine to medium or coarse. According to Newman (1997), in an example of the pigment he studied 44% of particles were <5 μm, while the largest was 45 μm; further analysis showed these in fact to be aggregates of smaller particles of the order of 0.07–0.12 μm long by 0.01 μm wide. Mactaggart also observes that in all samples he has examined that there is a small proportion of very fine, doubly terminated, crystals; the composition of these appears to be unknown.

Under crossed polars, chromium oxide hydrate has high birefringence, but this is masked by the body colour and the particles appear bright blue green.

Distinctively, the particles have a marked sweeping extinction, revealing them (as noted above) to be aggregates of finely fibrous crystals. Newman describes this as 'irregular undulose extinction', also reporting an erroneous attribution of this phenomenon to strain in the particles caused during cooling in manufacture. With the sensitive tint plate most particles seem to be composed of more than one crystal, these having grown together in different orientations (Mactaggart). The fibrous particles are length slow.

When viewed through the Chelsea filter, chromium oxide hydrate transmits a very weak red, which is visible only in the larger, more intensely coloured particles.

The optical and physical properties of chromium oxide and hydrated chromium oxide greens have been reviewed by Newman (1997).

Lit.: Mactaggart (2002); McCrone *et al.* (1973–80); Newman (1997)

The Pigment Compendium

CHROMIUM PHOSPHATE HEXAHYDRATE

CrPO$_4$.6H$_2$O

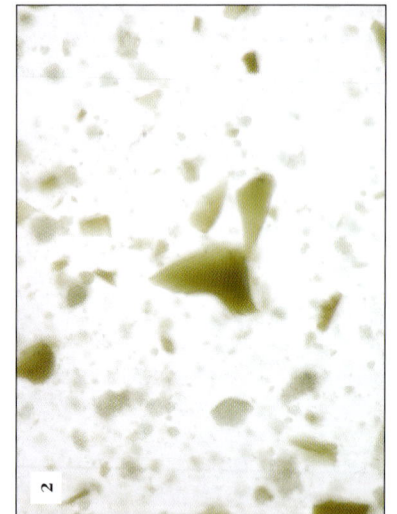

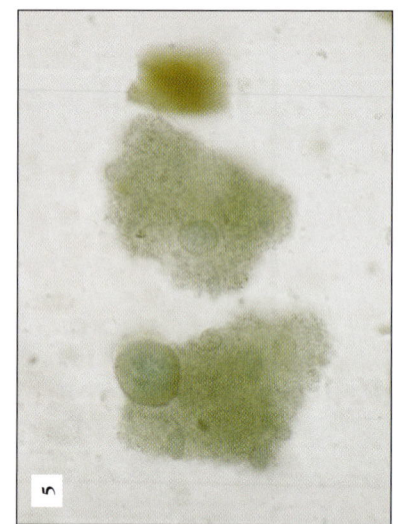

Key:

1. P1524: PPL/40×. Typical sample of chromium phosphate hexahydrate showing the two particle morphologies; granular particles and smooth plate-like particles.

2. P1524: PPL/100×. Olive green platy angular shards of chromium phosphate hexahydrate.

3. P1524: PPL/100×. Olive green platy crystal of chromium phosphate hexahydrate.

4. P1524: PPL/100×. Crumb-like particle morphology.

5. P1524: PPL/100×. Crumb-like particle morphology.

Crystallinity	Optic sign	2V	$n_\alpha/n_\beta/n_\gamma$	δ
Monoclinic	[unknown]	[unknown]	<1.662	

(data from ICSD, 2003/authors)

In modern examples of chromium phosphate hexahydrate examined by the authors, this compound clearly appeared to contain two separate phases, present in approximately equal amounts. One phase comprises translucent, olive green to brown particles, with low relief and RI just less than that of the medium. These form angular shards, with smooth surfaces and a conchoidal fracture giving curved grain boundaries. The grain size of these particles is medium–coarse. The second phase consists of rounded to crumb-like granules which are translucent, pale green in colour, again with low relief and RI just less than that of the medium. Particle surfaces are rough and finely pitted, while the particle size distribution of this latter phase is broad and ranges from fine to coarse.

Under crossed polars, both phases appear isotropic.

HEXACYANOFERRATE(II) COMPOUNDS [1]

$Fe_4[Fe(CN)_6]_3 \cdot xH_2O$ where $x = 14$–16 (and substitutions)

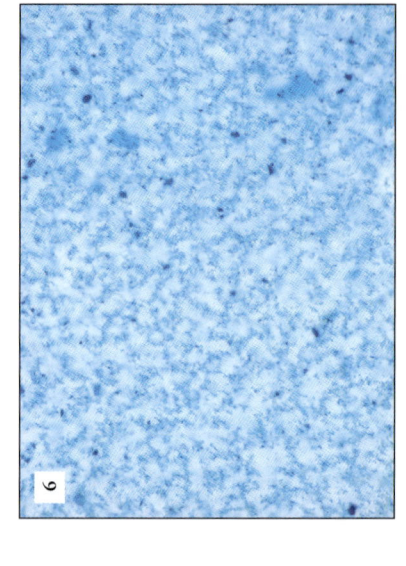

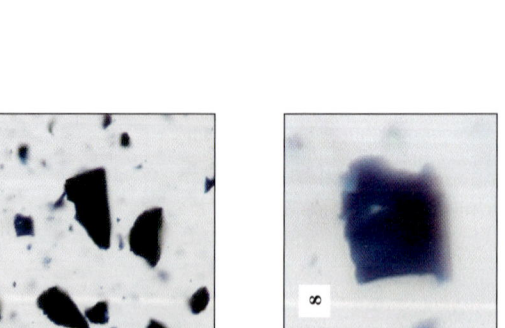

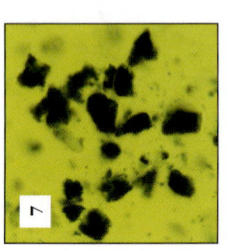

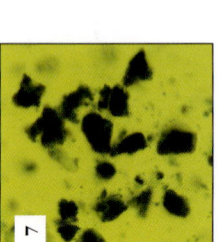

Key:

1. P0939: PPL/40×. Antimony-substituted form; particles with a colourless substrate.

2. P0939: PPL/100×. Antimony-substituted form; particles with a colourless substrate.

3. P0939: RPL/100×. The field of view as in Fig. 2. Under crossed polars, the substrate is weakly birefringent.

4. P0936: PPL/40×. Sodium form; note the angular particle morphology.

5. P0936: RPL/40×. The field of view as in Fig. 4 in reflected light. Note the very weak, orange reflections.

6. P0204: PPL/100×. Well-dispersed particles of a very fine modern commercial pigment.

7. P0939: PPL/CF. Antimony-substituted form; particles observed through the Chelsea filter.

8. 0936: PPL. Detail of particle morphology from Fig. 4.

9. P0204: PPL. Detail of particle morphology from Fig. 6.

Crystallinity	n	δ
Cubic	1.56	Isotropic

(data for arbitrary compound from McCrone et al., 1973–80)

A series of blue hexacyanoferrate(II) compounds exists, commonly known under the names 'Prussian blue', 'Berlin blue' or 'Iron blue'. Various substitutions into the structure are possible; however, the different varieties of hexacyanoferrate(II) compounds are generally impossible to distinguish optically. Combined with the fact that particle size of these pigments is extremely fine (0.01–0.2 μm) and therefore below the resolution of the optical microscope, characterisation is very difficult beyond group level. Berrie (1997), though reports that particle shape in fact varies depending on manufacturing process.

In plane-polarised light the various types of hexacyanoferrate compound form aggregates of the fine particles noted above, with individual particles not visible. Particle aggregates vary in colour and diaphaneity from mid-blue and moderately translucent through deep blue with low translucency to opaque. Relief is low and the RI is less than that of the medium. Aggregate sizes vary from very fine to coarse while aggregate shape is typically of angular shards, commonly with curved margins; from samples examined it would seem that the sodium and ammonia-treated forms commonly adopt this habit. The occurrence of angular fragments has also been noted by Welsh (1988) in material prepared according to traditional recipes involving blood and similar samples prepared by the authors confirmed this. Such morphology has also

been noted in a sample from the palette of J.M.W. Turner (Townsend, 1993). Examples of both are shown here. On the other hand zinc hexacyanoferrates examined tended to form seemingly granular, irregular, crumb-like particles. Other preparations may adopt subrounded or rounded habits. It is also not uncommon to encounter aggregates that have become smeared in sample preparation showing as a paler 'wash' of colour. Their high tinting strength and small particle size also means that in practice only very small quantities may be present in a sample that otherwise appears blue.

Under crossed polars, these compounds are isotropic and most varieties become completely extinguished. In reflected light the pigment is normally blue, though the reflectance is often weak; occasionally the reflective colour is a dark dull orange, probably due to the same phenomenon as that known as 'bronzing' in the bulk pigment.

The blue hexacyanoferrates appear black when viewed through the Chelsea filter since the transmission in the red/near-infrared of this pigment is low. While the superficially similar indigo ($q.v$) has a high red transmission, this occurs sufficiently far into the red/infrared that it is not commonly visible to the eye with a Chelsea filter and the two pigments therefore cannot necessarily be reliably differentiated on this basis.

HEXACYANOFERRATE(II) COMPOUNDS [2]

$Fe_4[Fe(CN)_6]_3 \cdot xH_2O$ where $x = 14-16$ (and substitutions)

Key:

1. P0965: PPL/100×/H. Rounded agglomerated mass of a nineteenth century sample of a hexacyanoferrate pigment.

2. P0101: PPL/100×. Paler masses of a commercial hexacyanoferrate pigment.

3. P0946: PPL/100×. General morphology found in a modern 'soluble' hexacyanoferrate.

4. P0943: PPL/100×. Iron hexacyanoferrate(II) prepared from blood.

5. P0935: PPL/100×. Appearance of an ammonia-treated hexacyanoferrate pigment.

6. P0965: PPL/H. Detail of particle morphology from Fig. 1.

7. P0946: PPL. Detail of particle morphology from Fig. 3.

8. P0943: PPL. Detail of grain boundary from Fig. 4.

Crystallinity
Cubic

n	δ
1.56	Isotropic

(data for arbitrary compound from McCrone et al., 1973–80)

These pigments have been commonly mixed with numerous other compounds, notably a wide variety of white 'extender' compounds. However, combinations with yellow chromates such as the lead, barium and strontium chromates have formed the basis of one group of the so-called 'chrome greens'; particle morphologies appropriate to these phases may predominate in such material.

The optical and chemical characteristics of Prussian blues have been reviewed by Berrie (1997).

Lit.: Berrie (1997); Townsend (1993); Welsh (1988)

INDIGO

$C_{16}H_{10}N_2O_2$

Crystallinity

Anisotropic	n	δ
	>1.662	Low
		(data from authors)

In plane-polarised light, indigo forms weakly translucent, dark blue particles. Relief is moderate and RI is greater than that of the medium, but this can be difficult to tell in practice. Mactaggart (2002) further notes that under very strong illumination the particles can be seen to have variable relief. Particle size is very fine and close to that resolvable by the optical microscope. Therefore, grains appear rounded or bacterioid in shape. Aggregates of particles have convoluted surfaces. Natural and synthetic forms of indigo are difficult to differentiate optically; synthetic pigments may be expected to be purer, but these may also be adulterated with other phases.

Under crossed polars, as McCrone et al. (1973–80) put it, 'much disappears'. Indigo is weakly birefringent but interference colours are strongly masked by the body colour. Particles appear blue.

Using the Chelsea filter, indigo can appear deep red due to the high transmission at that end of the spectrum commencing at about 690 nm. However, this may be difficult to see.

Indigo may be adulterated or extended with various phases including aluminium hydroxide, calcite and clays. Howard (2003) additionally notes in a mediaeval English context the use of lead white, chalk, crushed marble, powdered eggshell and gypsum. Some preparations of indigo strongly resemble Prussian blue; however, the two compounds may be differentiated by the presence of indigo's weak birefringence and high red transmission.

So-called Maya blue, a structural composite of indigo and palygorskite, is discussed in the entry for palygorskite.

The optical and physical properties of indigo and woad have been described by Schweppe (1997).

Lit.: Howard (2003); Mactaggart (2002); McCrone et al. (1973–80); Schweppe (1997)

The Pigment Compendium

INDIGO, BROMINATED AND INDIGO, SULFONATED

$C_{16}H_8N_2O_2Br_2$ and $C_{16}H_8N_2Na_2O_8S_2$

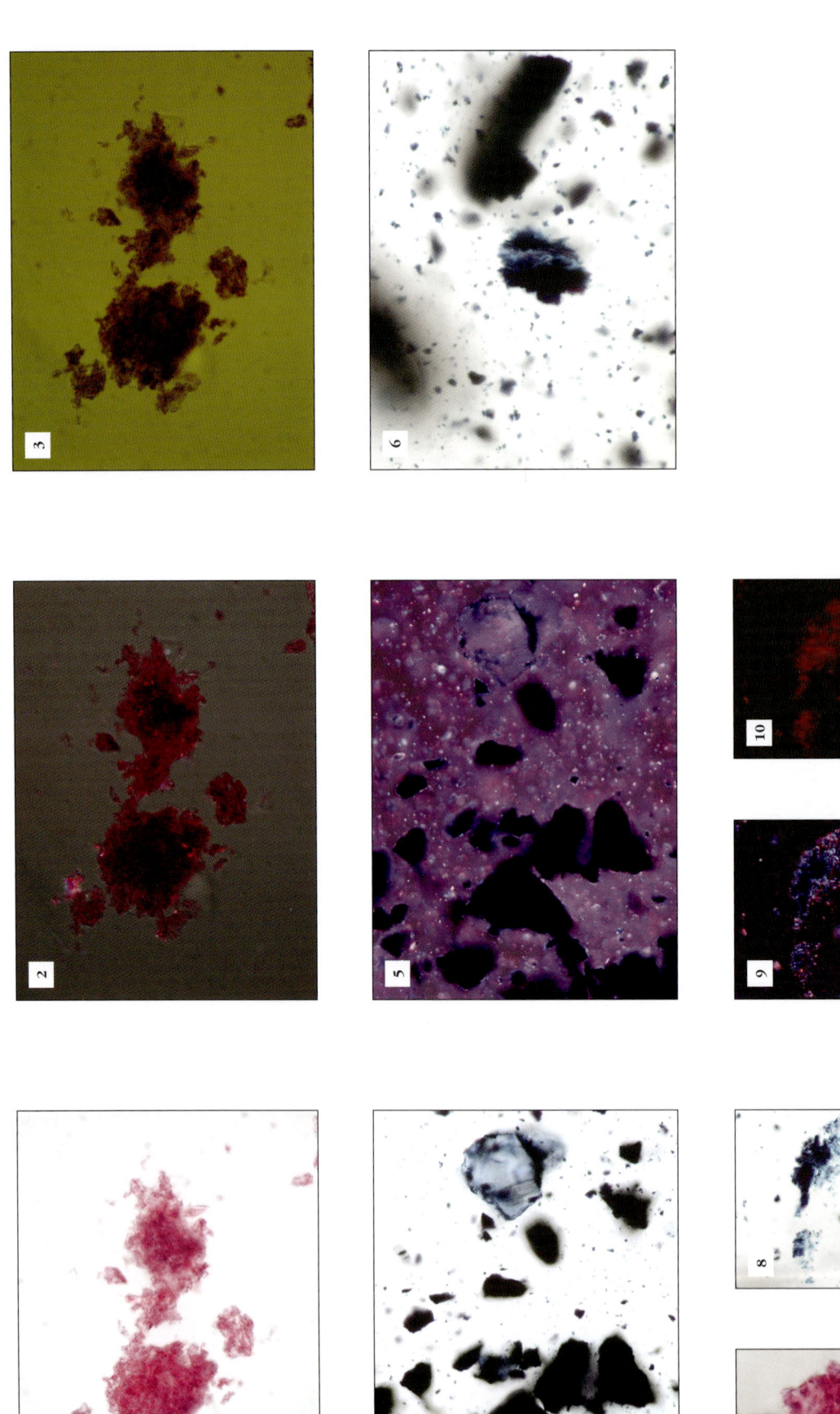

Key:

1. P0178: PPL/100×. Polycrystalline aggregates of brominated indigo.

2. P0178: ~XPL/100×. The same field of view as in Fig. 1 under crossed polars.

3. P0178: PPL/CF/100×. The same field of view as in Fig. 1 through the Chelsea filter.

4. P1520: PPL/40×. Opaque aggregates of sulfonated indigo with translucent quartz grain.

5. P1520: XPL/40×. ○ The same view as in Fig. 4 under crossed polars, note the anomalous pink birefringence of indigo particles.

6. P1520: PPL/100×. Fibrous particle morphology of sulfonated indigo.

7. P0178: PPL. Enlargement of particles of brominated indigo.

8. P1520: PPL. Aggregate of fibrous particles of sulfonated indigo.

9. P1520: XPL. The same view as in Fig. 8 under crossed polars.

10. P1520: RPL. The same view as in Fig. 8 in reflected light.

Crystallinity

	n	δ
Brominated indigo: Anisotropic	>1.662	Low
Sulfonated indigo: Anisotropic	<1.662	Low
		(data from authors)

In plane-polarised light, brominated indigo ('Tyrian purple') forms translucent to weakly translucent, pink-purple particles. Relief is moderate and RI is greater than that of the medium. Particle habits are of irregular, ragged flakes with rough, pitted and convoluted grain surfaces. Particle size varies from fine to very coarse.

Under crossed polars, the particles are observed to be polycrystalline and weakly anisotropic. Birefringence is moderate to low and interference colours are strongly masked by the body colour. Particles appear pink. Complete extinction is not obtained and particles twinkle as the stage is rotated.

Brominated indigo appears bright red when viewed through the Chelsea filter.

In plane-polarised light, sulfonated indigo ('indigo carmine') forms very weakly translucent to opaque dark blue grains; a few particles are red. Relief is moderate and RI is less than that of the medium. Particles appear as angular flakes. Particle size distribution is broad, with grains ranging in size from very fine to very coarse.

Under crossed polars, coarse, opaque indigo carmine is isotropic. However, fine-grained particles are birefringent and have anomalous pink and blue interference colours. Extinction is sweeping or undulose.

Sulfonated indigo appears a dull dark red when viewed through the Chelsea filter.

Lit.: Schweppe (1997)

PHTHALOCYANINE GROUP [1]

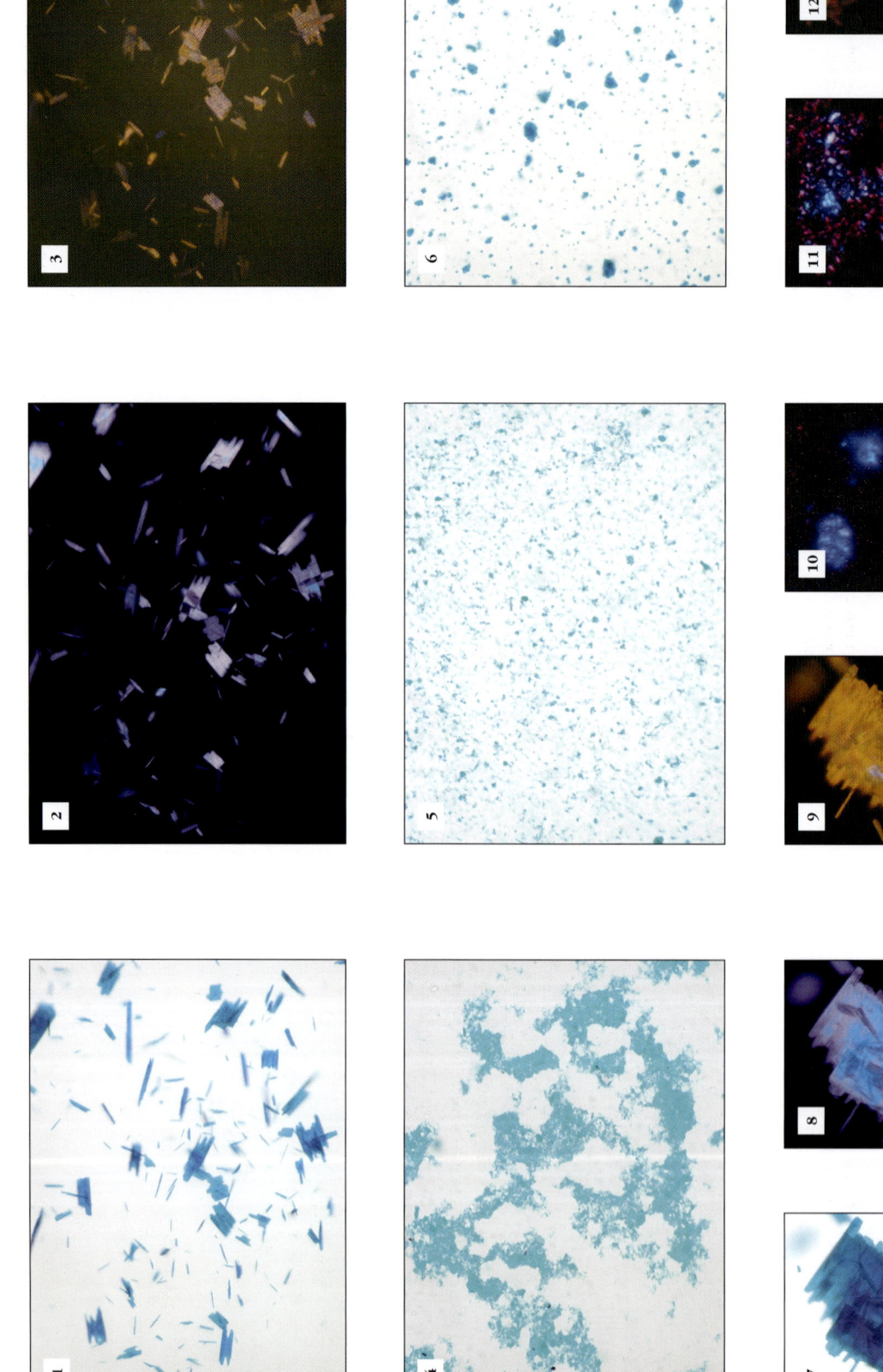

Key:

1. P0856: PPL/40×. Bladed and tabular crystals.

2. P0856: XPL/40×. View as in Fig. 1 under crossed polars.

3. 0856: RPL/40×. View as in Fig. 1 in reflected light.

4. P0197: PPL/100×. Finely crystalline turquoise blue particles with particle morphology difficult to determine.

5. P0749: PPL/100×/H. Finely crystalline turquoise phthalocyanine blue grains.

6. P0826: PPL/100×. Turquoise blue aggregates and individual particles of phthalocyanine blue.

7. P0856: PPL. Enlargement of particle from sample in Fig. 1.

8. P0856: XPL. Particle as in Fig. 7 under crossed polars.

9. P0856: RPL. Particle as in Fig. 7 in reflected light.

10. P0826: XPL. Enlarged area of Fig. 6 under crossed polars.

11. P0197: XPL. Enlarged area of Fig. 4 under crossed polars, note anomalous red and blue interference colours.

12. P0197: RPL. The same view as in Fig. 11 in reflected light.

Crystallinity
Anisotropic

n	δ
<1.662	Weak
	(data from authors)

The phthalocyanines most likely to be encountered as pigments are copper phthalocyanine ('CuPc') and its halogenated derivatives. Unsubstituted CuPc is polymorphous with five known crystal modifications (α, β, γ, δ, ε), though these have different stabilities. It is the phase-stabilised α- and the β-form which have found most commercial use. These various forms fall within the *Colour Index* (1971) designation CI Pigment Blue 15. In the copper polyhalophthalocyanines it is the degree of substitution that affects the colour, with only one crystal modification having been reported (similar to the α blue form). The bluish chlorinated form is CI Pigment Green 7, the yellowish bromochloro form CI Pigment Green 36 (Herbst and Hunger, 1997). As a result of the range of possible forms, confirmation of the presence of specific species of phthalocyanine compound normally requires a range of analytical tools in addition to optical microscopy. However, some examples of those that may be encountered are described below.

In plane-polarised light, phthalocyanine blue I forms weakly translucent to almost opaque, bright turquoise blue particles. Particles have three distinct morphologies. Some particles are fine-grained euhedral, acicular or lath-like crystals. Others are medium-grained, rounded aggregates of fine particles, and finally there is a very fine-grained groundmass, with particle size close to that resolvable by the optical microscope; however, some of these can be observed to be finely acicular. All particles have low relief, with RI just less than that of the medium.

In crossed polars, phthalocyanine blue I has moderate birefringence, but interference colours are highly anomalous. Acicular and rounded polycrystalline particles are bright blue and laths show complete, straight extinction. In contrast the fine-grained material is anomalous deep pink-red. The overall effect is for the blue particles to appear to stand proud of the red background. Phthalocyanine blue I appears deep red when viewed through the Chelsea filter.

In plane-polarised light, phthalocyanine blue II forms translucent to weakly translucent, intense blue particles. Crystals have low relief and RI just less than that of the medium. Particle shape is entirely of bladed and tabular crystals. Particle surfaces are striated parallel to the long axes of the crystals. Particle size is uniform and coarse.

Under crossed polars, phthalocyanine blue II is strongly birefringent, but interference colours are masked by the strong body colour and particles appear bright blue. Particles have complete and straight extinction. Phthalocyanine blue II appears black when viewed through the Chelsea filter.

Lit.: Herbst & Hunger (1997)

The Pigment Compendium

PHTHALOCYANINE GROUP [2]

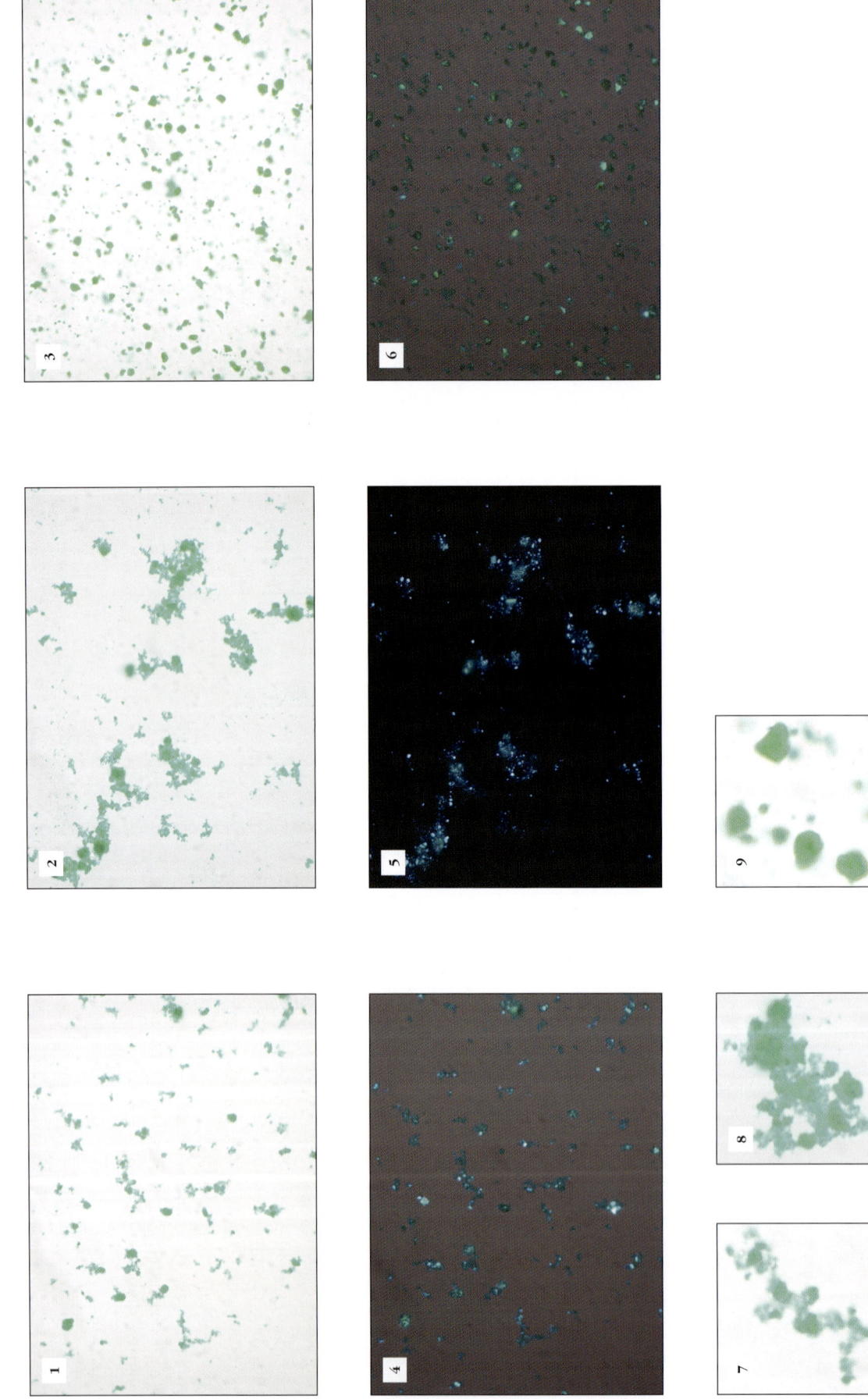

Key:

1. P0035: PPL/100×. Blue-green subrounded crystals and aggregates of phthalocyanine green.

2. P0035: PPL/100×. Aggregates of blue-green particles of phthalocyanine green.

3. P0036: PPL/100×. Subrounded green platy particles of phthalocyanine green.

4. P0035: ~XPL/100×. Phthalocyanine green crystals in partially crossed polars with interference colours masked by the body colour. Same field of view as in Fig. 1.

5. P0035: XPL/100×/○. Anomalous blue to green interference colours of phthalocyanine green grains in crossed polars. Same view as in Fig. 2.

6. P0036: ~XPL/100×. Weakly birefringent phthalocyanine grains with interference colours masked by the green body colour. Same view as in Fig. 3.

7. P0035: PPL. Blue-green subrounded crystals and aggregates of phthalocyanine green. Enlargement of part of Fig. 1.

8. P0035: PPL. Aggregates of blue-green particles of phthalocyanine green. Enlargement of part of Fig. 2.

9. P0036: PPL. Subrounded green plates of phthalocyanine green. Enlargement of part of Fig. 3.

Crystallinity

Anisotropic	n	δ
	<1.662	Weak
		(data from authors)

In plane-polarised light, phthalocyanine green I forms weakly translucent, intense blue green particles. Relief is low and RI is just less than that of the medium. Particle morphologies are predominantly rounded and subrounded plates, but a few bladed euhedral crystals occur. Particle surfaces are smooth. Particle size ranges from fine to medium.

In crossed polars, phthalocyanine green I has high birefringence, but interference colours are highly anomalous and masked by the body colour. Coarser-grained particles appear bright blue-green, whereas the finest particles are anomalous pink-red. Extinction is undulose.

In plane-polarised light, phthalocyanine green II forms weakly translucent, intense green particles. Relief is low and RI is just less than that of the medium. Particle morphologies are of rounded and subrounded plates. Particle surfaces are smooth. Particle size ranges from fine to medium.

Under crossed polars, some phthalocyanine green II particles are weakly birefringent and show low order interference colours, strongly masked by the body colour. Particles are polycrystalline and have sweeping or undulose extinction. Others particles are isotropic.

The appearance of phthalocyanine green is unaffected when viewed through the Chelsea filter.

Lit.: *Colour Index* (1971); Herbst & Hunger (1997)

ANTIMONY(III) SULFIDE, AMORPHOUS

$Sb_2S_3(am)$

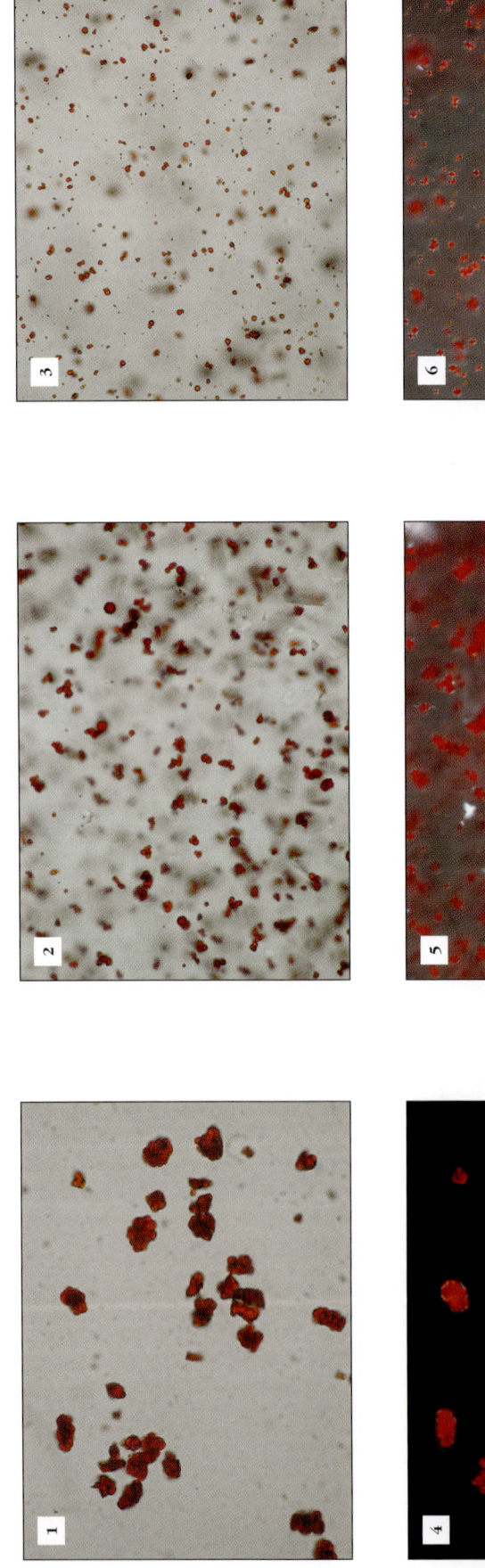

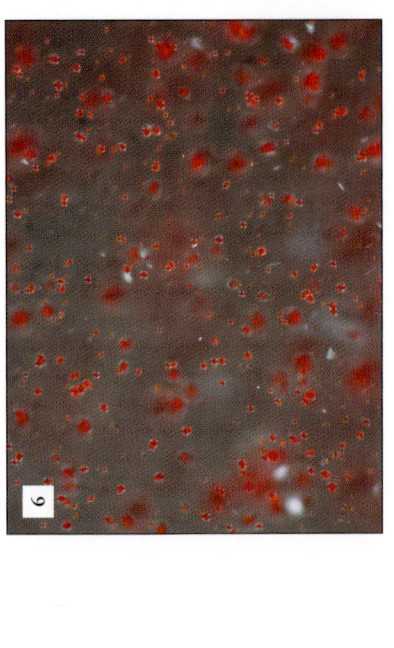

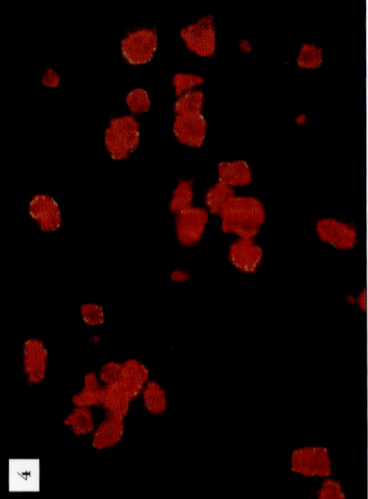

Key:

1. P1087: PPL/100×. ○ Particles of antimony(III) sulfide.

4. P1087: XPL/100×. ○ The field of view as in Fig. 1 under crossed polars.

2. P1094: PPL/100×. Particles of antimony(III) sulfide.

5. P1094: ~XPL/100×. The field of view as in Fig. 2 under partially crossed polars.

3. P1094: PPL/100×. Translucent plates of antimony(III) sulfide.

6. P1094: ~XPL/100×. The field of view as in Fig. 3 under crossed polars; the particles are weakly anisotropic at their margins, and extinction crosses may be observed.

Crystallinity
Amorphous

n	$\hat{\delta}$
>>1.662	Isotropic
	(data from authors)

A strong red powder is produced from this compound, which is generally considered to be an amorphous form of the otherwise normally black antimony(III) sulfide mineral stibnite (*q.v.*).

Under plane-polarised light this compound appears as orange translucent particles. The colour varies from orange to orange-red, although the individual crystals are not pleochroic. Relief is high and RI greater than that of the medium. Some internal and/or surface features are visible, giving the particles a pitted appearance. Particles are anhedral and angular to subangular, to bacterioid for the smallest size fraction. Particle size ranges from fine to very fine.

Under cross-polarised light the compound is isotropic, but with strong internal reflections that cause the crystals to glow with a blood-red colour.

ANTIMONY(III) SULFIDE, AMORPHOUS

ORPIMENT AND ARSENIC SULFIDE, ORPIMENT TYPE

As_2S_3

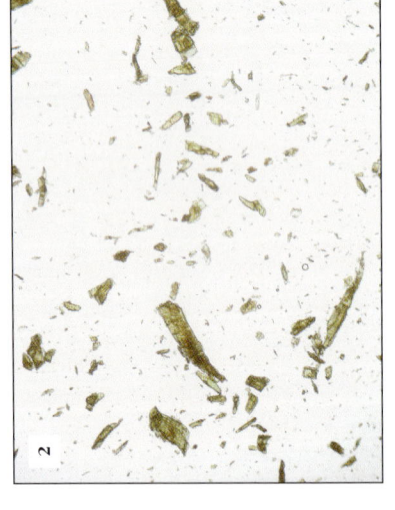

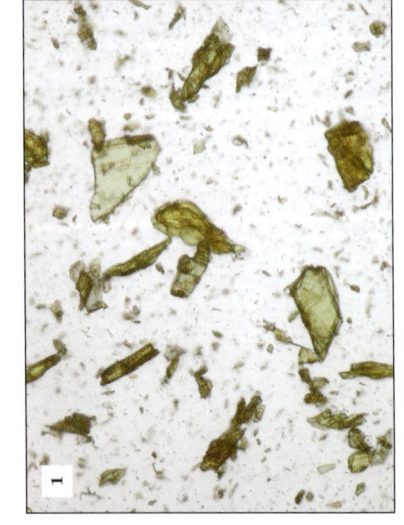

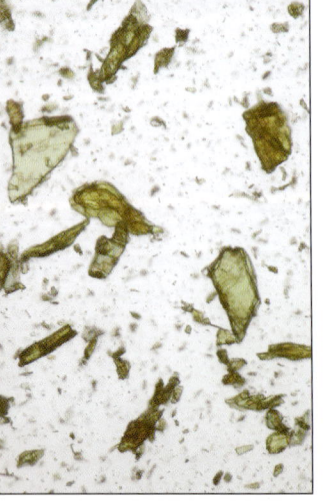

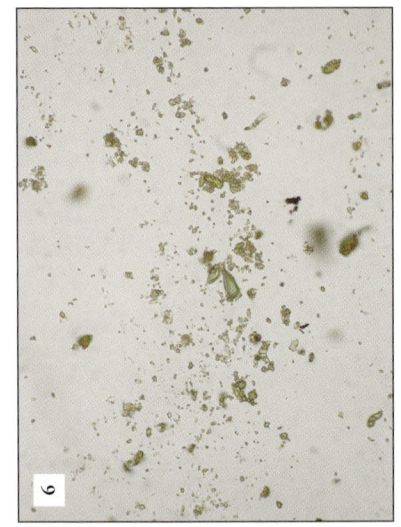

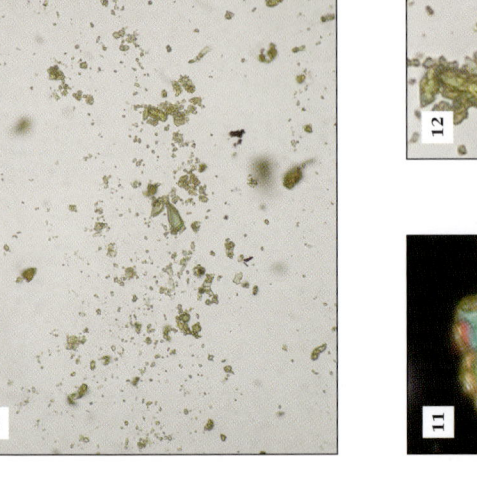

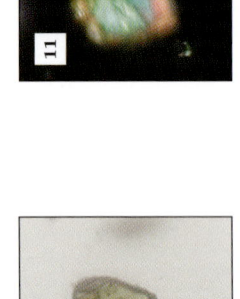

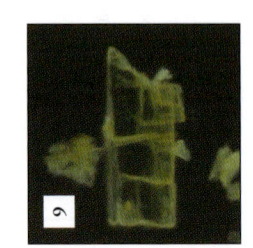

Key:

1. P0333: PPL/40×/H. Angular crystals of orpiment.
2. P0333: PPL/10×/H. Splinter-like particles of orpiment.
3. P0333: XPL/40×/H. Orpiment crystals under crossed polars.
4. P0333: XPL/40×/H. The field of view as in Fig. 1 under crossed polars.
5. P0333: XPL/10×/H. The field of view as in Fig. 2 under crossed polars.
6. P0332: PPL/100×/H. Fine- to medium-grained angular particles of orpiment.
7. P0011: PPL. Bladed crystal of orpiment.
8. P0011: PPL. The crystal as in Fig. 7 with the polariser rotated through 90° showing change in body colour due to pleochroism.
9. P0011: UV. The crystal as in Fig. 7 in reflected light.
10. P0658: PPL/H. An anhedral crystal of orpiment.
11. P0658: XPL/H. The crystal as in Fig. 10 showing third order interference colours.
12. P0332: PPL/H. Detail of particle morphology from Fig. 6.

Crystallinity	Optic sign	2V	n_α	n_β	n_γ	δ
Monoclinic	−ve	30°–76°	2.40	2.81	3.02	0.6200

(data for mineral from Mason, 1968; cf. Webmineral, 2003)

In plane-polarised light, the most striking characteristics of the arsenic sulfide mineral orpiment are its very high relief and its straw-yellow colour. Dispersion is relatively strong. Additionally, it frequently forms coarse-grained, foliated, platy and micaceous-looking crystals that, coupled with its perfect cleavage, give the internal structures of the particles a crosshatched appearance. This phenomenon can be so pronounced as to render the crystals almost opaque. Other crystal habits include fibrous and acicular-prismatic forms and earthy to granular aggregates. With reflected light orpiment appears grey.

Under crossed polars orpiment is strongly birefringent, typically showing bright pink and green interference colours; this is a diagnostically important feature. Orpiment has straight extinction; particles are length fast.

Synthetic orpiment, produced by a 'dry' sublimation process, is difficult to differentiate from the naturally occurring mineral. However, as a rule of thumb synthetic orpiment is likely to have a fine to medium particle size. In such instances the extremely high relief all but obliterates the internal features of the crystals in both plane-polarised and cross-polarised light. Arsenic(III) oxide is also likely to be present in synthetic material. Electron micrographs of synthetic orpiment have been published by Wallert (1984).

Orpiment is often derived from geothermal and volcanic environments and consequently generally occurs with other arsenic sulfide minerals, notably realgar and pararealgar. Orpiment is also a common trace mineral found in association with zones of hydrothermal antimony mineralization. In addition to phases such as stibnite, minerals including calcite, baryte and gypsum are also commonly associated with orpiment-bearing assemblages. However, according to FitzHugh (1997), mineral orpiment found in artefacts is always remarkably pure while the synthetic analogue is often contaminated with arsenic(III) oxide, a starting material in the common manufacturing process.

An amorphous synthetic arsenic sulfide prepared by aqueous precipitation has been described by FitzHugh; it is unclear whether this ever found commercial use; however, the example documented by that author had a particle size of about 1 μm with a rounded morphology. It is, of course, isotropic under crossed polars.

The chemical and optical properties of orpiment have been reviewed by FitzHugh (1997).

Lit.: FitzHugh (1997); Wallert (1984)

The Pigment Compendium

REALGAR

As$_2$S$_2$

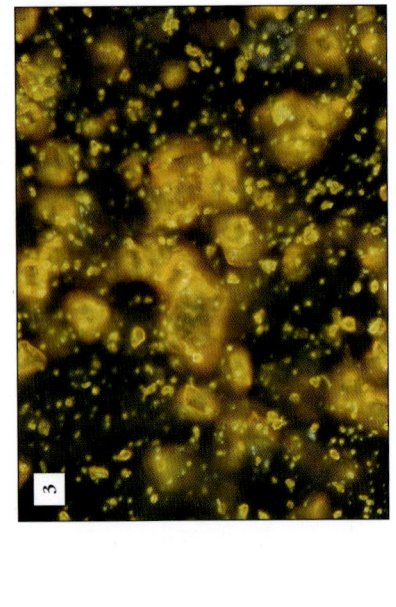

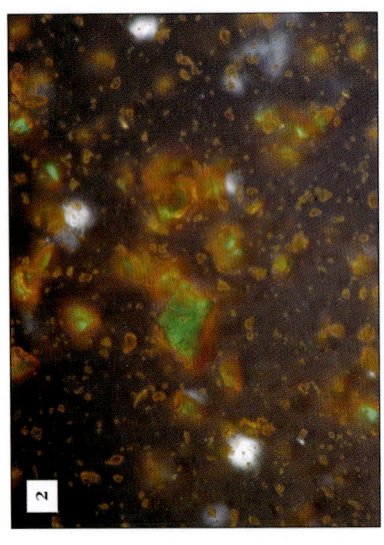

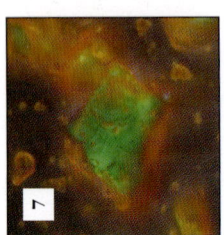

Key:

1. P1474: PPL/100×. Fine-grained realgar sample with typical rhombic grain morphology.

2. P1474: XPL/100×. Typical green interference colours in crossed polars. Same field of view as in Fig. 1.

3. P1474: RPL/100×. Golden yellow appearance of realgar crystals in reflected light . Same field of view as in Fig. 1.

4. P1312: PPL/40×/H. Coarse realgar grains with conchoidal fracture and sets of parallel cleavages.

5. P1312: XPL/40×/H. Isotropic example with strong orange reflections at grain edges. Same field of view as in Fig. 4.

6. P1312: RPL/40×/H. Typical golden yellow colour in reflected light from grain edges. Same field of view as in Fig. 4.

7. P1474: XPL. Rhombic crystal showing typical green interference colours in crossed polars.

8. P1312: PPL/H. Coarse realgar grains with conchoidal fracture.

9. P1312: PPL/H. Common rhombic crystal habit formed after grinding.

10. P1286: PPL. Conchoidal fracture and evidence of cleavage.

Crystallinity	Optic sign	2V	n_α	n_β	n_γ	δ
Monoclinic	−ve	38°–40°	2.538	2.684	2.704	0.1660

(data for mineral from Sinkankas, 1966; cf. Webmineral, 2003)

Realgar is often found in association with orpiment (*q.v.*), and may be present as traces in samples of orpiment. However, this orange arsenic sulfide has been used as a pigment in its own right.

In plane-polarised light, realgar occurs as yellow-orange particles. Some, but not all, samples exhibit pleochroism, with colours that range from orange-red to orange-yellow. The crystals have extremely high relief but are on the whole more translucent and have less internal structure than the similar orpiment. Fluid inclusions may be present. Dispersion is relatively strong. Particle size is likely to be medium to coarse and size distribution is often broad. The process of crushing and the one perfect cleavage generally dictate particle morphology, commonly resulting in straight-edged rhombic forms. A weak conchoidal fracture also gives shard like particles with curved boundaries.

Under crossed polars, realgar has distinctive and strong orange internal reflections. It has high birefringence, although the interference colours are masked by the body colour and the high relief; however, bright blues and greens are typical. Realgar has straight extinction.

Both FitzHugh (1997) and Short (1940) have noted that realgar is isotropic and some polymorphs of the compound are indeed amorphous (see Lengke and Tempel, 2003). In such cases, a good conchoidal fracture is developed and strong orange internal reflections occur.

Realgar may be found in association with orpiment. It may also occur with cinnabar.

The chemical and optical properties of realgar have been reviewed by FitzHugh (1997).

Lit.: FitzHugh (1997); Lengke & Tempel (2003); Short (1940); Sinkankas (1966)

The Pigment Compendium

PARAREALGAR

AsS

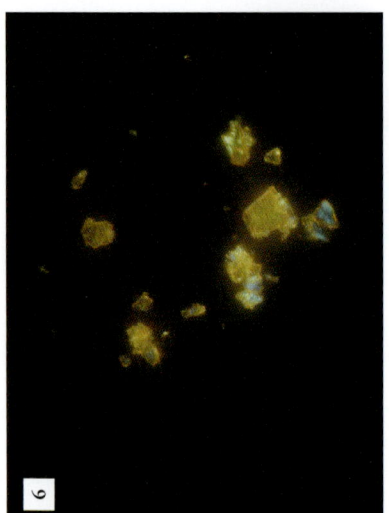

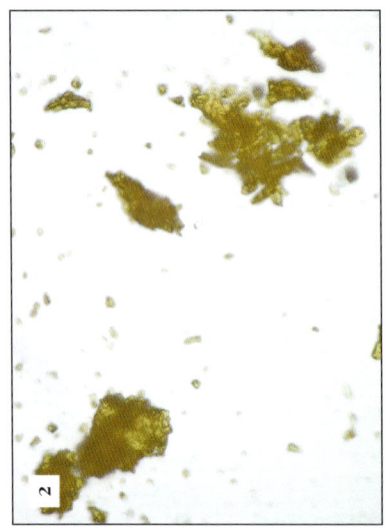

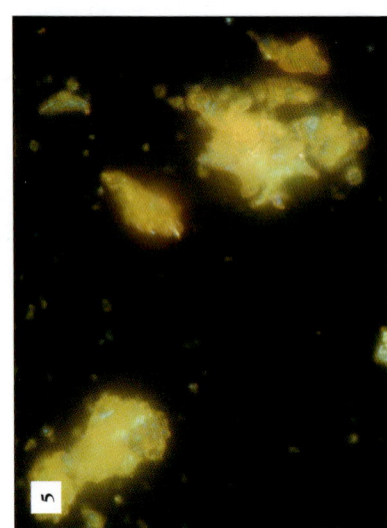

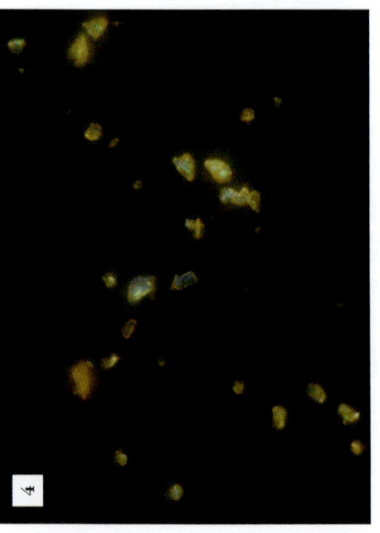

Key:

1. P1514: PPL/100×. Angular crystals of pararealgar.
2. P1514: PPL/100×. Polycrystalline particles of pararealgar.
3. P1514: PPL/100×. Angular crystals of pararealgar.
4. P1514: XPL/100×. The field of view as in Fig. 1 under crossed polars.
5. P1514: XPL/100×. The field of view as in Fig. 2 in crossed polars.
6. P1514: XPL/100×. The field of view as in Fig. 3 in crossed polars.
7. P1514: PPL. Detail of crystal from Fig. 3.
8. P1514: XPL. The crystal in Fig. 7 under crossed polars.

Crystallinity	Optic sign	2V	$n_\alpha/n_\beta/n_\gamma$	δ
Monoclinic	[unknown]	[unknown]	>2.02	High
				(data from Roberts et al., 1980)

Pararealgar most commonly occurs as a light-induced alteration product of realgar (q.v.), passing through an intermediate phase that is often also present.

Corbeil and Helwig (1995) describe examples of pararealgar derived from alteration as appearing as coarse (≥10 μm) yellow particles with irregular faces and a pebbled surface. Relief is very high, the RI being substantially greater than that of the mounting medium. Pararealgar exhibits gold to orange-red internal reflections.

Under crossed polars pararealgar particles exhibit green or blue anomalous interference colours, each particle showing only one colour. Extinction is complete.

Pararealgar is most likely to be found in association with realgar.

Lit.: Corbeil & Helwig (1995); Roberts et al. (1980)

CADMIUM SULFIDE, GREENOCKITE TYPE

CdS

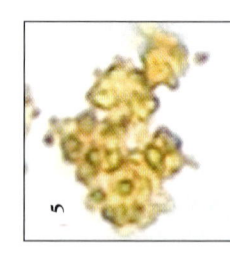

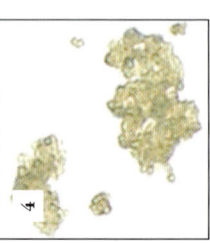

Key:

1 P1305: PPL/100×/H. Aggregates of rounded grains.

2. P1305: XPL/100×/H. Same field of view as Fig. 1 in crossed polars.

3. P1305: RPL/100×/H. Same field of view as Fig. 1 in reflected light.

4. P0030: PPL. Aggregates of rounded yellow particles.

5. P1305: PPL/H. grains and aggregates of finely crystalline particles.

Crystallinity	Optic sign	n_ω	n_ϵ	δ
Hexagonal	−ve	2.529	2.506	0.0230
			(data for mineral from Ford, 1932; cf. Webmineral, 2003)	

Under plane-polarised light, the synthetic cadmium sulfide analogue of the mineral greenockite may be seen as very fine-grained particles, of 1 μm or less in the samples examined; typical modern pigments might have particle sizes as low as 0.1 μm, ranging to 3–4 μm, with narrow size distribution (Fiedler and Bayard, 1986). Consequently particle shape is difficult to determine, though they appear to have spherical to bacterioid habits. Rare coarser-grained material (fine particle size) can be seen to adopt cubic shapes. Material produced by the 'dry' process, which involves sublimation, can show filaments and needles, stubby prisms and twins (Fiedler and Bayard). Fiedler and Bayard also document rarer pseudo-spherulitic (fused agglomerates of particles with no specific orientation but rosette-like morphology) and true spherulitic forms. The particles are translucent, but strongly coloured and colour may vary from sample to sample from slightly greenish yellow to orange yellow depending on shade and grain size. The particles cannot normally be seen to exhibit pleochroism although cadmium sulfide has been reported to be weakly so (Dana, 1944). Relief is high, with the RI greater than that of the medium.

Strikingly, under crossed polars this form of cadmium sulfide has very strong acid yellow to yellow-green internal reflections that cause the sample to glow; Fiedler and Bayard additionally note that anomalous polarisation effects in the edges of particles can lead to only the edges of grains being visible. The birefringence is moderate but strongly masked by the body colour, the relief and the reflections. The finely crystalline particles can be seen to twinkle as the stage is rotated, though individual particles may not necessarily exhibit extinction to a degree that varies from sample to sample. Agglomeration of very fine particles may also lead to poorly visible extinction. It is yellow in reflected light.

Cadmium sulfide also notably fluoresces (de la Rie, 1982) although the strength of this varies and in some cases pigments may have no visible fluorescence at all. Emission wavelengths peak in the red part of the spectrum.

Cadmium lithopones, cadmium sulfide co-precipitated with barium sulfate, are known; these are difficult to characterise using optical microscopy (Fiedler and Bayard).

Lit.: Dana (1944); de la Rie (1982); Fiedler & Bayard (1986)

CADMIUM SULFIDE, HAWLEYITE TYPE AND CADMIUM SULFIDE, AMORPHOUS TYPE

CdS

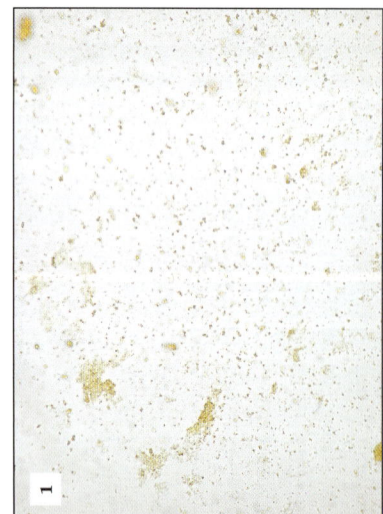

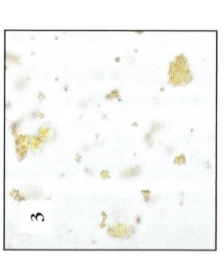

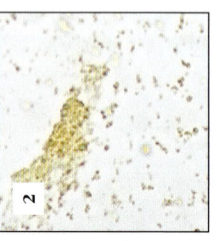

Crystallinity
Cubic

n
>2.0

δ
Isotropic
(data from Fiedler and Bayard, 1986)

Under plane-polarised light the synthetic cadmium sulfide analogue of the mineral *hawleyite* may be seen as very fine-grained particles, of 1 μm or less in the samples examined. Consequently particle shape is difficult to determine, though they appear to have spherical to bacterioid habits. Rare coarser-grained material (fine particle size) can be seen to adopt cubic shapes. The particles are translucent to transparent, but strongly coloured bright yellow; in reflected light the body colour of the pigment can be seen as weakly yellow to greenish yellow. The particles do not exhibit pleochroism. Relief is high, with RI greater than the medium.

Under crossed polars this form of cadmium sulfide can be seen to be isotropic and, unlike the greenockite type, has very low or no internal reflections. It is yellow in reflected light.

Earlier examples of this form of cadmium sulfide may occur with a proportion of cadmium carbonate (*otavite*) present, this compound being used to prepare paler shades; sample P1307 illustrated above is an example of this. Fiedler and Bayard (1986) also note that the example of the synthetic hawleyite form that they studied also contained a proportion of the hexagonal greenockite type of cadmium sulfide; this is not unexpected since this latter compound was usually roasted to prepare the former.

An amorphous form of cadmium sulfide also exists. According to Fiedler and Bayard, this pigment appears as composed of individual particles or agglomerates which often contain centres with a fused appearance; individual areas of fused material are also observed. Small amounts of crystalline cadmium sulfide are also reported to be present.

The optical and chemical properties of cadmium yellows, oranges and reds have been reviewed by Fiedler and Bayard (1986).

Lit.: de la Rie (1982); Fiedler & Bayard (1986)

CADMIUM SULFIDE, HAWLEYITE TYPE AND CADMIUM SULFIDE, AMORPHOUS TYPE

CADMIUM SELENIDE SULFIDE AND CADMIUM SELENIDE

Cd(Se,S) and CdSe

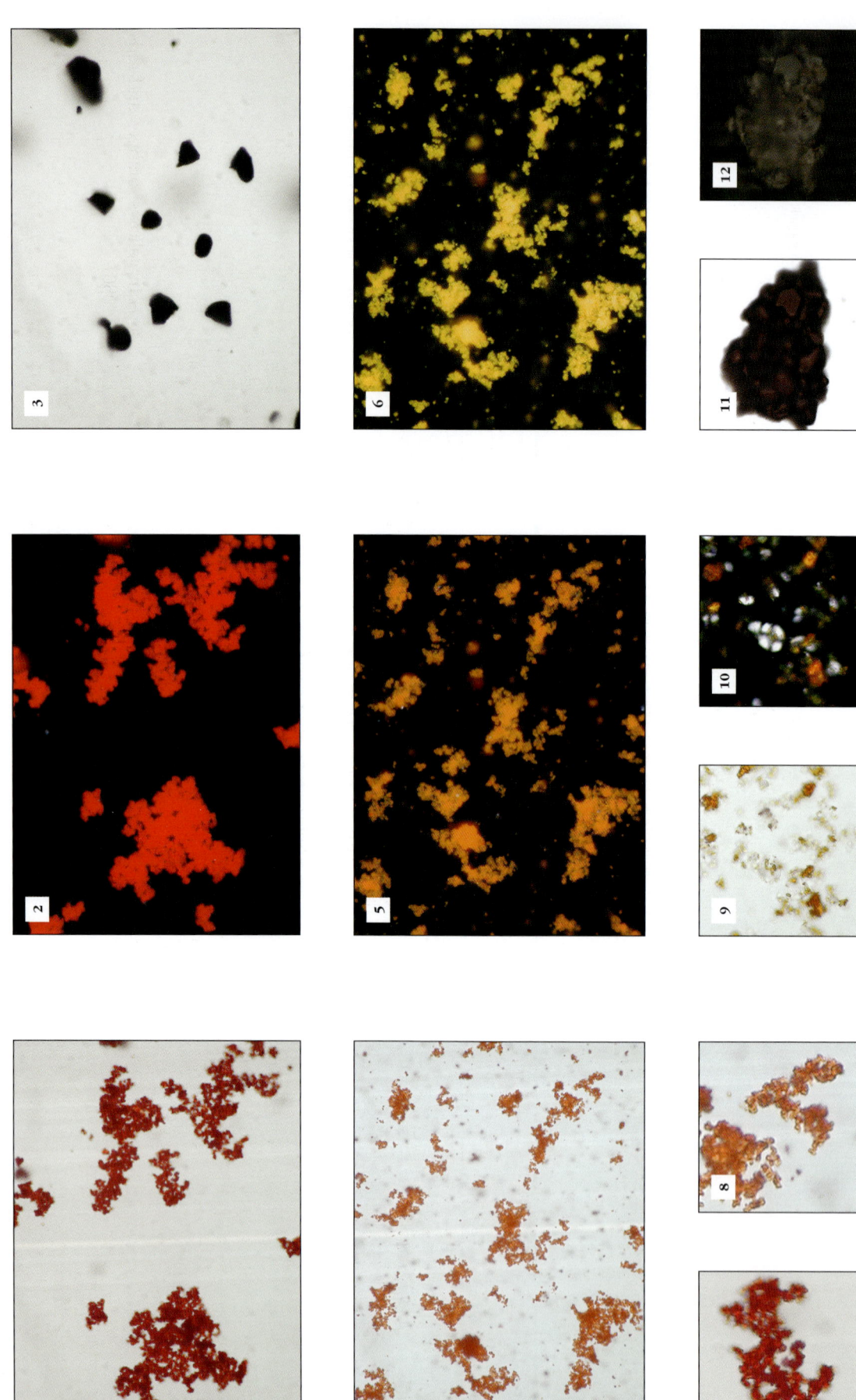

Key:

1. P0033: PPL/100×. Fine-grained rounded grains of cadmium red.

2. P0033: XPL/100× ● Strong internal reflections and masking by the body colour of the interference colours of cadmium red. Same field of view as in Fig. 1.

3. P1526: PPL/100×. Opaque particles of pure cadmium selenide.

4. P0244: PPL/100×. Fine-grained rounded particles of cadmium orange.

5. P0244: XPL/100×. Field of view as in Fig. 4 under crossed polars with bright orange internal reflections.

6. P0244: RPL/100×. Field of view as in Fig. 4 in reflected light, note strong yellow internal reflections.

7. P0033: PPL. Fine-grained aggregate of rounded grains of cadmium red.

8. P0244: PPL. Fine-grained aggregates of rounded particles of cadmium orange.

9. P0245: PPL. Cadmium orange mixed with chalk.

10. P0245: XPL. The field view as in Fig. 9 in crossed polars; note coccolith with standing extinction cross.

11. P1526: PPL/ ● Weakly translucent particle of cadmium selenide.

12. P1526: RPL. The particle as in Fig. 11 in reflected light.

Crystallinity	Optic sign	n_ω	n_ε	δ
Hexagonal	–ve	~2.53	~2.51	~0.02
				(data derived from Fiedler and Bayard, 1986)

Substitution of selenium for sulfur in cadmium sulfide produces a range of colours, from the yellow cadmium sulfide to red and orange cadmium sulfide selenides to red cadmium selenide. However, the greenockite crystal structure persists and therefore other optical properties such as refractive index and birefringence are not greatly dissimilar to those of cadmium sulfide itself.

Under plane-polarised light, cadmium selenide sulfides form very fine-grained, spherical to bacterioid particles. In colour they are translucent and bright orange, with high relief, RI is greater than that of the medium. According to McCrone *et al.* (1973–80) the RI varies such that for 'bright red' $n = 2.64$ and 'deep red' $n = 2.77$; this presumably reflects the increasing substitution of selenium for sulfur. With reflected light the body colour of the pigment is more apparent, seen as yellow or orange to red.

Under crossed polars, the most distinctive phenomenon observed is the very strong, bright orange internal reflections. The particles are anisotropic and have moderate birefringence, but interference colours are masked by the strong reflections and the body colour. The fine particle size makes extinction angle determinations difficult to analyse.

Under plane-polarised light, *cadmium selenide* forms very fine-grained, spherical to bacterioid particles. In colour the pigment appears as translucent to opaque and of a deep red colour. The particles have high relief, with the RI greater than that of the medium. Colour in reflected light is red.

Under crossed polars, the most distinctive phenomenon observed is the very strong, red internal reflections. The particles are anisotropic and have moderate birefringence, but interference colours are masked by the strong reflections and the body colour. The fine grain size of the particles makes extinction angle determinations difficult to analyse.

The cadmium sulfide selenides notably fluoresce (de la Rie, 1982) although the strength of this varies and in some cases pigments may have no visible fluorescence at all. Emission wavelengths peak in the red to near infrared part of the spectrum.

Lit.: de la Rie (1982); Fiedler and Bayard (1986); McCrone *et al.* (1973–80)

CADMIUM MERCURY SULFIDE AND CADMIUM ZINC SULFIDE

$(Cd_{1-x}Hg_x)S$ and $(Cd_{1-x}Zn_x)S$

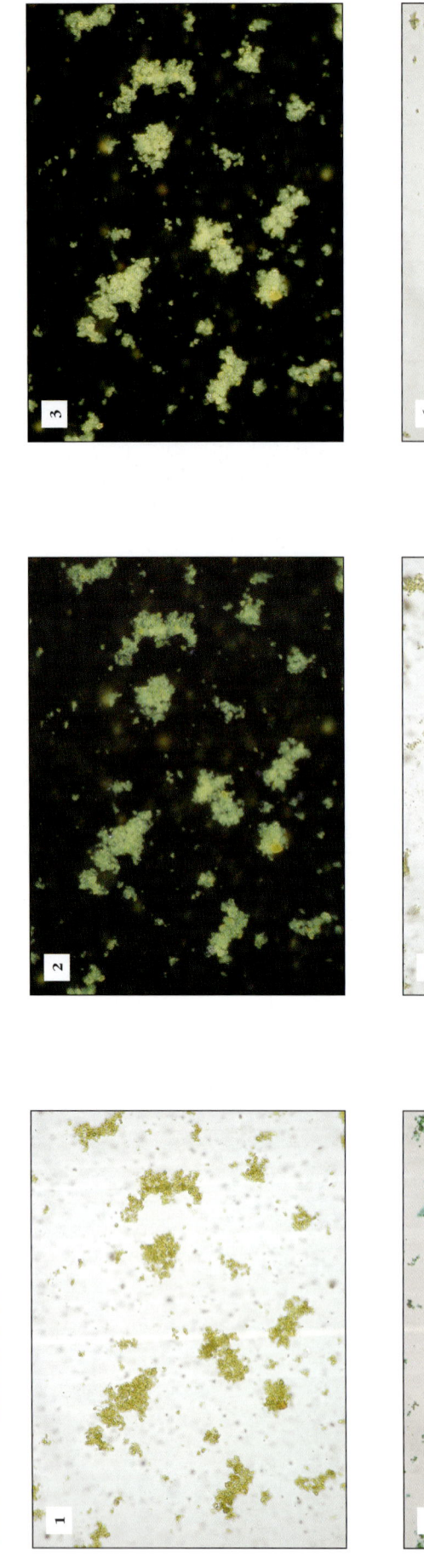
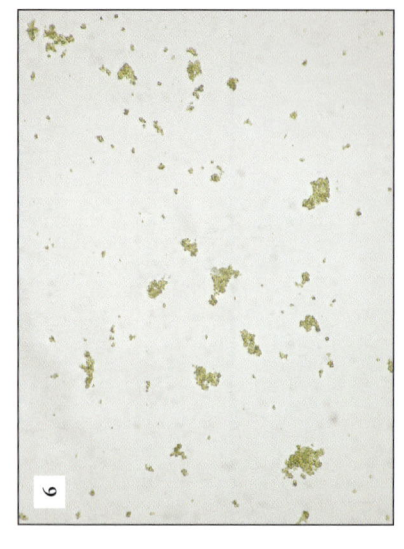

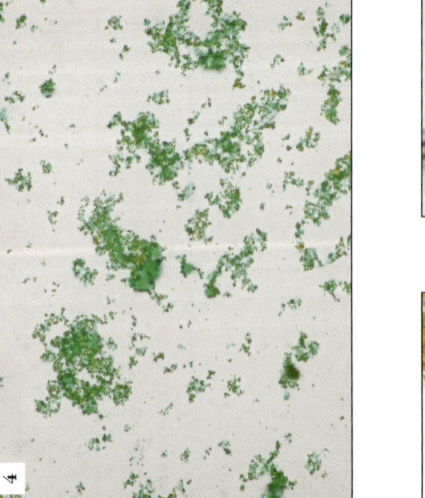

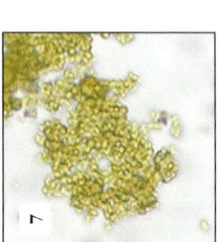

1. P0233: PPL/100×. Fine-grained particles of cadmium zinc sulfide.

2. P0233: XPL/100×. The field of view as in Fig. 1 under crossed polars.

3. P0233: RPL/100×. The field of view as in Fig. 1 under reflected light.

4. P0097: PPL/100×. Cadmium zinc sulfide mixed with a green phthalocyanine pigment.

5. P0236: PPL/100×. Fine-grained particles of cadmium zinc sulfide.

6. P0030: PPL/100×. Fine-grained particles of cadmium zinc sulfide.

7. P0233: PPL. Detail of particle morphology from Fig. 1.

8. P0097: PPL. Detail of particle morphology from Fig. 4.

Crystallinity	Optic sign	n_ω	n_ε	δ
Hexagonal	[unknown]	>2.529	>2.506	>0.0230
Hexagonal	−ve	2.378–2.529	2.356–2.506	~0.0230
			(data from Moore, 1973 and Fiedler and Bayard, 1986)	

Solid solutions can be produced that form cadmium-mercury and cadmium-zinc sulfide series (($Cd_{1-x}Hg_x$)S and ($Cd_{1-x}Zn_x$)S). According to Fiedler and Bayard (1986, citing Huckle et al., 1966, and Moore, 1973), the cadmium mercury sulfides were produced by initially co-precipitating cadmium and mercury sulfides in solution; the precipitate was then calcined in an inert atmosphere, a process that converts the cubic crystal structure to a hexagonal one. A compositional range of 11–26% HgS in the solid solution gives rise to colours from orange through to maroon. Forming paler yellow shades, the cadmium zinc sulfides were originally prepared by calcining a mixture of cadmium and zinc sulfides with either zinc or magnesium oxide (Laurie, 1914; cf. Fiedler and Bayard, 1986). However, the principal modern approach has been to form cadmium zinc sulfide where cadmium and zinc salts with the same anion were used, with up to about 25% zinc content (Huckle et al.; Moore).

Particle size of commercial cadmium mercury sulfide pigments falls into the 0.1–1.0 μm range (Moore, 1973). Relief is very high and the RI greater than

the mounting medium. Refractive indices of cadmium mercury sulfides will in fact deviate from the lower end-point of trigonal-hexagonal cadmium sulfide (2.529/2.506) towards that of the trigonal-trapezohedral mercury(II) sulfide, cinnabar type (2.905/3.256), the precise value depending on the extent of substitution; birefringence may also be significantly greater than that of pure cadmium sulfide. However, precise values over the normal range of substitution are not known.

According to Fiedler and Bayard (1986), the refractive index in cadmium zinc sulfides varies according to the proportion of zinc present between the values for hexagonal cadmium sulfide and a lower end point of $n_\varepsilon = 2.356$ and $n_\omega = 2.378$. Pure zinc sulfide is cubic, with a refractive index of 2.37.

Cadmium mercury sulfide pigments were also manufactured in lithopone varieties with barium sulfate present (Moore).

Lit.: Fiedler & Bayard (1986); Huckle et al. (1966); Laurie (1914); Moore (1973)

IRON OXIDE HYDROXIDE, GOETHITE TYPE

α-FeOOH

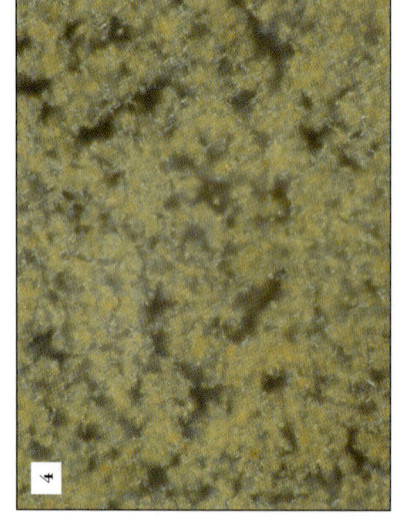

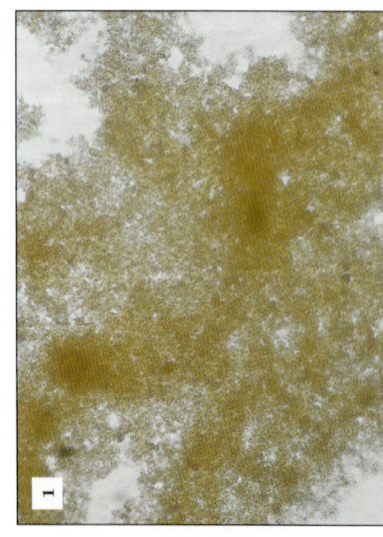

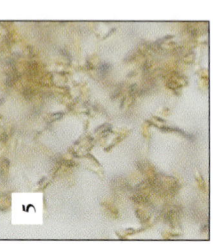

Key:

1. P0130: PPL/100×/ ⬤ Very fine-grained yellow particles of iron oxide hydroxide.

2. P0130: XPL/100×/ ⬤ Same field of view as in Fig. 1 under crossed polars.

3. P0226: PPL/100×. Acicular and rounded yellow particles of iron oxide hydroxide.

4. P0226: XPL/100×. Same field of view as in Fig. 3 under crossed polars, crystals are masked by the strong body colour and internal reflections.

5. P0226: PPL. Enlarged field from Fig. 3 showing acicular habit of crystals.

Crystallinity	**Optic sign**	**2V**	n_α	n_β	n_γ	**δ**
Orthorhombic	–ve	0°–27°	2.260–2.275	2.393–2.409	2.398–2.515	0.138–0.140
						(data for mineral from Deer *et al.*, 1992)

In plane-polarised light iron oxide hydroxide, the synthetic equivalent of the mineral goethite, generally forms translucent yellow particles of very fine size. As a consequence of the particles being at or near the resolution of the optical microscope they typically appear as finely fibrous, round or bacterioid in shape. According to Cornell and Schwertmann (1996), the principal habit for this compound is acicular, although stars (twins), hexagons, bipyramids, cubes and thin rods are other possible morphologies. In the acicular form reported particle sizes are from a few tens of nm to several microns or more. The acicular habit is further promoted by rapid growth conditions during manufacture. A broad particle size distribution is also apparently characteristic. Buxbaum (1998) has published electron micrographs of this pigment clearly illustrating the acicular and spheroidal habits possible.

Relief of this compound is moderate with RI greater than that of the medium. Iron oxide hydroxide has strong dispersion and therefore refractive index will vary depending on the light source.

Under crossed polars, iron oxide hydroxide is strongly birefringent, with high order interference colours that are strongly masked by the body colour with the result that anomalous bright greens are typically produced. The compound also exhibits moderate mustard yellow-coloured internal reflections. Fibrous particles are length slow.

Synthetic iron oxide hydroxide may be generally distinguished from the naturally occurring material (goethite and yellow ochre) by the uniformly fine particle size and high purity.

Lit.: Buxbaum (1998) 92; Cornell & Schwertmann (1996); Deer *et al.* (1992)

GOETHITE

α-FeOOH

Key:

1. P1200: PPL/40×. Yellow-brown crystals of goethite with a variety of morphologies and a broad distribution of particle size.

2. P0356: PPL/100×/H. Rounded platy and finely disseminated particles of yellow goethite.

3. P0785: PPL/40×. Coarse-grained plate-like particles on orange goethite.

4. P1200: XPL/40×. Field of view as in Fig. 1 under crossed polars.

5. P0271: RPL/100×. The field of view as in Fig. 2 in reflected light.

6. P0785: XPL/40×. The field of view as in Fig. 3 under crossed polars.

7. P1200: PPL. Prismatic crystal of goethite with cleavage planes parallel to the long axis of the grain.

8. P1200: XPL. The crystal in Fig. 7 under crossed polars.

9. P0356: PPL/H. Enlargement of a particle from Fig. 2.

10. P0785: PPL. An irregular plate of orange goethite.

11. P0785: XPL/ ● The particle in Fig. 10 under crossed polars.

12. P1324: PPL/H. Plate-like particles of yellow goethite.

Crystallinity	Optic sign	2V	n_α	n_β	n_γ	δ
Orthorhombic	−ve	0°–27°	2.260–2.275	2.393–2.409	2.398–2.515	0.138–0.140

(data from Deer et al. 1993)

Goethite is the iron oxide hydroxide mineral that is the primary constituent of yellow ochre.

Under plane-polarised light, goethite forms translucent yellow crystals often with a fibrous habit. These may occur as elongate sheaves or as individual fine needles. Goethite has perfect cleavage, although this is generally difficult to see in the fibrous crystals. Particle size distribution is therefore typically very broad from very fine to coarse. Relief is moderate with RI greater than that of the medium. Goethite has strong dispersion and therefore refractive index will vary depending on the light source.

Under crossed polars, goethite has high birefringence, with the interference colours strongly masked by the body colour. Typically red, yellow and orange colours occur. Extinction is straight, but sweeping across the fibrous particle aggregates. Fibrous particles are length slow.

Natural assemblages containing goethite are discussed under yellow ochre.

Lit.: Cornell & Schwertmann (1996); Deer et al. (1992)

The Pigment Compendium

IRON(III) OXIDE, HEMATITE TYPE

α-Fe$_2$O$_3$

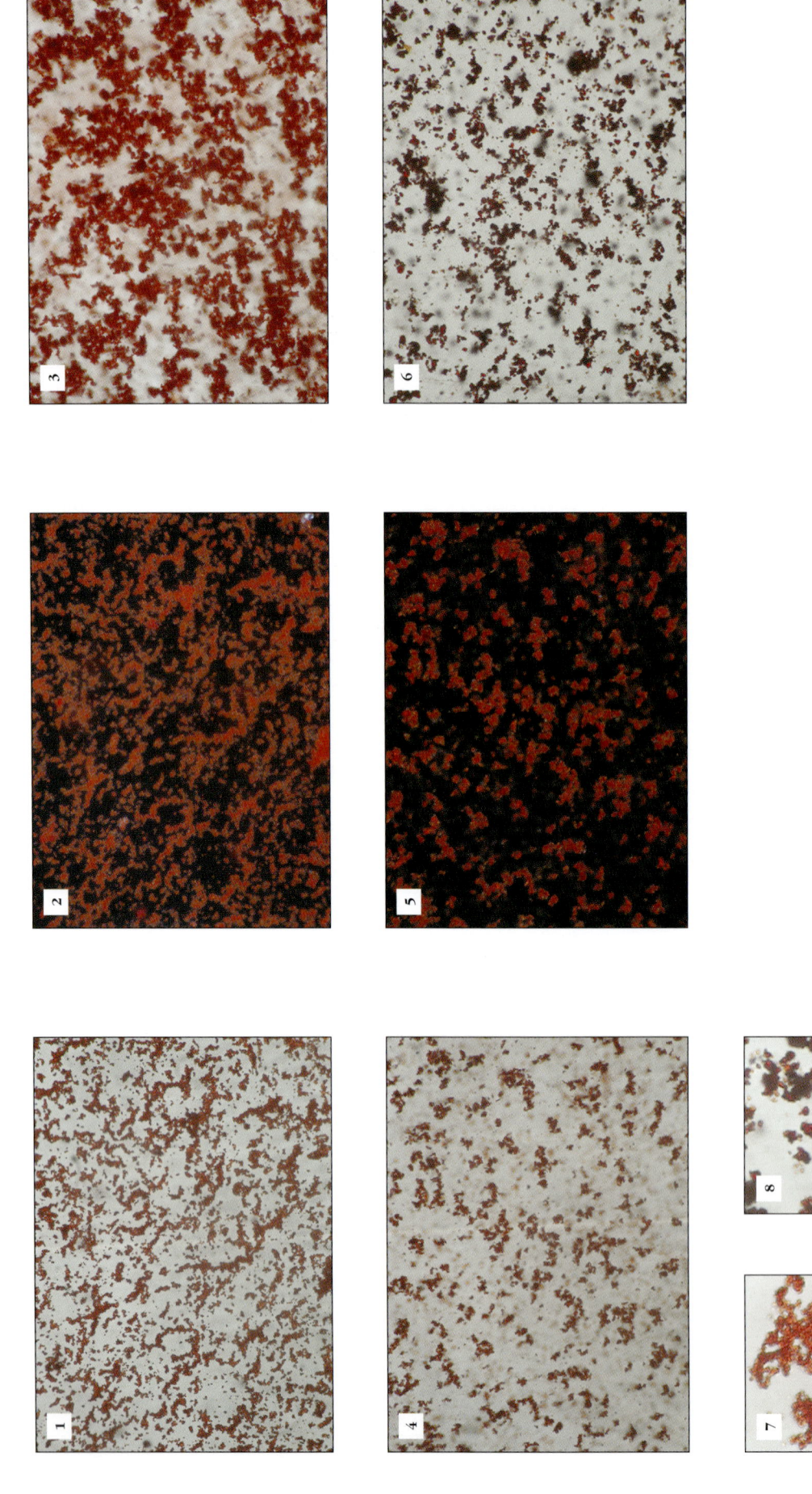

Key:

1. P0132: PPL/100×. Very fine-grained red and orange-red particles of iron(III) oxide.

2. P0132: XPL/100×. Interference colours of iron(III) oxide masked by the body colour and strong internal reflections. Same field of view as in Fig. 1 under crossed polars.

3. P0257: PPL/100×. Aggregates of red and red-orange particles of iron(III) oxide.

4. P0258: PPL/100×. Aggregates of red and red-orange particles of iron(III) oxide.

5. P0258: XPL/100×/ ○. Interference colours of red iron(III) oxide particles masked in crossed polars by the particle body colour and strong internal reflections. Same field of view as in Fig. 4.

6. P0194: PPL/100×. Aggregates of red-orange and opaque 'caput mortuum' particles.

7. P0253: PPL. Fine-grained rounded red to orange-red grains of iron(III) oxide.

8. P0194: PPL. Enlarged portion of Fig. 6 to show particle morphology.

Crystallinity	Optic sign	n_ω	n_ε	δ
Trigonal to Disordered	−ve	2.95–3.01	2.74–2.78	0.21–0.23

(data from McCrone et al., 1973–80)

Iron(III) oxide, hematite type is the synthetic analogue of the mineral hematite. In plane-polarised light, this iron oxide forms translucent to almost opaque red to orange-red particles of very fine grain size. Particles are often at the resolution of the optical microscope and they consequently appear as round or bacterioid grains. According to Cornell and Schwertmann (1996), the principal habits for this compound are hexagonal plates and rhombohedra; the plates vary in thickness and can be round, hexagonal or irregular in shape. Spindles, rods, ellipsoids, cubes, discs, spheres, double ellipsoids, stars and bipyramids may also occur.

Relief is seen as high with the RI greater than that of the medium. Iron(III) oxide has strong dispersion and therefore refractive index will vary depending on the light source.

Under crossed polars, iron(III) oxide has high birefringence, although interference colours are strongly masked by the body colour and are typically bright red or orange. Iron(III) oxide also has strong, red internal reflections.

The synthetic compound may be differentiated from the mineral and red ochres by its purity and uniform particle size.

Lit.: Cornell & Schwertmann (1996); McCrone et al. (1973–80)

HEMATITE

α-Fe$_2$O$_3$

Key:

1. P1208: PPL/40×. Deep red translucent crystals of hematite and opaque martite.

2. P1097: PPL/100×. Finely crystalline particles of hematite.

3. P0137: PPL/40×/●. Aggregates of orange, red-orange and opaque particles of hematite.

4. P1208: XPL/40×/●. The field of view as in Fig. 1 under crossed polars.

5. P0735: PPL/40×/H. Finely disseminated hematite particles coating sand.

6. P1104: PPL/40×. Finely crystalline particles and plates of hematite plus other impurities.

7. P1208: PPL. An enlarged image of a red translucent hematite particle from Fig. 1.

8. P1208: PPL. A euhedral prismatic crystal of hematite.

9. P1097: PPL. An enlarged section of Fig. 2 showing particle morphologies.

10. P0735: PPL/H. Quartz crystal coated in fine-grained red hematite particles.

11. P0735: XPL/●/H. The grain in Fig. 10 under crossed polars.

12. P0638: PPL/H. Fine particles of hematite coating euhedral gypsum crystals from a litharenite.

Crystallinity	Optic sign	n_ω	n_ε	δ
Trigonal to Disordered	–ve	3.15–3.22	2.87–2.94	0.28

(data from Deer et al., 1992)

Hematite is the commonest red iron oxide mineral. It is used as a pigment on its own and is also the primary constituent of red ochres. Hematite forms in a wide variety of geological environments. Massive deposits of hematite are associated with the oxidising zones of iron ore deposits. However, sedimentary rocks may also be hematite rich, with the iron oxide finely disseminated throughout the rock. Ochres (q.v.) are an extreme form of iron oxide-rich sediment, they have earthy textures and are discussed elsewhere.

Under plane-polarised light, hematite forms translucent to opaque, red-orange-brown particles. Particle size and shape is extremely variable and may range from very fine particles with a rounded appearance, which have a size at the limits of the resolution of the microscope, through medium and coarse-grained crystals, which have either a platy to tabular appearance ('micaceous hematite'), or more frequently aggregates of fibrous crystals. Hematite from sedimentary rather than ore deposit sources often occurs as films coating sand grains. Under the microscope this is observed as finely particulate hematite clinging to particles of quartz and other minerals. Relief is high with an RI greater than that of the medium. Hematite, like the other iron oxide minerals, has strong dispersion, and the refractive index varies with the wavelength of the light source. Hematite has no cleavage, but does

possess a weak conchoidal fracture, which is evident as curved particle boundaries.

Under crossed polars, hematite has very high birefringence, but the interference colours are typically masked by the strong body colour and bright reds and yellows are produced. Additionally, hematite has strong red internal reflections. Hematite has straight extinction, although this is generally sweeping across fibrous particle aggregates. Elongation in fibrous particles could not be determined in examples studied.

Isotropic crystals associated with hematite are likely to be pseudomorphs after magnetite, a variety of hematite known as 'martite'.

Natural hematite may be differentiated from synthetic iron(III) oxides by the variability of particle shape and the presence of impurities. These may include quartz, calcite, baryte for samples derived mineralised veins. For ochres, a wide variety of minerals may be present.

Lit.: Cornell & Schwertmann (1996); Deer et al. (1992)

IRON OXIDE HYDROXIDE, LEPIDOCROCITE TYPE AND LEPIDOCROCITE

γ-FeOOH

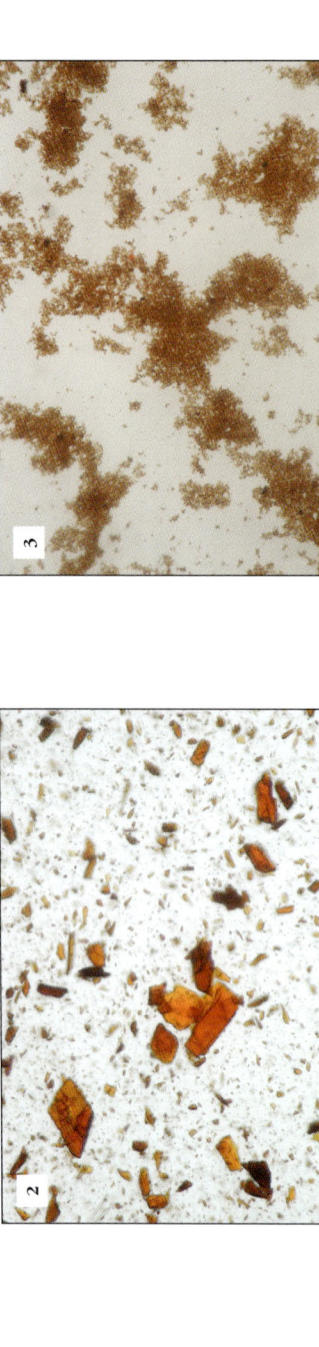

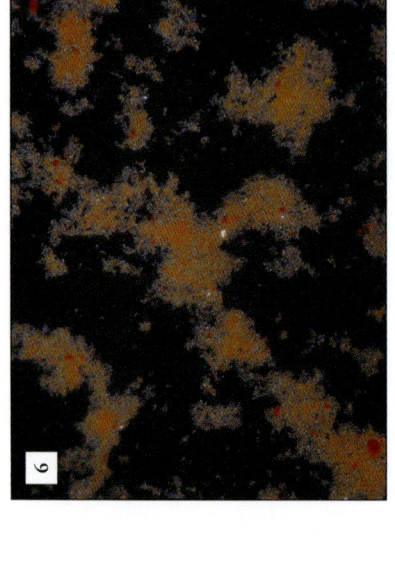

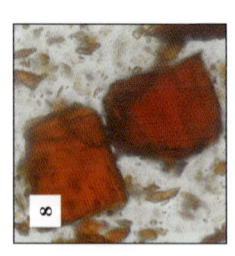

1. P1220: PPL/40×/ **O** Lepidocrocite crystals with fibrous, tabular and equant morphologies.

2. P1220: PPL/40×. Crystals of lepidocrocite.

3. P0131: PPL/40×/ **O** A mixture of synthetic iron oxides and hydroxides including lepidocrocite type.

4. P1220: XPL/40×/ **O** Highly birefringent lepidocrocite particles with interference colours obscured by the body colour.

5. P1220: XPL/40×. The field of view as in Fig. 2 under crossed polars.

6. P0131: XPL/40×/ **O** The view as in Fig. 3 under crossed polars with minor highly birefringent impurities.

7. P1220: PPL/ **O** Tabular lepidocrocite crystals.

8. P1220: PPL/ **O** The crystals in Fig. 7 with the polariser rotated through 90° showing change in body colour due to pleochroism.

9. P1220: PPL. Rhombic crystal of lepidocrocite.

10. P1220: XPL. The crystal in Fig. 9 under crossed polars.

Crystallinity	Optic sign	2V	n_α	n_β	n_γ	δ
Orthorhombic	−ve	83°	1.94	2.20	2.51	0.57
					(data for mineral from Deer *et al.*, 1992)	

In plane-polarised light, lepidocrocite and its synthetic analogue form translucent orange crystals, weakly pleochroic from dark to pale orange. Relief is weakly variable and high, RI is greater than that of the medium. Particle shapes for the mineral are fibrous, occurring in sheaves, or as platy to tabular crystals. When these are euhedral they have hexagonal, polyhedral habits. Both forms may be found in the same sample. Particle surfaces for the fibrous crystals appear finely striated. For tabular crystals, particle surfaces may be smooth or stepped. Lepidocrocite has a perfect cleavage that influences particle shape on crushing; this is commonly visible in the resulting particles. Particle size has a broad distribution from fine to very coarse. In the synthetic analogue the basic morphologies encountered are lath-like or tabular, though massive varieties may exhibit micaceous and fibrous textures and aggregated scales (Cornell and Schwertmann, 1996); 'grassy', 'hedgehog' type spherulites have also been observed.

Under crossed polars lepidocrocite has high birefringence with its interference colours masked by the strong body colour. Typically, however, bright orange reds, pinks and greens are observed. Lepidocrocite has straight extinction and is length slow.

Lepidocrocite forms as an alteration product of other iron minerals (especially goethite, *q.v.*) and is a common constituent of yellow and brown ochres as well as so-called 'limonites'. Consequently it may be encountered in the full range of these contexts.

Lit.: Cornell & Schwertmann (1996); Deer *et al.* (1992)

The Pigment Compendium

JAROSITE AND NATROJAROSITE

$KFe_3(SO_4)_2(OH)_6$ and $NaFe_3(SO_4)_2(OH)_6$

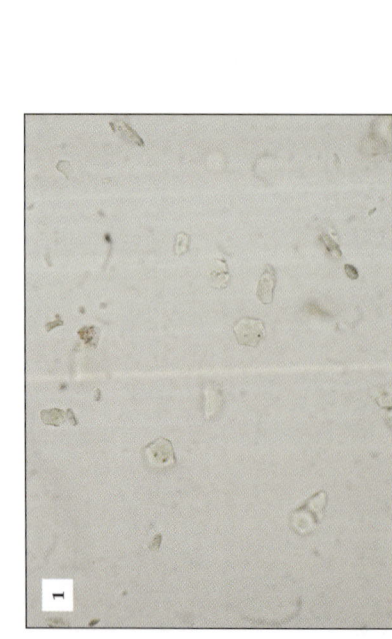

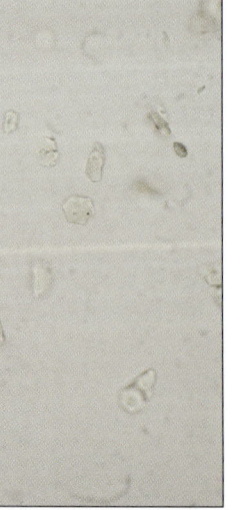

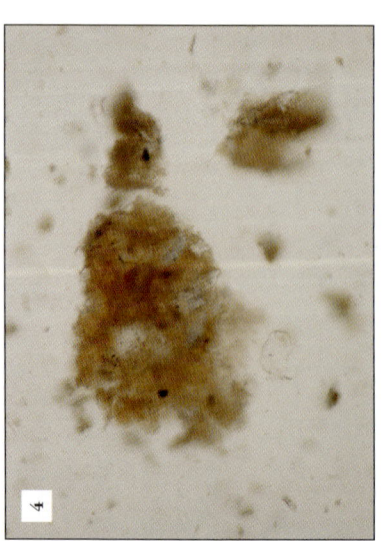

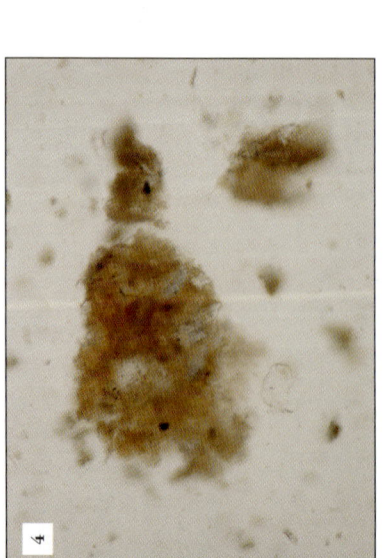

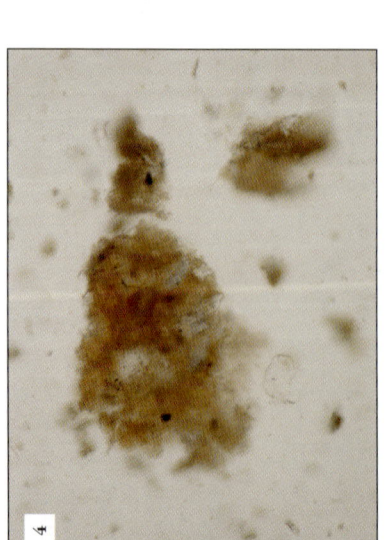

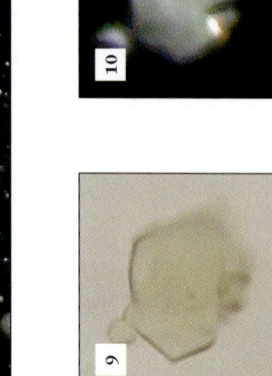

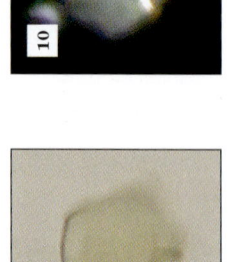

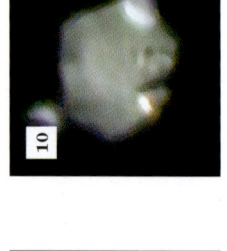

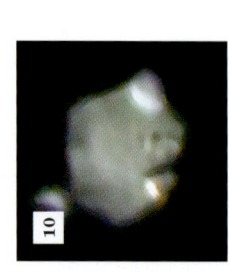

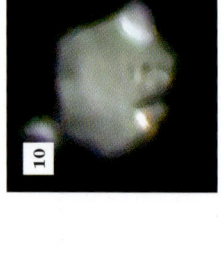

Key:

1. P1106: PPL/100X/ ◐ Euhedral hexagonal plates of jarosite.

2. P1234: PPL/10X. Cryptocrystalline aggregates of jarosite.

3. P1234: XPL/10X/ ◐ The field of view as in Fig. 2 in crossed polars.

4. P1234: PPL/40X. Cryptocrystalline aggregates of natrojarosite.

5. P1234: XPL/40X. The same field of view as in Fig. 4 under crossed polars.

6. P1234: UV/40X. The same field of view as in Fig. 4 with UV fluorescence.

7. P1106: PPL/ ◐ Euhedral hexagonal plate of jarosite.

8. P1106: PPL/ ◐ Euhedral hexagonal plate of jarosite.

9. P1106: PPL/ ◐ Euhedral hexagonal plate of jarosite.

10. P1106: XPL/ ◐ The crystal seen in Fig. 9 under crossed polars.

Crystallinity	Optic sign	n_ω	n_ε	δ
Trigonal	−ve	1.815–1.820	1.713–1.715	0.102–0.105

(data for jarosite from Heinrich, 1965; cf. Webmineral, 2003)

Under plane-polarised light, jarosite is commonly observed to form a variety of habits, of which the end-members are euhedral hexagonal plates and aggregates of cryptocrystalline, crumb-like particles. A variety of particle forms can occur between these end-members, through subhedral to anhedral plates to varying degrees of crystallinity and particle size. Euhedral and subhedral forms are distinctive and not uncommon. They occur as transparent, pale yellow-green coloured plates. Particle surfaces are smooth although inclusions are often visible within the particles. Particle size varies from fine to medium. Less well-developed forms appear as aggregates of fine-grained plates. Cryptocrystalline particles are also common. These form rounded, crumb-like aggregates, with a dirty yellow colour and rough, pitted particle surfaces. For this morphology, particle size ranges from very fine to coarse-grained particles. For all forms, relief is moderate and RI is greater than that of the medium.

Under crossed polars, jarosite has high birefringence, though the thin plates or finely cryptocrystalline masses typically show less than the maximum birefringence. For euhedral forms, plates tend to align lying on their basal sections and therefore show the lowest birefringence in this orientation. Interference colours range from first order greys and whites, and sometimes slightly anomalous blues to high third and fourth order colours. Jarosite can be seen to have straight extinction when in this form. Higher order, brighter birefringence is generally seen in polycrystalline masses. Bright yellows are typically transmitted. These particles have sweeping or undulose extinction. Cryptocrystalline aggregates twinkle as the stage is rotated.

Natrojarosite is optically indistinguishable from jarosite and typically occurs as crumb-like polycrystalline aggregates.

As naturally occurring minerals, jarosite and natrojarosite commonly contain impurities of phases such as quartz, feldspar and iron oxide minerals.

LEAD CHLORIDE OXIDE

PbCl$_2$.5-7PbO

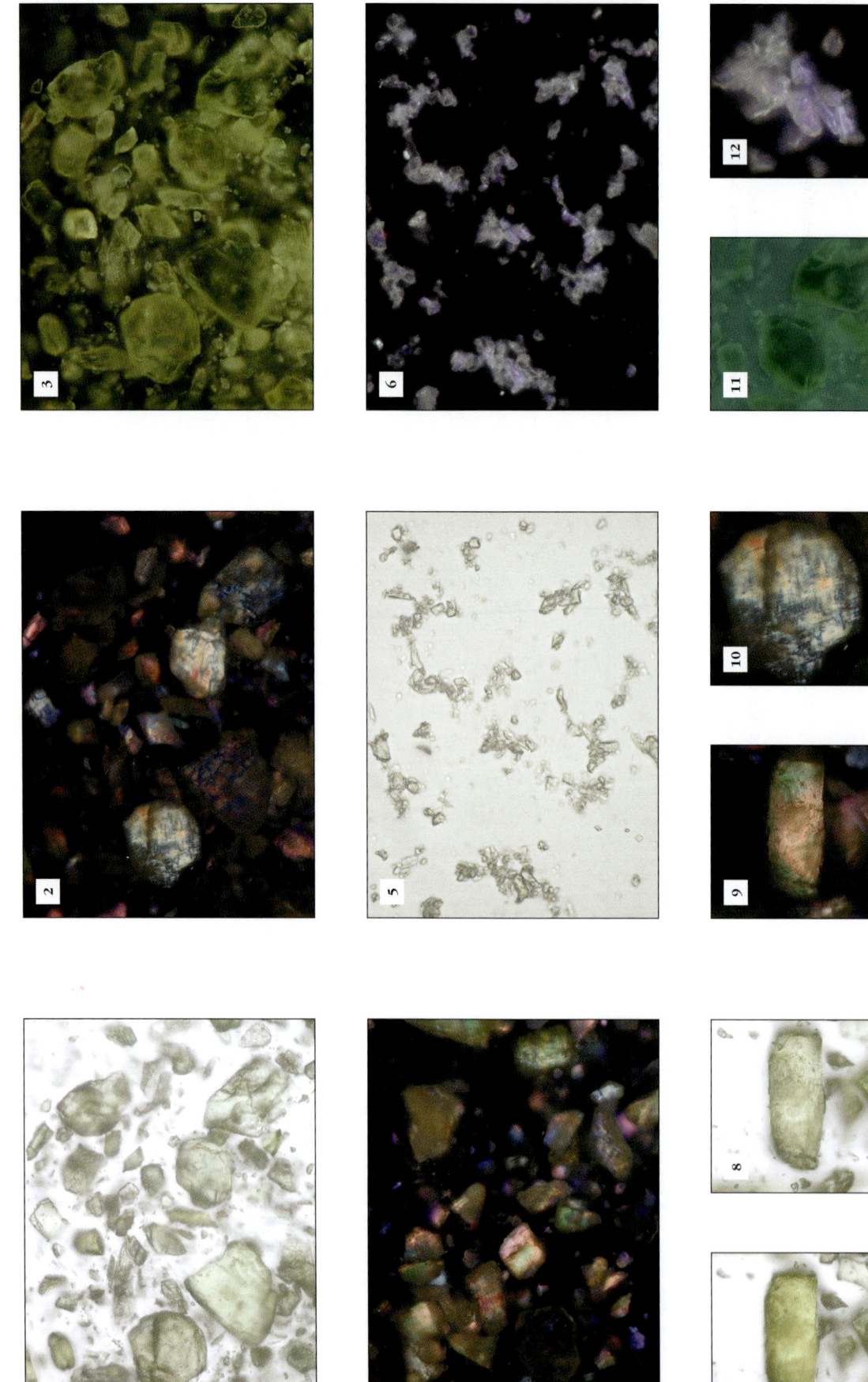

Key:

1. P1528: PPL/40×. Coarse-grained crystals of lead chloride oxide.

2. P1528: XPL/40×. The field of view as in Fig. 1 under crossed polars.

3. P1528: RPL/40×. The field of view as in Fig. 1 in reflected light.

4. P1528: XPL/40×/●. Lead chloride oxide under crossed polars, note the fine-grained colours show nomalous purples, awhile the coarse-grained particles show third order interference colours.

5. P1528: PPL/100×. Medium-grained particles of lead chloride oxide.

6. P1528: XPL/100×/●. The field of view as in Fig. 5 with particles showing anomalous purple interference colours.

7. P1528: PPL. Subhedral crystal of lead chloride oxide.

8. P1528: PPL. The crystal in Fig. 7 with the polariser rotated through 90° showing change in colour due to pleochroism.

9. P1528: XPL/●. The crystal in Fig. 7 in crossed polars showing third order interference colours.

10. P1528: XPL. Detail of crystals from Fig. 2 showing first and second order and anomalous blue interference colours.

11. P1528: UV. Detail of crystals from Fig. 1 under UV illumination.

12. P1528: XPL/●. Detail of particles from Fig. 6 showing anomalous purple interference colours.

Crystallinity

Anisotropic

n	δ
$\gg 1.662$	High

(data from authors)

In plane-polarised light lead chloride oxide (known historically under such names as 'Turner's yellow', 'Patent yellow' or 'mineral yellow') may be observed as translucent lemon yellow crystals with moderate to high relief. The RI is greater than that of the medium. Crystal shape is of crushed angular shards, although the habit is strongly influenced by the perfect cleavage shown by this compound. Commonly the particle surfaces are seen as striated, this being the manifestation of the cleavage; flakes that have fallen with the cleavage plane parallel to the slide may then be observed to have a stepped surface. Particles frequently exhibit conchoidal fracture. Particle size distribution is broad ranging from fine to very coarse.

Under crossed polars, the particles have high birefringence with up to third order interference colours present. Particles that exhibit first order and low second order colours often also show anomalous blue interference colours. Striking anomalous purple colours are typical of grains showing higher birefringence, a phenomenon particularly evident in medium- to fine-grained particles. Extinction is straight and complete; particles are length fast.

With UV excitation the crystals show a weak to moderate green fluorescence at grain boundaries. In reflected light the grains show strong lemon yellow reflections, mainly associated with grain boundaries.

The Pigment Compendium

LEAD IODIDE

PbI₂

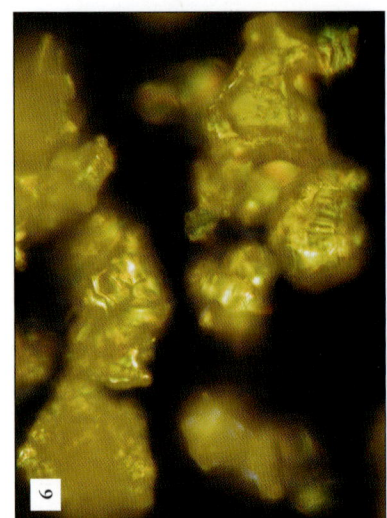

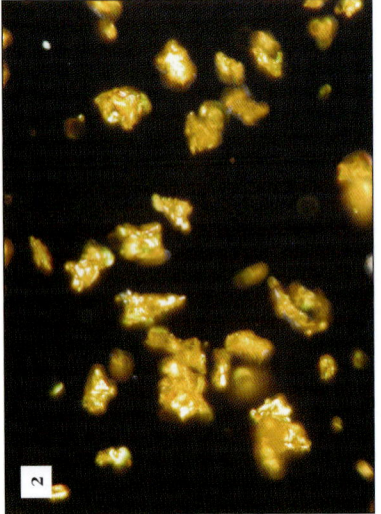

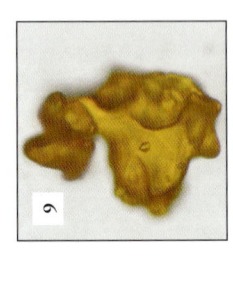

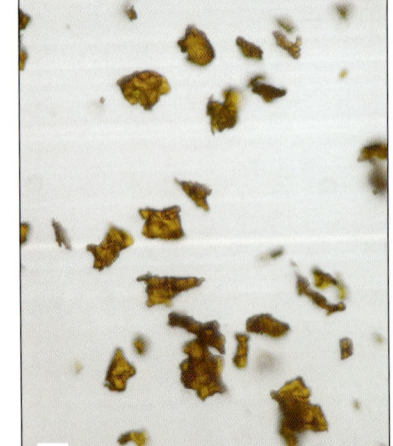

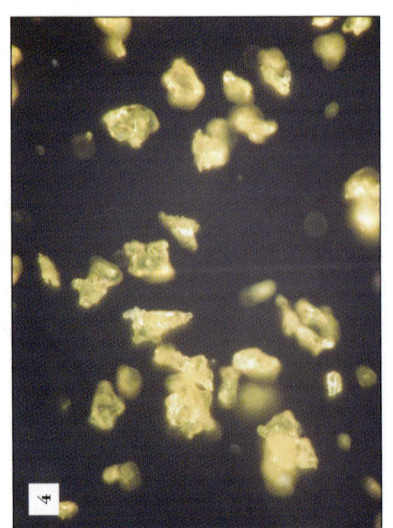

Key: **1.** P1522: PPL/40×. Crystals of lead iodide.

2. P1522: XPL/40× ○. The field of view as in Fig. 1 under crossed polars.

3. P1522: UV/40×. The field of view as in Fig. 1 in UV.

4. P1522: RPL/40×. The field of view as in Fig. 1 in reflected light.

5. P1522: PPL/100×. Anhedral embayed crystals of lead iodide.

6. P1522: XPL/100× ○. The field of view as in Fig. 5 under crossed polars.

7. P1522: PPL. Subhedral hexagonal plates of lead iodide.

8. P1522: XPL/ ○. The crystal in Fig. 7 under crossed polars.

9. P1522: PPL. Highly irregular anhedral crystal of lead iodide.

Crystallinity	Optic sign	n_ω	n_ε	δ
Trigonal		$\gg$1.662		High
				(data from ICSD, 2003/authors)

In plane-polarised light lead(II) iodide is seen as translucent to very weakly translucent, strongly yellow crystals that are pleochroic from yellow to dark yellow. Relief is very high and the RI is much greater than that of the medium. Particle shape is predominantly of rounded and subhedral rounded grains although some particles can be seen to take euhedral form and occur as hexagonal, rhombic or bladed plates. Grain surfaces are uneven and fractured and appear domed. Particle size is likely to be medium to coarse.

Under crossed polars, lead(II) iodide has strong birefringence but the high third order and greater interference colours are strongly masked by the body colour. Anomalous yellow, greens and oranges are common. Extinction is straight. Lead(II) iodide has strong yellow internal reflections. The birefringence is too high to determine elongation.

The Pigment Compendium

LEAD OXIDE, LITHARGE TYPE AND LITHARGE

α-PbO

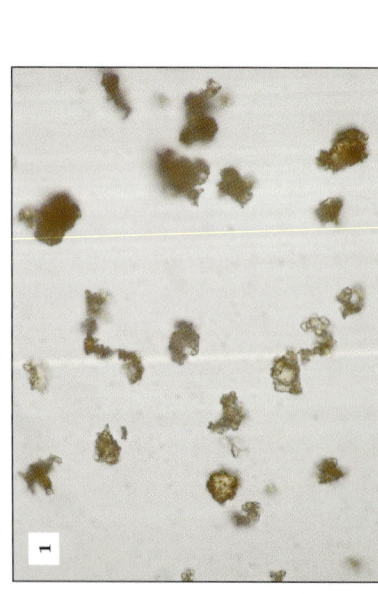

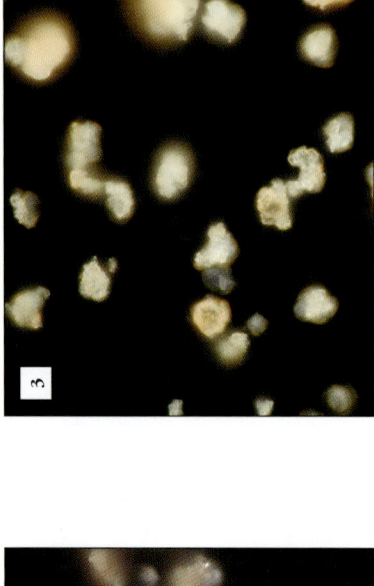

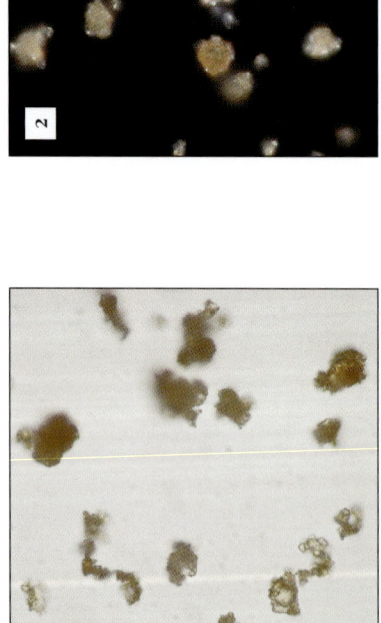

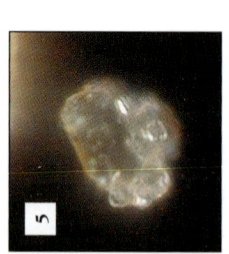

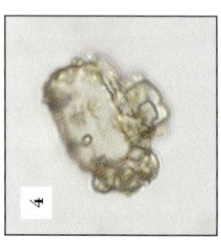

1. P1432: PPL/100×. Particles of lead(II) oxide, litharge type.

2. P1432: XPL/100×/**○**. The field of view as in Fig. 1 under crossed polars.

3. P1432: RPL/100×. The field of view as in Fig. 1 in reflected light.

4. P1432: PPL. Detail of particles from the sample in Fig. 1.

5. P1432: XPL. The particle as in Fig. 4 under crossed polars.

Crystallinity	Optic sign	n_ω	n_ε	δ
Tetragonal	−ve	2.665	2.535	0.1300
			(data for mineral from Dana, 1997; cf. Webmineral, 2003)	

A tetragonal α-lead(II) oxide and its mineral analogue litharge exist in addition to the β-, massicot, type discussed below, but appear to be relatively unstable forms and are therefore unlikely to be encountered. However, in plane-polarised light litharge can be seen as pale yellow, translucent crystals or semi-opaque aggregates of dark red-brown crystals. Relief is high and RI is much greater than that of the medium; some differences in RI values appear in the literature – in addition to those cited above, Winchell (1929) gives $n_\omega = 2.64$ and $n_\varepsilon = 2.51$. Particle size was fine in the synthetic sample studied by the authors, with many particles that were close to the resolution of the microscope and therefore appearing rounded or bacterioid in shape. However, some particles, particularly the pale yellow translucent crystals show square, rectangular or acicular euhedral crystals. The semi-opaque

particles appear as aggregates of finely crystalline grains. The natural mineral is likely to exhibit a much broader range of particle size and a more varied morphology.

Under crossed polars, this compound has high birefringence and third order interference colours are typical. Crystals have straight extinction and prism sections are length fast. Polycrystalline grains do not achieve complete extinction and twinkle as the stage is rotated.

Lit.: Winchell (1929)

The Pigment Compendium

LEAD OXIDE, MASSICOT TYPE AND MASSICOT

β-PbO

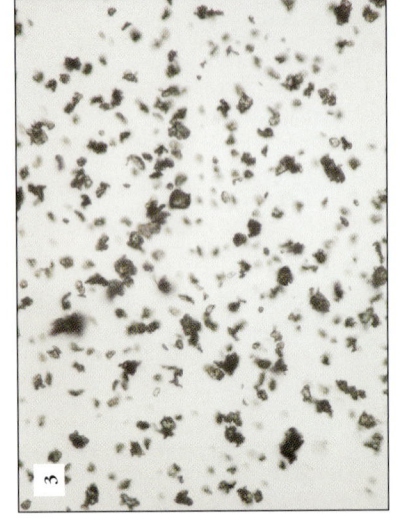

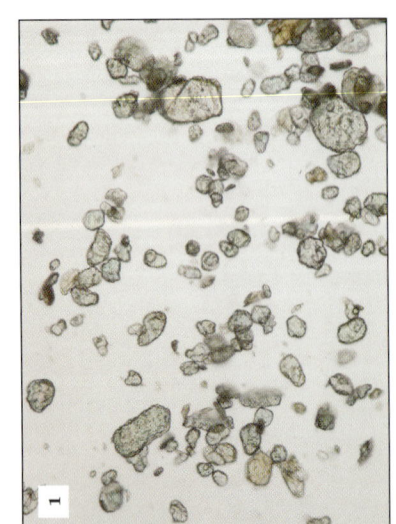

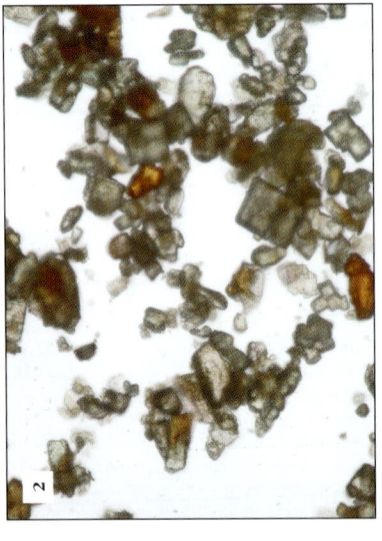

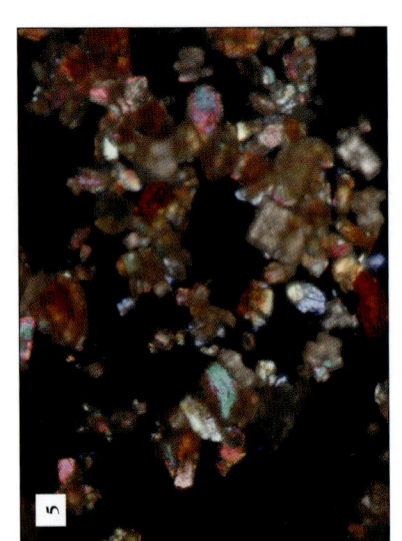

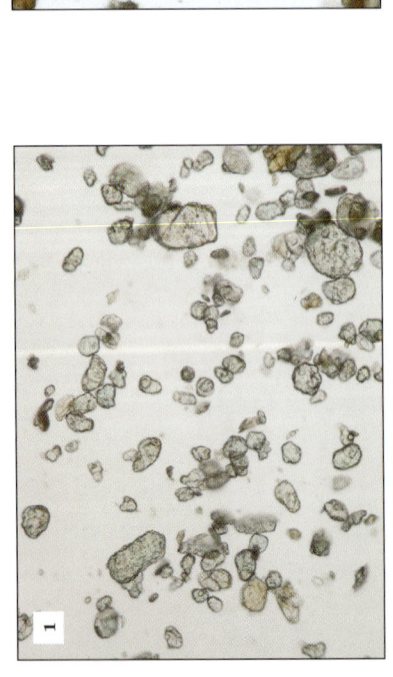

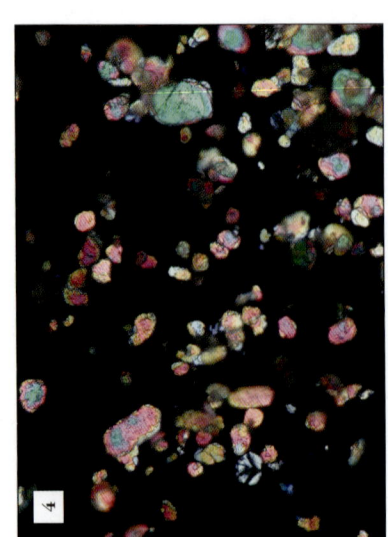

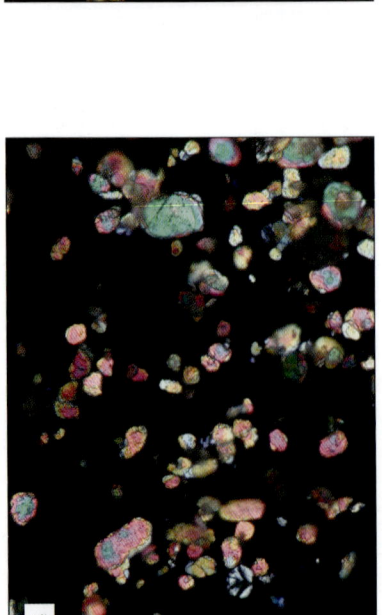

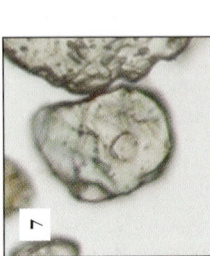

Key:

1. P0073: PPL/40×. Rounded, subhedral particles of lead oxide, massicot type.

2. P0242: PPL/40×. Euhedral and anhedral particles of massicot, showing wide variation in colour.

3. P1428: PPL/40×. Fine-grained angular particles of massicot.

4. P0073: XPL/40×/ **O**. The field of view as in Fig. 1 under crossed polars.

5. P0242: XPL/40×. The field of view as in Fig. 2 under crossed polars.

6. P1428: XPL/40×/ **O**. The field of view as in Fig. 3 under crossed polars.

7. P0073: PPL. Detail of particle morphology from Fig. 1.

8. P0073: XPL/ **O**. The particle in Fig. 7 in crossed polars, showing third order interference colours.

Crystallinity	Optic sign	2V	n_α	n_β	n_γ	δ
Orthorhombic	+ve	~90°	2.71	2.61	2.51	0.20
					(data for mineral from Winchell, 1929)	

In plane-polarised light β-lead(II) oxide and its mineral analogue massicot form translucent, pale yellow to yellow-brown crystals. Crystal shape is typically subhedral to euhedral, with prismatic forms with pyramidal terminations and rectangular basal sections common. Relief is very high and RI is much greater than that of the medium. Particle surfaces appear rough and pitted. Growth zones are visible in some crystals.

Under crossed polars massicot can be seen to have high birefringence, showing bright third order interference colours, typically pinks, turquoises, yellows and greens. The crystals have straight extinction, which is typically complete or slightly undulose. The particles also show moderate brown-yellow internal reflections.

Lit.: Winchell (1929)

LEAD ANTIMONY OXIDE, BINDHEIMITE TYPE

$Pb_2Sb_2O_7$

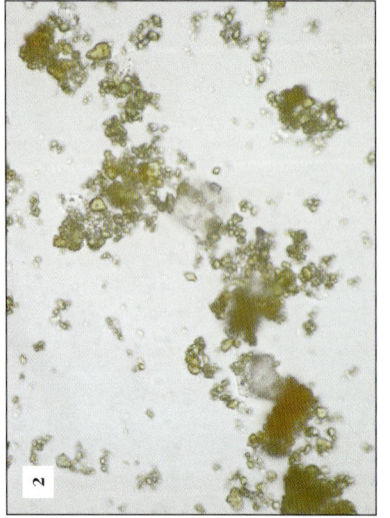

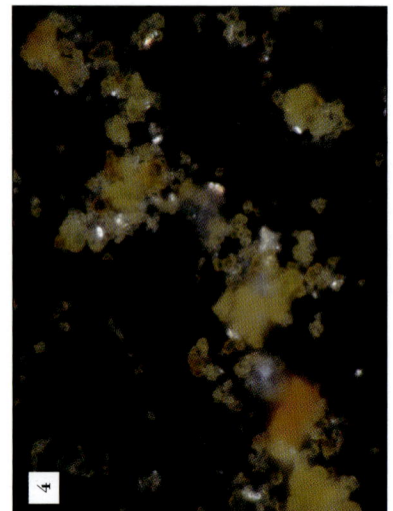

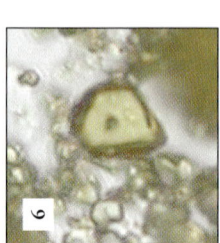

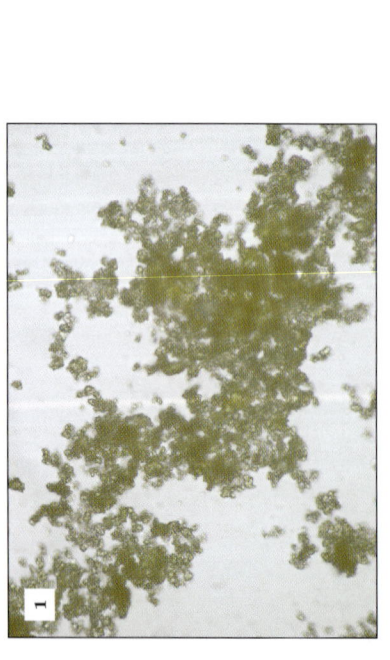

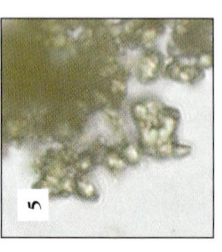

Key:

1. P0074: PPL/100×. Yellow particles of, bindheimite type lead antimony oxide.

2. P0075: PPL/100×. Coarse-grained particles of lead antimony oxide, bindheimite type.

3. P0074: XPL/100×. The field of view as in Fig. 1 in crossed polars.

4. P0075: XPL/100×/ ● . The field of view as in Fig. 2 in crossed polars.

5. P0074: PPL. Detail of particle morphology from Fig. 1.

6. P0075: PPL. Detail of particle morphology from Fig. 2.

Crystallinity

Cubic

n

2.01–2.28

δ

Isotropic

(data for synthetic compound from Gettens & Stout, 1966)

In plane-polarised light lead antimony oxide, the synthetic analogue of the rare mineral bindheimite, forms pale yellow particles with high relief and an RI much greater than that of the medium. Since the particles are often small, the relief effects make the crystals appear almost opaque; however, closer examination shows the particles to be pale yellow in colour. Particle morphologies range from rounded or bacterioid for those particles close to the resolution of the optical microscope, to anhedral, angular crystals. A few euhedral cubic plates may be present. Particle size is very fine to medium, although most particles are fine or very fine.

As a cubic compound, under crossed polars lead antimony oxide is isotropic. However, the particles have strong internal reflections and particle boundaries glow a strong, bright yellow.

Many samples have trace amounts of fine, euhedral, lath-shaped crystals that are yellow in plane-polarised light and high fourth order interference colours under crossed polars; these are the rosiaite type of lead antimony oxide (*q.v.*).

Examples of this pigment examined by the authors appeared a lime yellow colour in reflected light. In one modern example studied a minor amount of free antimony oxide (senarmontite form) was visible as bireflecting laths.

The optical, physical and chemical properties of lead antimonate yellows have been reviewed by Wainwright *et al.* (1986). This compound yellow is difficult to distinguish from lead tin silicon oxide using optical microscopy, though the presence of free tin(IV) oxide (p. 314) would indicate that the compound is lead tin yellow.

Lit.: Gettens & Stout (1966); Wainwright *et al.* (1986)

The Pigment Compendium

LEAD ANTIMONY OXIDE, ROSIAITE TYPE

$PbSb_2O_6$

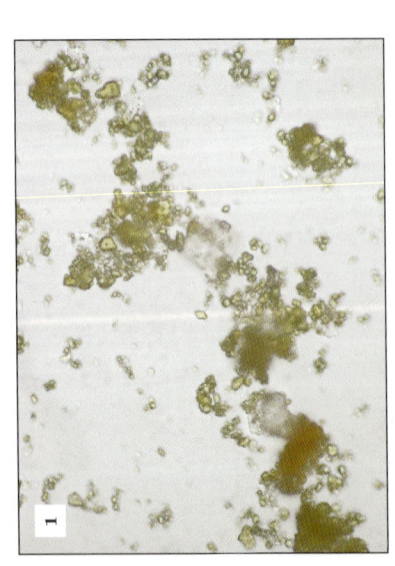

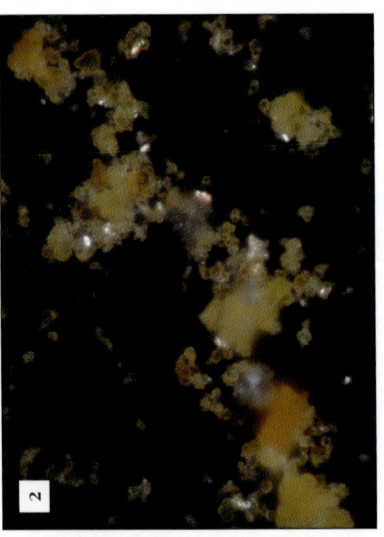

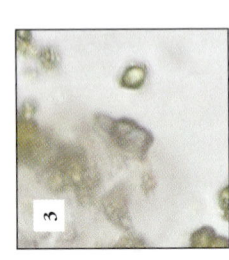

Key:

1. P0075: PPL/100×. This sample predominantly contains the yellow bindheimite type of lead antimony oxide; however, it also contains rosiaite type as colourless impurities.

2. P0075: XPL/100×/ ●. The field of view as in Fig. 1 in crossed polars.

3. P0075: PPL. Detail of Fig. 1, showing colourless lath of lead antimony oxide, rosiaite type.

Crystallinity	Optic sign	n_ω	n_ε	δ
Trigonal-Hexagonal	−ve	2.092	1.920	0.1720

(data for mineral from Webmineral, 2003)

The rosiaite form of lead antimony oxide is unlikely to be found in isolation from the bindheimite type (*q.v.*); usually it occurs as a minor phase in samples of the latter. It has, however, been shown during a study of lead antimony tin oxide that under certain conditions (notably where tin and antimony are mixed in equal proportions) significant amounts of $PbSb_2O_6$ are formed (Cascales *et al.*, 1986).

Lead antimony oxide, rosiaite type can be observed as fine, euhedral, lath-shaped crystals, yellow in plane-polarised light although rare examples may demonstrate euhedral hexagonal habits (see Wainwright *et al.*, 1986). Relief is high, the RI much greater than that of the medium.

Under crossed polars, this compound shows high fourth order interference colours.

As implied above, this compound is likely to be accompanied by a variety of other lead-antimony-tin compounds.

Lit.: Cascales *et al.* (1986); Wainwright *et al.* (1986)

LEAD ANTIMONY TIN OXIDE AND LEAD ANTIMONY ZINC OXIDE   ~Pb$_2$(Sb,Sn)$_2$O$_7$ and Pb$_2$(Sb,Zn)$_2$O$_7$?

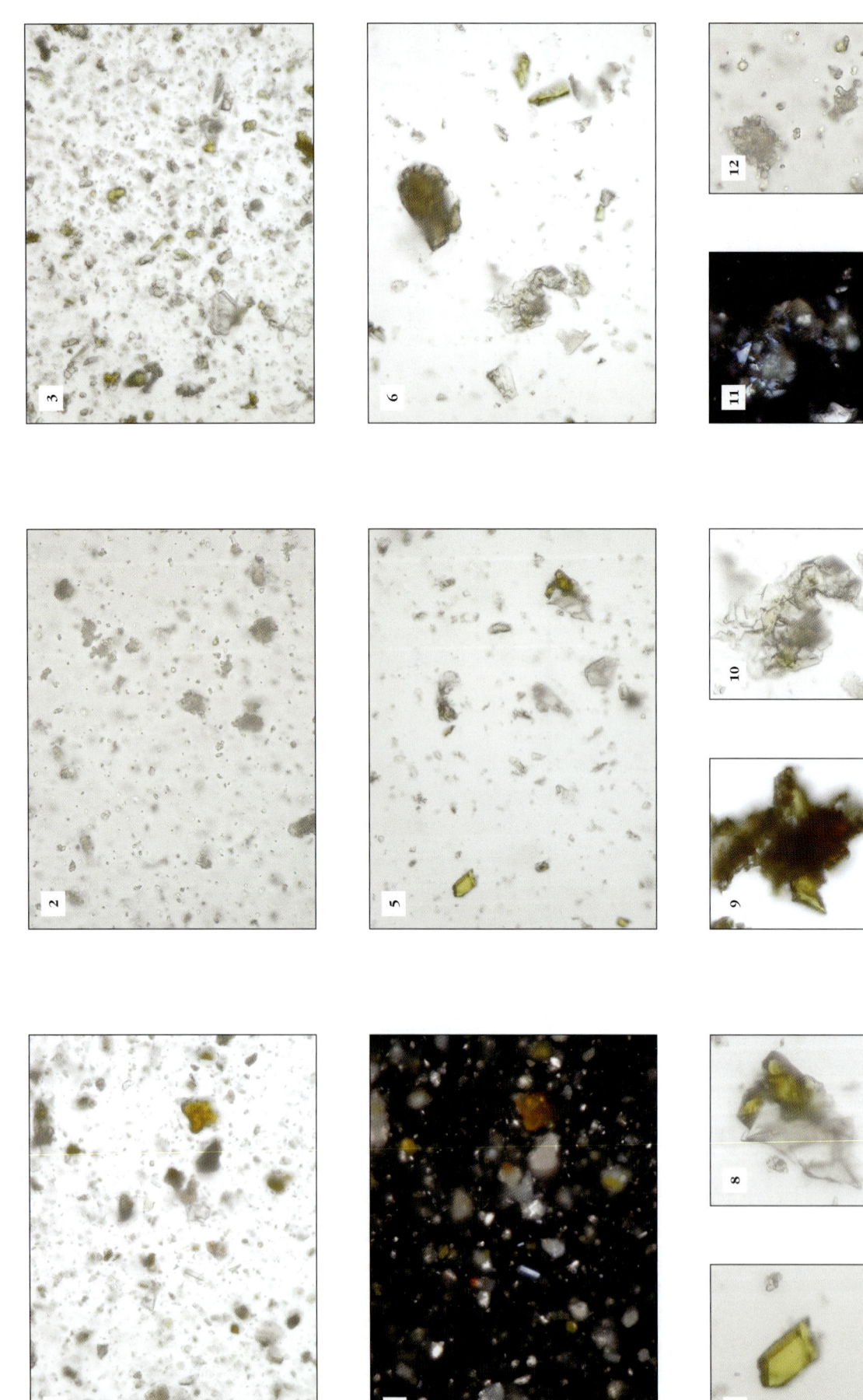

Crystallinity

Cubic	n >2.0	δ Isotropic

The samples of lead antimony tin oxide and lead antimony zinc oxide that were examined by the authors are modern facsimiles of recipes in, respectively, Kunckel (1689) and various seventeenth century Italian manuscripts (Buzzegoli *et al.*, 2000; Hermens, 2002). In all cases the resulting pigment is a complex mixture of a number of phases.

According to Roy and Berrie (1998), the bulk lead antimony tin oxide pigment is a pure yellow colour, warmer-hued than the lead-tin yellows. Microscopically the particle size is in the range of 10μm to 40μm or more. Relief is very high, the RI much greater than that of the mounting medium. Particle morphology appears similar to that of lead tin silicon oxide. However, samples of lead-antimony-tin pigments prepared following historical recipes contained three phases. In plane-polarised light, phase 1 appears as colourless, translucent to transparent crystals with very high relief and RI greater than that of the medium. Crystal habit is of subhedral to euhedral rectangular particles to angular anhedral fragments of these. Particle surfaces are domed and striated. In crossed polars these particles have high first order interference colours and straight, complete extinction and are length fast. Particle size distribution was bimodal with fine- and coarse-grained populations. Phase 2 appears in plane-polarised light as weakly translucent to opaque particles, with moderate relief and RI greater than that of the medium. Particle habit is of anhedral crumbs and grain surfaces are rough,

irregular and pitted. Under crossed polars, the particles have strong grey-yellow internal reflections which strongly mask the moderate interference colours. Particles are, however, clearly polycrystalline and complete extinction is not achieved. The rarer phase 3 is composed of dark yellow, translucent particles with angular shards of a glassy appearance, high relief and RI greater than that of the medium. Under crossed polars, this phase is isotropic and has weak, yellow internal reflections. Roy and Berrie have also noted samples where the compound is embedded in a lead calcium silicate matrix.

The lead-antimony-zinc pigments studied were more complex still, with up to five phases present in a single sample. In purer samples the appearance under plane-polarised light was of weakly translucent, dirty grey-yellow particles with moderate relief and RI greater than that of the medium. Particle shape was very irregular, with ragged crumb-like aggregates and individual grains; fine inclusions could also be present with intergrowths of golden yellow translucent, inclusion-free particles. Particle size ranges from very fine to coarse. For the most part particles appear isotropic, but a few have weak first order grey interference colours, largely masked by the reflections. The particles can also have strong grey-yellow internal reflections.

Lit.: Buzzegoli *et al.* (2000); Hermens (2002); Kunckel (1689); Roy & Berrie (1998)

LEAD ANTIMONY TIN OXIDE AND LEAD ANTIMONY ZINC OXIDE

LEAD TIN OXIDE

Pb_2SnO_4

1. P1016: PPL/100×. Fine-grained pale yellow particles.

4. P1018: PPL/40×. Fine- and coarse-grained dark yellow particles.

7. P1041: PPL. Enlarged view of dark yellow particles.

2. P1016: XPL/100×. The same view as in Fig. 1 under crossed polars.

5. P1018: XPL/40×/ ○. The same view as in Fig. 4 under crossed polars.

8. P1041: XPL. The same view as in Fig. 7 under crossed polars.

3. P1016: RPL/100×. The same view as in Fig. 1 in reflected light.

6. P1018: RPL/40×. The same view as in Fig. 4 in reflected light.

9. P1041: RPL. The same view as in Fig. 7 in reflected light.

Crystallinity	**Optic sign**	n_ω	n_ε	δ
Tetragonal	+ve	2.29	2.31	0.02
				(data from Verbitskaya and Burakova, 1965)

Under plane-polarised light, lead tin oxide (lead tin yellow type I') is observed as very fine-grained particles, and dense aggregates of these, with an angular, irregular, crumb-like morphology. Particle surfaces are rough and pitted. Colour ranges from translucent pale yellow in individual particles to dull yellow-brown weakly translucent, almost opaque in the aggregates. In reflected light the aggregates appear extremely pale yellow to white. Individual particle size is very fine to fine and at the resolution of the optical microscope, thus particles appear to have a rounded or bacterioid morphology; analytical studies of this pigment indicate particle size is typically in the range <1 μm–3 μm with a narrow distribution (Eastaugh, 1988). Relief is high, with RI greater than that of the medium.

Under crossed polars particles are birefringent, but the interference colours are strongly masked by the body colour and hard to see due to the high

relief and very strong lemon yellow internal reflections produced by the particles. However, individual particles and the aggregates are seen to twinkle as the stage is rotated.

Trace impurities of tin oxide may be present. These are distinctive under crossed polars as they have bright third and fourth order birefringence colours.

The chemical, physical and optical characteristics of this pigment have been reviewed by Kühn (1993a).

Lit.: Eastaugh (1988); Kühn (1993a); Verbitskaya & Burakova (1965)

The Pigment Compendium

LEAD TIN SILICON OXIDE

$Pb(Si_x,Sn_{1-x})O_3$

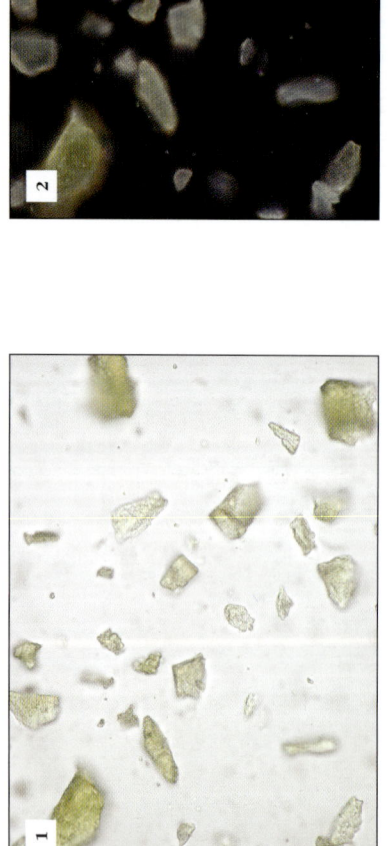

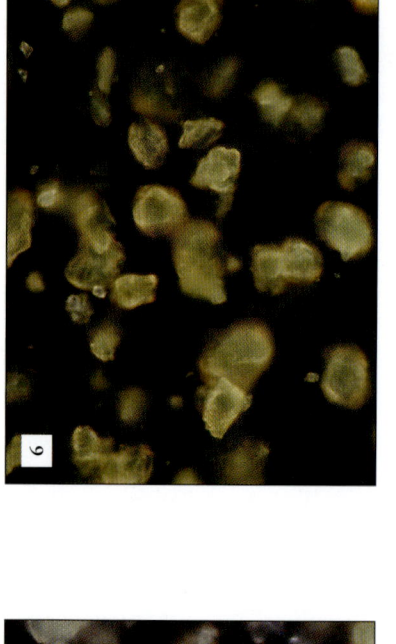

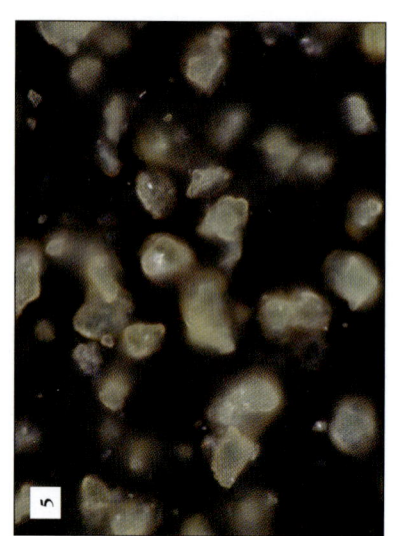

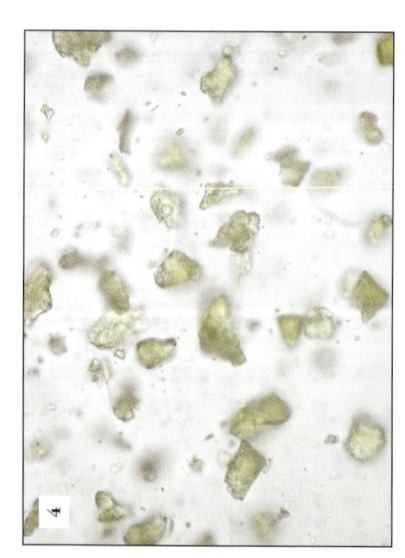

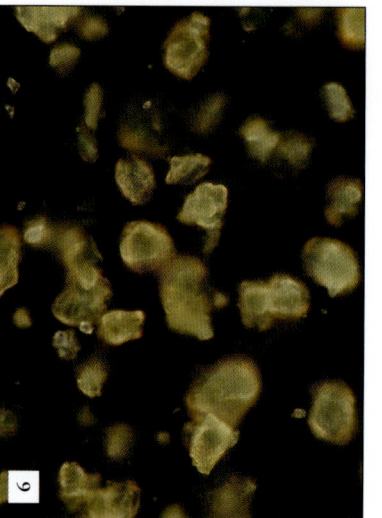

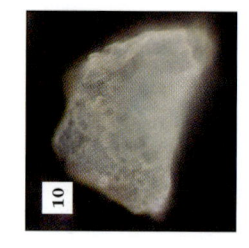

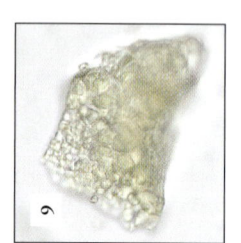

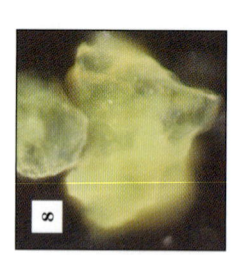

Key:

1. P1045: PPL/100X. Subhedral bladed crystals.

2. P1045: XPL/100X. The field of view as in Fig. 1 under crossed polars.

3. P1045: RPL/100X. The field of view as in Fig. 1 in reflected light.

4. P1044: PPL/100X. Subhedral bladed and anhedral irregular crystals.

5. P1044: XPL/100X/○. The field of view as in Fig. 4 under crossed polars.

6. P1044: RPL/100X. The field of view as in Fig. 4 in reflected light.

7. P1005: PPL. A yellow-brown particle.

8. P1005: XPL. Same field of view as in Fig. 7 under crossed polars.

9. P1046: PPL. Framboidal particle.

10. P1046: XPL. The same particle as in Fig. 9 under crossed polars.

11. P1046: RPL. The same grain as in Fig. 9 in reflected light.

12. P1044: PPL. Prismatic forms from Fig. 4.

Crystallinity
Cubic

n	δ
~2.3	0.0

(data from Kühn, 1993)

Under plane-polarised light, lead tin silicon oxide (lead tin yellow type II') forms medium-grained crystals. The particles are translucent and pale yellow to orange yellow in colour, colour intensity increasing with particle size; in reflected light the particles are clearly yellow and translucent, in contrast to lead tin oxide. The particles have high relief, with the RI much greater than that of the medium; although Kühn (1993) gives a value of ~2.3 for the RI, other sources (notably from extrapolation of values in Verbitskaya and Burakova, 1965) suggest higher values, around 2.47. Particle surfaces have a rough, etched appearance. Particle shape is likely to be predominantly of subhedral bladed and tabular forms. However, a range of compositions can be achieved by varying the silicon content where multiphase mixtures of lead tin silicon oxide, lead tin oxide and tin oxide occur with corresponding changes in habit. Low silicon varieties have framboidal morphology while higher silicon varieties adopt progressively euhedral prismatic forms (Eastaugh, 1988). Some particles demonstrate a weak conchoidal fracture, with curved particle boundaries.

Under crossed polars, the particles are isotropic. However, the particles have strong lemon yellow internal reflections and these are distinctive, particularly outlining the particle boundaries.

Lead tin silicon oxide is often found with free tin(IV) oxide ($q.v.$), which forms moderate relief particles that are readily distinguished from the lead tin silicon oxide particles under crossed polars by their bright, high third and fourth order interference colours.

A weak, dull yellow fluorescence is observable in this pigment with UV excitation.

The chemical, physical and optical characteristics of this pigment have been reviewed by Kühn (1993a).

Lit.: Eastaugh (1988); Kühn (1993a); Verbitskaya & Burakova (1965)

The Pigment Compendium

LEAD CHROMATE, CROCOITE TYPE

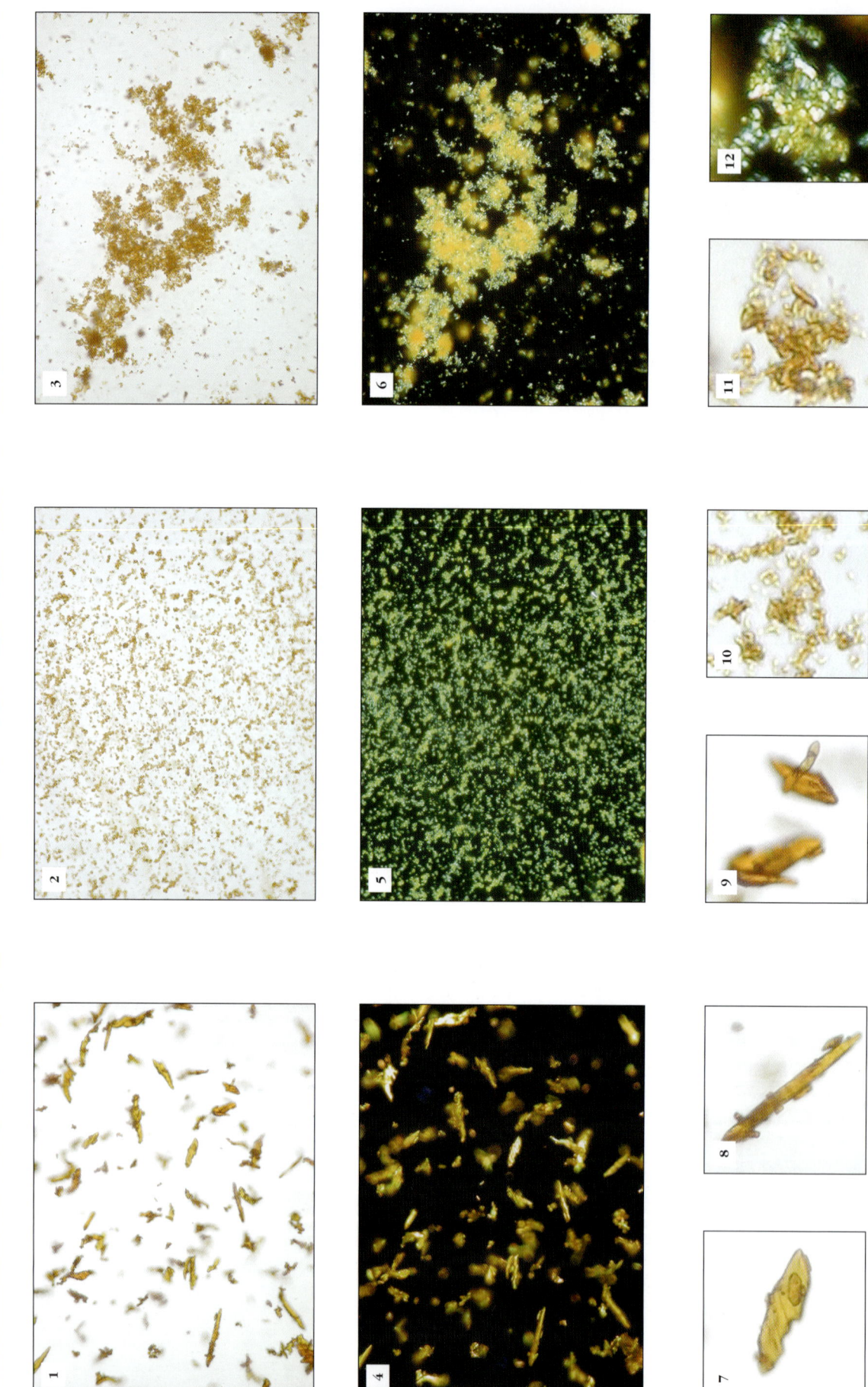

Key:

1. P0898: PPL/40×. Coarse-grained lead chromate with terminated acicular crystals.

2. P0083: PPL/100×. Fine-grained lead chromate.

3. P0224: PPL/100×. A mixture of lead chromate and lead chromate oxide.

4. P0898: XPL/40×. The field of view as in Fig. 1 under crossed polars.

5. P0083: XPL/100×. The field of view as in Fig. 2 under crossed polars, note anomalous green interference colours.

6. P0224: XPL/100×. The field of view as in Fig. 3, the lead chromate shows anomalous greens, while the lead chromate oxide has strong yellow reflections.

7. P0898: PPL. Crystal morphology from Fig. 1.

8. P0898: PPL. Crystal morphology from Fig. 1.

9. P0898: PPL. Crystal morphology from Fig. 1.

10. P0083: PPL. Detail of particle morphology from Fig. 2.

11. P0224: PPL. Detail of particle morphology from Fig. 3.

12. P0224: XPL. The field of view in Fig. 11 under crossed polars.

Crystallinity	Optic sign	2V	n_α	n_β	n_γ	δ
Monoclinic	+ve	54°–57°	2.31	2.37	2.66	0.3500
					(data for mineral from Ford, 1932; cf. Webmineral, 2003)	

Under plane-polarised light, lead chromate appears as acicular to lath-shaped crystals with high relief and RI greater than that of the medium. The crystals are translucent and strongly coloured yellow and exhibit pleochroic colours from yellow to golden yellow (though this may be difficult to observe in small particles). Crystal surfaces are smooth or striated, with the striations forming at an angle oblique to the long axes, and crystals have a domed appearance. Particle size ranges from very fine to medium, size distribution being typically narrow. Burnstock *et al.* (2003) have observed rounded particles of the order of 0.3–0.5 μm mixed with rods 0.8–2.0 μm in proportions that vary from sample to sample among a group of nineteenth century pigments as well as 0.05–0.6 μm aggregates of subrounded rhombs in an example of the 1950s.

Under crossed polars, the crystals are strongly birefringent, with high third and fourth order interference colours, masked by the yellow body colour.

McCrone *et al.* (1973–80) note grey to bright yellow colours and that agglomerates may appear dull orange-yellow. Moderate yellow internal reflections are also present. The particles have inclined extinction and elongation in acicular particles can be seen to be length slow.

Kühn and Curran (1986) list a number of extenders that can be found with lead chromate including barium sulfate, calcium sulfate, kaolinite ('china clay') and diatomite ('diatomaceous earth'). Lead chromate may also be found with other yellow chromates such as lead chromate oxide (phoenicochroite) and strontium chromate.

Lit.: Burnstock *et al.* (2003); Kühn & Curran (1986); McCrone *et al.* (1973–80)

CROCOITE

PbCrO₄

Key:

1. P1182: PPL/40×. Shows variation in particle morphology and colour.

2. P1182: PPL/40×. ● Subhedral prismatic particle with prism-parallel cleavage.

3. P1182: PPL/40×. ● The particle in Fig. 2 rotated through 90° to show pleochroism.

4. P1182: XPL/40×. ● The same field of view as in Fig. 1 under crossed polars.

5. P1182: PPL/40×. Crystal of crocoite.

6. P1182: PPL/40×. The particle in Fig. 5 rotated through 90° to show pleochroism.

Crystallinity	Optic sign	2V	n_α	n_β	n_γ	δ
Monoclinic	+ve	54°–57°	2.31	2.37	2.66	0.3500
					(data from Ford, 1932; cf. Webmineral, 2003)	

Under plane-polarised light, crocoite forms translucent crystals which are strongly pleochroic from acid yellow to golden yellow to orange (though other authors describe the colours as orange-red to blood red; see: Kühn and Curran, 1986). It has very high relief, with RI much greater than that of the medium. There is no dispersion. Typically, the crystals have acicular to prismatic forms and possess perfect cleavage, parallel to the long axes of the particles, and a weaker cleavage parallel to the basal section. When crushed, the particle shapes are strongly influenced by these features, and produce angular, elongated particles. Surfaces and particle boundaries may additionally show a weak conchoidal fracture. Particle surfaces appear glassy and striated. Clay and Watson (1944, 1948) and Wagner *et al.* (1933) have published optical photomicrographs of various prismatic and rhombic forms of lead chromate.

Under crossed polars, crocoite has high birefringence, with third and fourth order interference colours, strongly masked by the body colour, and generally producing bright oranges, yellows, greens and pinks. Crocoite has straight extinction and moderate yellow-orange internal reflections. Crystals are length slow.

Crocoite appears green when viewed through the Chelsea filter, all red and yellow light being absorbed. Samples examined with UV excitation gave a weak orange fluorescence.

Lit.: Clay & Watson (1944, 1948); Kühn & Curran (1986); Wagner *et al.* (1933)

The Pigment Compendium

LEAD CHROMATE OXIDE, PHOENICOCHROITE TYPE

$Pb_2(CrO_4)O$

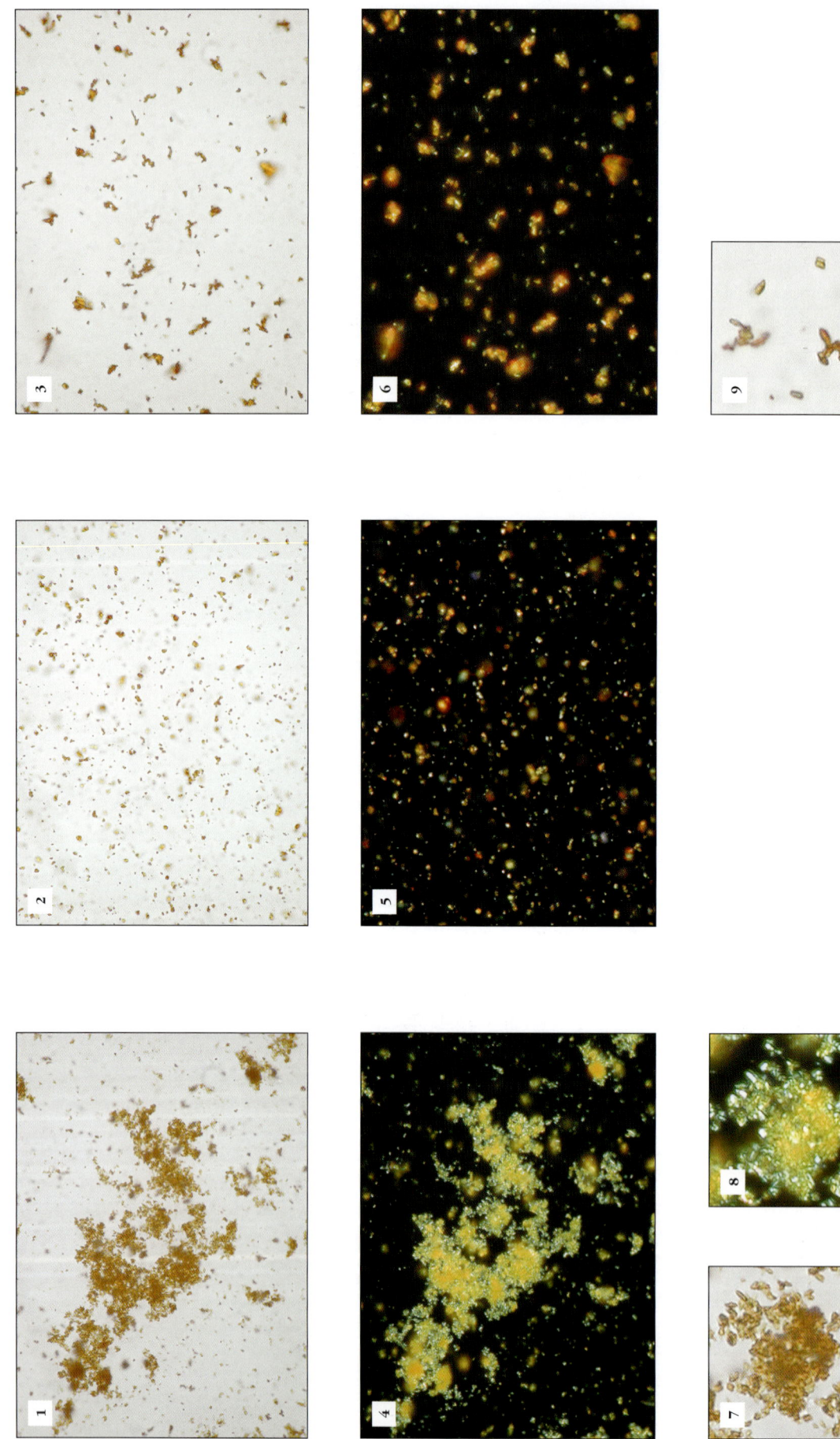

Key:

1. P0224: PPL/100×. Lath-shaped particles of phoenicochroite with crocoite.

2. P1310: PPL/100×/H. Fine-grained particles of phoenicochroite.

3. P0723: PPL/100×/H. Bladed and complex form crystals of phoenicochroite.

4. P0224: XPL/100×. The field of view as in Fig. 1 under crossed polars.

5. P1310: XPL/100×/H. The field of view in Fig. 2 under crossed polars.

6. P0723: XPL/100×/H. The field of view as in Fig. 3 under crossed polars.

7. P0224: PPL. Enlarged view of particle morphologies from Fig. 1.

8. P0224: XPL. The field of view in Fig. 7 under crossed polars.

9. P0723: PPL/H. Enlargement of crystals from Fig. 3.

Crystallinity	Optic sign	2V	n_α	n_β	n_γ	δ
Monoclinic	+ve	58°–62°	2.38	2.44	2.65	0.2700
					(data for mineral from Webmineral, 2003)	

In plane-polarised light lead chromate oxide, the synthetic analogue of the mineral phoenicochroite, can be seen to form as translucent, yellow-orange crystals of high relief and RI greater than that of the medium. McCrone et al. (1973–80) report the RI of this compound as $n_\alpha = 2.42$, $n_\beta = 2.7$ and $n_\gamma > 2.7$. Particle shapes are of acicular to bladed laths, some of which show clear pyramidal terminations. Particle surfaces are smooth. Particle size is fine with a narrow size distribution; McCrone et al. (1973–80) give sizes as typically 1 μm, with rods <1 μm wide and 2–3 μm long.

Under crossed polars, lead chromate oxide has moderate birefringence, but the interference colours are masked by the strong body colour. Particles appear orange and grey-green in colour. Grains also display moderate, orange internal reflections. Lead chromate oxide has straight extinction and is length fast.

Burnstock et al. (2003) have noted the occurrence of this compound in pigments with the crocoite type of lead chromate, quartz, lead carbonate and gypsum.

Lit.: Burnstock et al. (2003); McCrone et al. (1973–80)

The Pigment Compendium

LEAD CHROMATE SULFATE

$PbCrO_4 \cdot xPbSO_4$

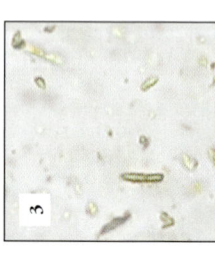

Key:

1. P0985: PPL/100×/H. Fine-grained acicular particles of lead chromate sulfate.

2. P0985: XPL/100×/H. The field of view as in Fig. 1 under crossed polars.

3. P0985: PPL/H. Enlargement of particle morphology from Fig. 1.

Crystallinity	Optic sign	2V	n_α	n_β	n_γ	δ
PbCrO$_4$: Monoclinic	+ve	54°–57°	2.31	2.37	2.66	0.3500
PbSO$_4$: Orthorhombic	+ve	68°–75.4°	1.878	1.883	1.895	0.0170

(data for crocoite and lanarkite from Webmineral, 2003)

Lead chromate sulfate, PbCrO$_4$.xPbSO$_4$, is a solid solution series of lead chromate(VI) and lead (II) sulfate (*qq.v.*). Lead chromate sulfates with a PbCrO$_4$ content of >30–35% have monoclinic symmetry, as does pure lead chromate(VI). However, in pigments with a lead chromate content of less than 10% it crystallises in the orthorhombic system, that of lead(II) sulfate. As a solid solution, this pigment will exhibit optical properties that are intermediate between those of the end-members.

However, modern commercial samples containing <10% lead chromate may be precipitated with the orthorhombic form if stabilisers are used.

Lit.: Kühn & Curran (1986)

The Pigment Compendium

BARIUM CHROMATE

$BaCr_2O_4$

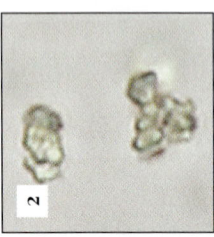

Key:

1. P0086: PPL/100×. Fine-grained particles of barium chromate.

2. P0086: PPL. Detail of particles from Fig. 1 showing particle morphology.

3. P0740: PPL/H. Detail of particles from Fig. 1 showing particle morphology.

Crystallinity	Optic sign	2V	$n_\alpha/n_\beta/n_\gamma$	δ
Orthorhombic	[unknown]	[unknown]	~1.94 and 1.98	Low
			(data from ICSD, 2003 and McCrone *et al.*, 1973–80)	

Under plane-polarised light barium chromate can be seen to form pale yellow translucent crystals. Particle size ranges from sample to sample but can be very fine to medium; older samples are typically coarser while modern preparations are very fine. In the examples illustrated here because particle size is close to that resolvable with the optical microscope particles appear rounded, spherical or bacterioid, though coarser particles form elongate laths. Irregular aggregates also occur: Burnstock *et al.* (2003) have also observed aggregates of angular crystalline particles in rosette form approximately 3–5 μm in diameter by electron microscopy in a modern sample of barium chromate, as well as ~1 μm rounded florets, ~0.5 μm subrounded particles and 2–5 μm plates in a single sample of the pigment dating to the earlier twentieth century. Particle surfaces were smooth in samples studied by the authors but Mactaggart (2002) has observed that some crystal surfaces have a reticulated pattern. Relief is high, with RI greater than that of the medium.

Under crossed polars, barium chromate is very weakly birefringent although Mactaggart has noted up to second order colours (however, see below).

Many particles appear isotropic; others show anomalous blue-grey colours. The compound shows moderate, grey-yellow internal reflections that mask the interference colours in aggregates of particles. Crystals are length slow.

Barium is also thought to substitute into the crocoite structure forming a barium crocoite (Burnstock *et al.*); crystal structure and therefore optical properties of this are more likely to accord to lead chromate. The higher birefringence observed for some samples by Mactaggart may be for this form of the pigment.

Microscopy of barium chromate has been reviewed by Kühn and Curran (1986) and Burnstock *et al.*

Lit.: Burnstock *et al.* (2003); Kühn & Curran (1986); Mactaggart (2002)

STRONTIUM CHROMATE

SrCrO$_4$

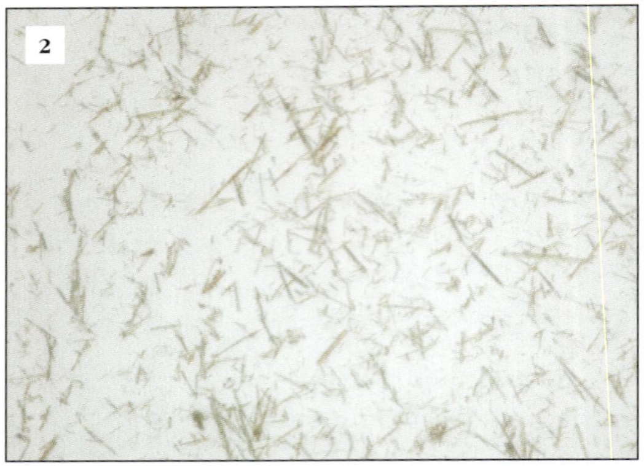

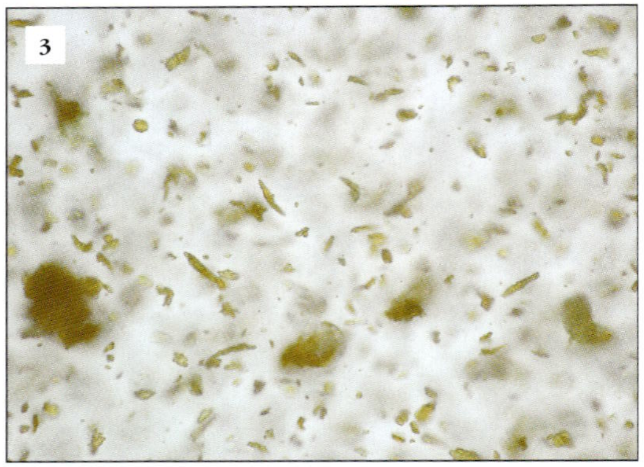

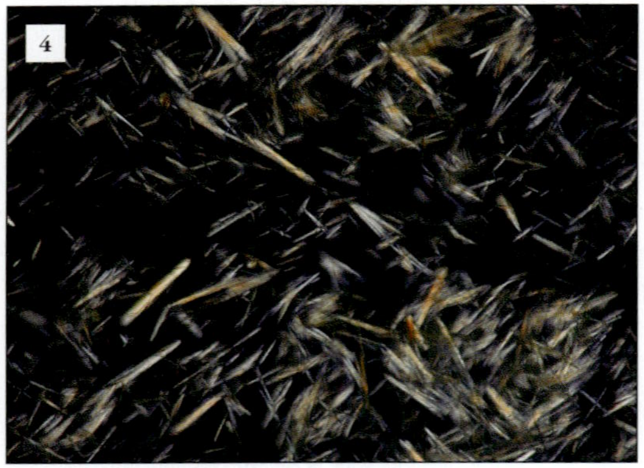

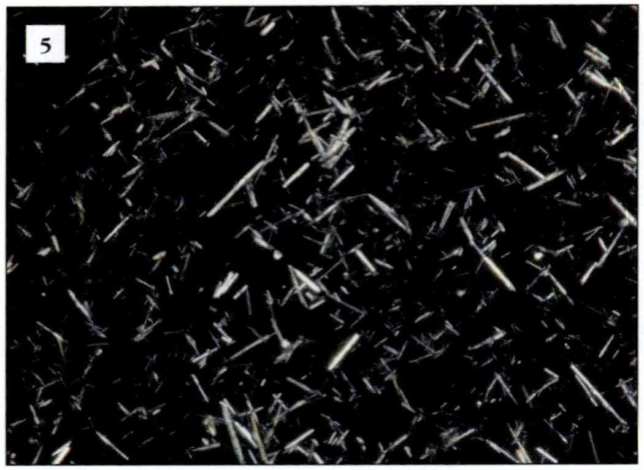

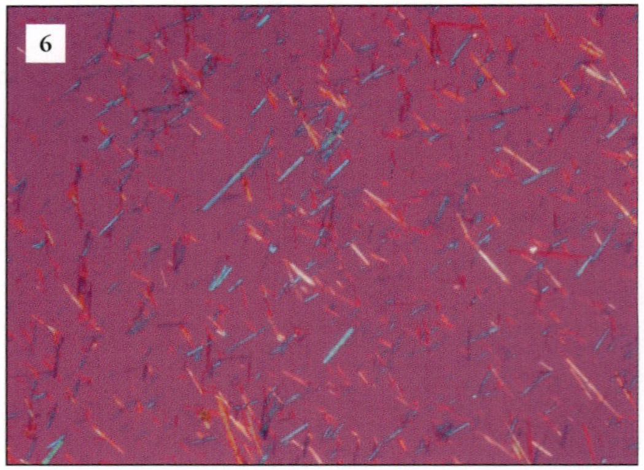

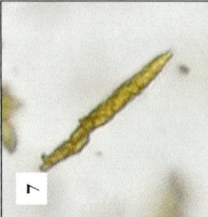

Key:

1. P1321: PPL/40×/H. Straw yellow acicular grains of strontium chromate.

2. P0232: PPL/40×. Acicular crystals of strontium chromate.

3. P0538: PPL/100×/H. Coarse-grained crystals of strontium chromate.

4. P1321: XPL/40×/○/H. The field of view as in Fig. 1 under crossed polars.

5. P0232: XPL/40×/○. The field of view as in Fig. 2 under crossed polars.

6. P0232: XPL/STP/40×. Strontium chromate with the sensitive tint plate inserted NE-SW showing the crystals to be length fast.

7. P0538: PPL/H. Crystal morphology of strontium chromate needles from Fig. 3.

Crystallinity	Optic sign	2V	$n_\alpha/n_\beta/n_\gamma$	δ
Monoclinic	[unknown]	[unknown]	>1.662	High

(data from ICSD, 2003/authors)

Under plane-polarised light, strontium chromate yellow is a distinctive pigment. It forms medium- to coarse-grained, acicular crystals, which are translucent and pale yellow in colour. Particles are weakly pleochroic. Crystals appear as individual needles and as mats of randomly orientated particles. Relief is moderate and RI is greater than that of the medium.

Under crossed polars, birefringence is high, and particles show fourth order and greater interference colours, appearing white. Larger crystals with stronger body colour mask the interference colours and bright yellows are seen. Strontium chromate has straight extinction and acicular particles can be seen to be length slow. Moderate yellow internal reflections also occur.

Microscopy of strontium chromate has been reviewed by Kühn and Curran (1986) and Burnstock et al. (2003).

Lit.: Burnstock et al. (2003); Kühn & Curran (1986)

The Pigment Compendium

LEAD CHROMATE MOLYBDATE

$xPbCrO_4 \cdot yPbMoO_4 \cdot zPbSO_4$, where $x + y + z = 1$

Key:

1. P0243: PPL/100×. Fine-grained particles of lead chromate molybdate.

2. P0243: XPL/100×/⊙. The field of view as in Fig. 1 under crossed polars.

3. P0243: RPL/100×. The field of view as in Fig. 1 in reflected light.

4. P1095: PPL/100×. Fine-grained platy crystals of lead chromate molybdate.

5. P1095: XPL/100×/⊙. The field of view as in Fig. 4 under crossed polars.

6. P1095: RPL/100×. The field of view as in Fig. 4 in reflected light.

7. P0243: PPL. Enlarged view of particle morphology from Fig. 1, note the yellow laths of lead chromate

8. P0243: XPL/⊙. The field of view as in Fig. 7 under crossed polars.

9. P1095: PPL. Detail of particle morphology from Fig. 4.

Crystallinity
Monoclinic or Tetragonal

n
>1.662

δ
High

(data from Ziobrowski, 1973 and authors)

Lead chromate molybdate appears to be a three-way solid solution between lead chromate(VI), lead molybdate and lead(II) sulfate, the latter of these not always being present.

In plane-polarised light lead chromate molybdate can be seen as translucent, orange-red particles, with deeper shades correlating with larger particles. Relief is moderate and RI is greater than that of the medium. Particle size is very fine, and particles appear with rounded or bacterioid habits. Particle surfaces are smooth and domed. According to Ziobrowski (1973), pigment grades have been tailored to meet a range of requirements; that author quotes particle sizes of 0.04 μm to 1.0 μm, with a tendency to form small aggregates.

Under crossed polars, particles are strongly birefringent, with high order interference colours, strongly masked by the body colour, so that particles

appear bright orange. It should be noted that, depending on the relative proportion of chromate to molybdate, the compound crystallographically assumes a tetragonal to monoclinic structure; compounds typical of pigments (90 chromate: 10 molybdate) are monoclinic (Ziobrowski).

The pigment appears orange in reflected light. Samples examined were also variably fluorescent with UV excitation, from no observable fluorescence through to a dull orange colour.

Free lead chromate is often present in samples of this pigment.

Lit.: Ziobrowski (1973)

CINNABAR AND MERCURY(II) SULFIDE, DRY-PROCESS TYPE

HgS

Key:

1. P0549: PPL/40×/H. Coarse grains showing variation in particle morphology, high relief and uneven fracture.

2. P0549: XPL/40×/H. Same field of view as in Fig. 1 showing birefringence masked by strong body colour and internal reflections; large crystal on right shows straight extinction.

3. P1342: PPL/100×/H. Anhedral crystals of cinnabar.

4. P1327: PPL/100×/H. Shows natural cinnabar with colourless impurities, predominantly carbonates.

5. P1327: XPL/100×/H. Same field of view as in Fig. 4, note the highly birefringent carbonate particles.

6. P1347: PPL/100×/H. Shows weak pleochroism and high relief in dry-process mercury(II) sulfide.

7. P0863: PPL. Crystal of cinnabar.

8. P0863: PPL. The crystal in Fig. 7 showing pleochroic change in body colour upon rotation of the polariser through 90°.

9. P1472: PPL. Striated crystal of cinnabar.

10. P1472: XPL/○. The crystal in Fig. 9 under crossed polars showing high birefringence.

11. P0010: PPL. Particles of cinnabar.

12. P0010: RPL. The crystals in Fig. 11 observed in reflected light.

Crystallinity	Optic sign	n_ω	n_ε	δ
Trigonal	+ve	2.905	3.256	0.3510
			(data for mineral from Mason, 1968; cf. Webmineral, 2003)	

The naturally occurring mercury sulfide mineral cinnabar and its synthetic analogue, mercury(II) sulfide ('vermilion') when produced by the dry-process method, are very difficult to distinguish optically. It has been remarked by Gettens *et al.* (1993c) that the natural variety may have colourless streaks within the crystals. Similarly, the presence of impurities is more likely to be a feature of pigments produced from the natural mineral. Otherwise differentiation is difficult, though it is commonly assumed on the basis of documentary sources and the absence of contrary evidence that most observations of this pigment in artefacts are of the synthetic form.

In plane-polarised light, the particles exhibit a strong red-orange body colour that is weakly pleochroic from pale to dark orange-red. The finer the particles, the more orange the hue will be. Most strikingly it has extremely high relief, since the RI is much greater than that of the medium. Cinnabar has perfect cleavage and the particles are unevenly to conchoidally fractured. These features along with the relief, render the crystals translucent rather than transparent, and in extreme cases in fine particles the particles appear almost opaque. Cinnabar is often coarse grained, with angular particles produced by crushing. However, particle size distribution is frequently

uneven. Crystals typically have a hexagonal platy or prismatic habit, but this is lost when the sample is ground. The cleavage causes particles to form as rectangular or columnar particles.

Under crossed polars the high birefringence of cinnabar produces strong interference colours. These are, however, masked by a combination of the body colour, relief and strong internal reflections, causing the crystal to glow a deep red. In instances where interference colours are visible anomalous red, orange and yellow colours are typically observed. Cinnabar and mercury(II) sulfide have straight extinction.

Natural cinnabar may be found in association with calcite or dolomite, quartz, orpiment, realgar (which may appear similar) and hematite.

The optical and physical properties of cinnabar and vermilion in pigment contexts have been reviewed by Gettens *et al.* (1993c).

Lit.: Gettens *et al.* (1993c)

MERCURY(II) SULFIDE, CINNABAR TYPE, WET-PROCESS TYPE

HgS

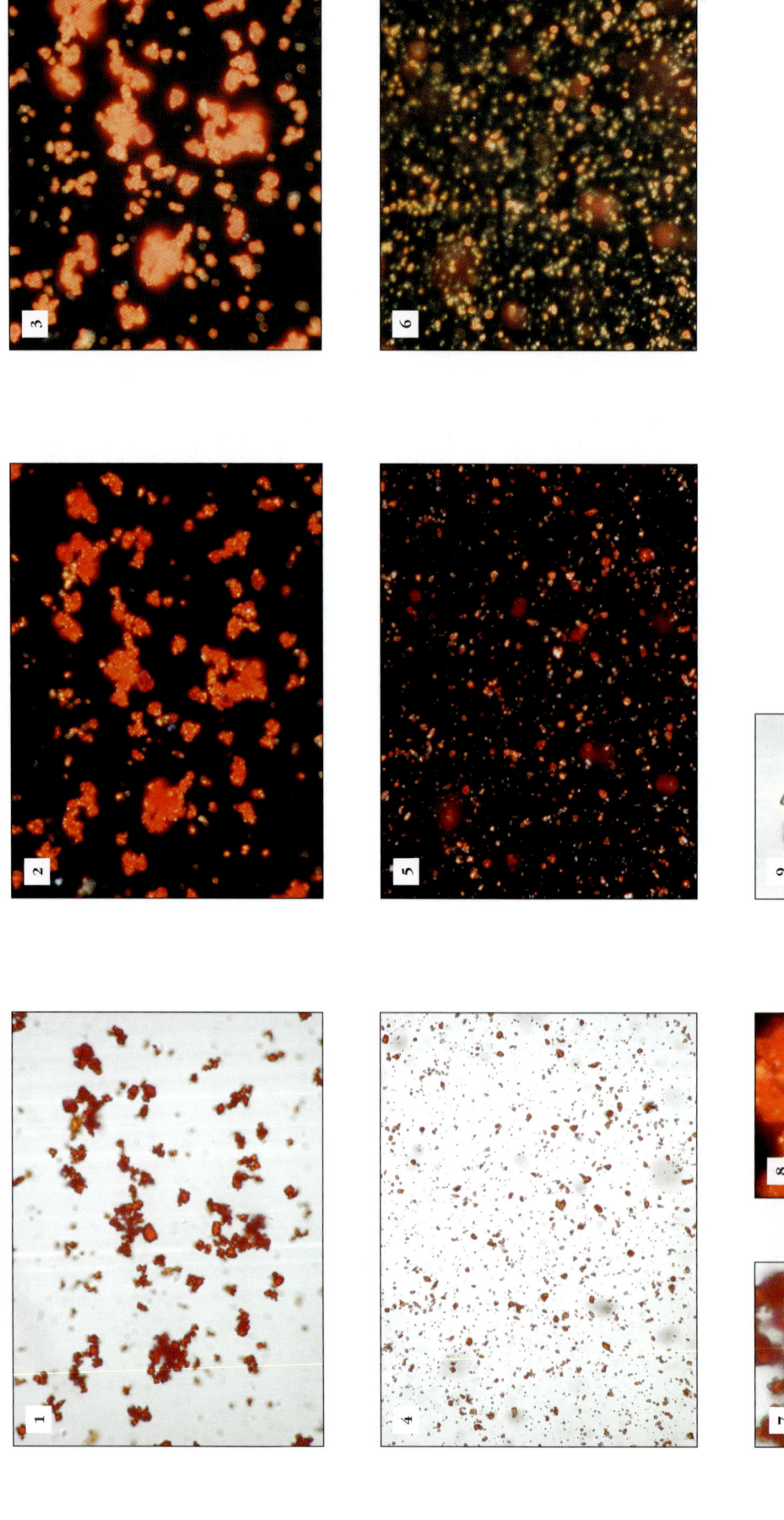

Key:

1. P0262: PPL/100×. Small particles of mercury(II) sulfide.

2. P0262: XPL/100×/●. Same field of view as in Fig. 1 under crossed polars.

3. P0262: RPL/100×. Same field of view as in Fig. 1 in reflected light.

4. P0553: PPL/100×/H. Fine-grained mercury(II) sulfide showing strongly coloured grains with high relief.

5. P0553: XPL/100×/●/H. The same field of view as in Fig. 4 under crossed polars.

6. P0553: RPL/100×/H. The same field of view as in Fig. 4 in reflected light.

7. P0262: PPL. Enlarged grains from Fig. 1 showing tabular morphology.

8. P0262: XPL/●. The grains in Fig. 7 under crossed polars.

9. P0551: PPL/●/H. Particles of mercury(II) sulfide showing striated, comb-like texture of crystals.

Crystallinity	Optic sign	n_ω	n_ϵ	δ
Trigonal	+ve	2.905	3.256	0.3510
				(data for mineral from Mason, 1968; cf. Webmineral, 2003)

This variety of synthetic mercury(II) sulfide, commonly known as 'wet' process vermilion, may be generally distinguished from the natural analogue cinnabar and mercury(II) sulfide produced via the 'dry' process by the fact that it is usually far more finely divided. Additionally, the high purity of the sample distinguishes the pigment from the naturally occurring variety.

Under plane-polarised light, wet-process vermilion is orange to orange-red, with high relief, and an RI greater than that of the medium. Particle size is typically very fine, with particles less than 1 μm and an even particle size distribution. Particle shape is often beyond the resolution of microscope and they thus appear spherical or bacterioid, or as larger aggregates.

Under crossed polars, wet-process vermilion has very strong red internal reflections that, along with the high relief and body colour, somewhat obscure the high birefringence.

The optical and physical properties of vermilion in pigment contexts have been reviewed by Gettens *et al.* (1993c).

Lit.: Gettens *et al.* (1993c)

The Pigment Compendium

METACINNABAR

HgS

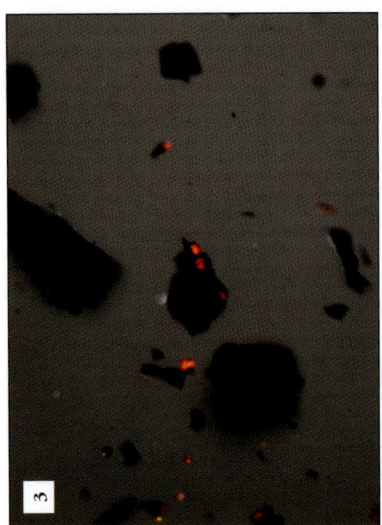

Key: **1.** P1517: PPL/10×. Opaque particles of metacinnabar.

2. P1517: PPL/40×/**○** Particles of metacinnabar showing unaltered patches of red cinnabar.

3. P1517: ~XPL/40×. The field of view as in Fig. 2 under crossed polars, note birefringent unaltered cinnabar.

Crystallinity

Cubic	*n*	Isotropic
	[Opaque]	δ

(data from Webmineral, 2003)

In plane-polarised light, metacinnabar is opaque. Particle shape closely resembles that of cinnabar ('vermilion') and its synthetic analogue mercury(II) sulfide, of which it occurs as an alteration product. Metacinnabar appears black in reflected light with resinous-looking surfaces; red-orange cinnabar particles may be seen in association.

Varying amounts of an unaltered cinnabar phase may be present, the optical properties of which have been described above and also by Gettens *et al.*

(1993). The blackening of mercury sulfides of certain compositions and in certain environmental conditions has been described by McCormack (2000).

Lit.: Gettens *et al.* (1993c); McCormack (2000)

(POTASSIUM, SODIUM) HEXANITROCOBALT(III)

$K_{3-x}Na_x[Co(NO_2)_6]\cdot nH_2O$

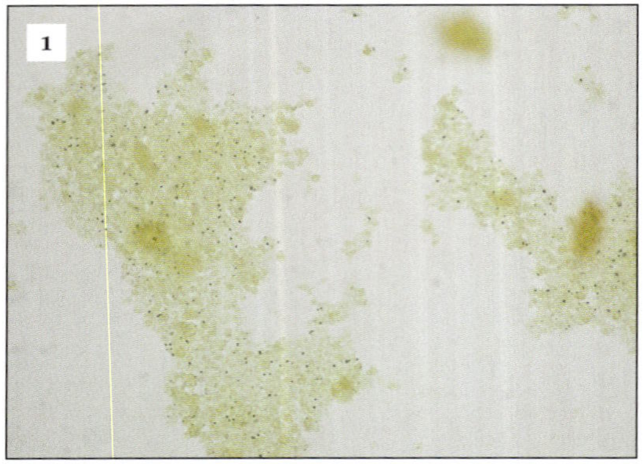

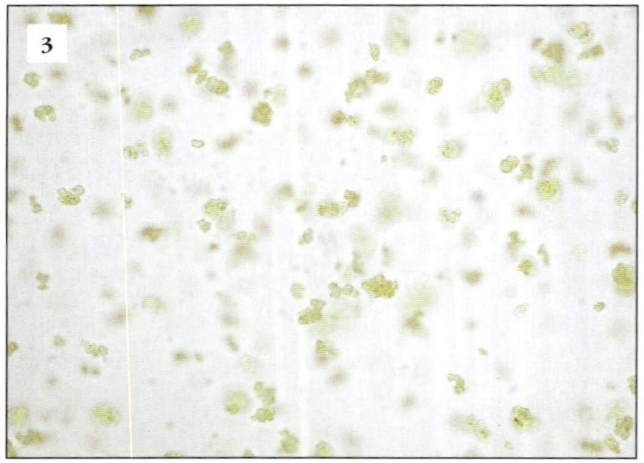

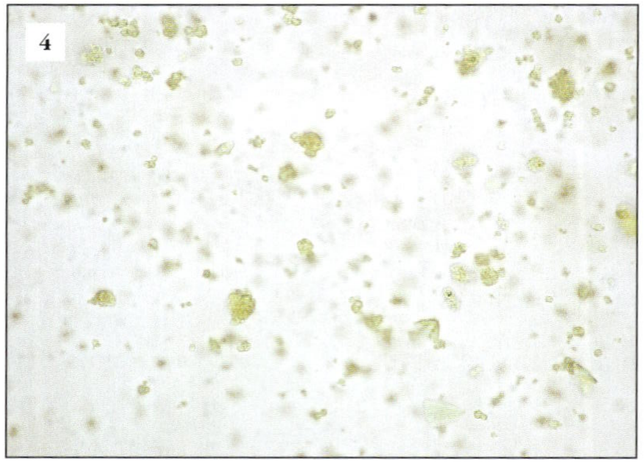

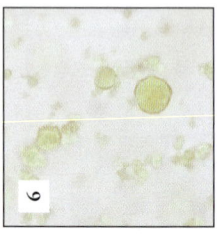

Key:

1. P0081: PPL/100×. Fine-grained rounded particles and larger pennate crystals.

2. P0081: RPL/100×. The field of view as in Fig. 1 in reflected light.

3. P1315: PPL/100×/⦿/H. Rosette-like clusters of particles.

4. P1316: PPL/100×/H. Rosettes of pennate particles.

5. P0846: PPL/100×. Spherical and platy particles.

6. P0846: PPL. Enlargement of spherulitic particles from Fig. 5.

7. P1316: PPL/H. Pennate particle morphology.

Crystallinity
Cubic

n
1.72–1.76

δ
Isotropic
(data from Cornman, 1986)

Under plane-polarised light (potassium, sodium) hexanitrocobalt(III), commonly known as 'cobalt yellow' or 'aureolin', forms translucent, yellow-coloured particles of moderate relief. The RI is greater than that of the medium and the compound is non-pleochroic. The particles are predominantly fine to very fine grained, bacterioid or spherulitic, either as isolated particles, or more commonly as aggregates. Coarse- examples show a rhombic habit and also rosettes of four rhombs. However, a distinctive second habit also occurs. These are medium-sized dendritic particles with pennate

habit that occur as individual particles and radiating aggregates, having the appearance of feathers or leaves. The occurrence of this latter shape is immediately diagnostic of this compound. McCrone *et al.* (1973–80) have also observed cubes and octahedral forms.

This pigment is isotropic.

Lit.: Cornman (1986); Gates (1995); McCrone *et al.* (1973–80)

POTASSIUM HEXACHLOROPLATINATE

K$_2$[PtCl$_6$]

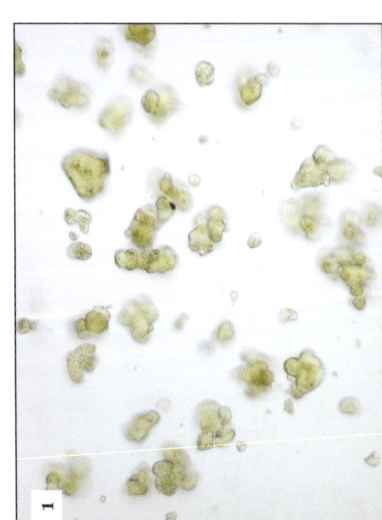

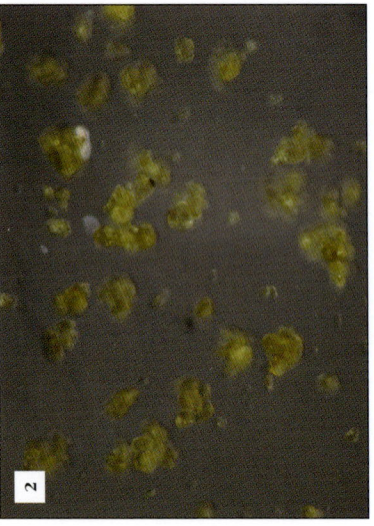

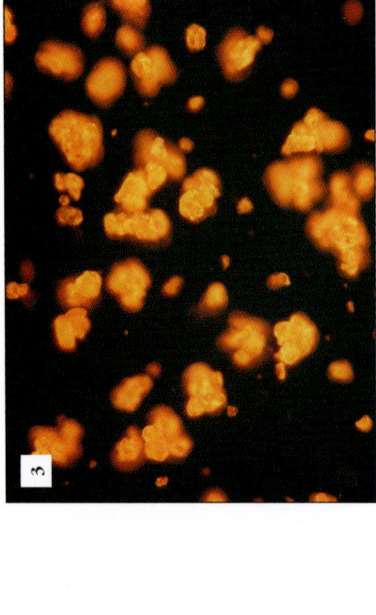

Key: **1.** P0914: PPL/40×. Subrounded equant particles of potassium hexachloroplatinate. **2.** P0914: XPL/40×. The same field of view as in Fig. 1 under crossed polars. Note the yellow internal reflections. **3.** P0914: UV/40×. The same field of view as in Fig. 1 in UV light; note the strong orange UV fluorescence.

Crystallinity	***n***	**δ**
Cubic	>1.662	Isotropic (data from authors)

In plane-polarised light, potassium hexachloroplatinate forms translucent yellow crystals with high relief and RI greater than that of the medium. Habit is of subrounded equant particles and framboidal aggregates of these, plus a few elongate, tabular grains. Particle surfaces are smooth and appear domed. Inclusions are visible, which appear as black spots inside the crystals. Grain size is medium to coarse.

Under crossed polars, potassium hexachloroplatinate is isotropic. However, it has moderate yellow internal reflections, which illuminate the grain boundaries.

Under UV illumination, the particles exhibit a strong orange fluorescence.

The Pigment Compendium

SULFUR

S(am) and S₈

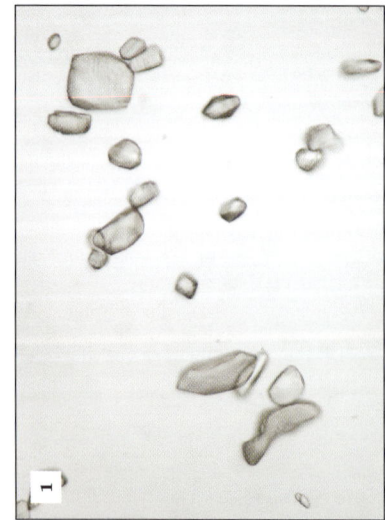

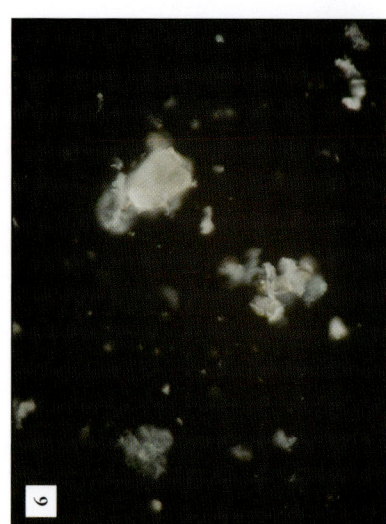

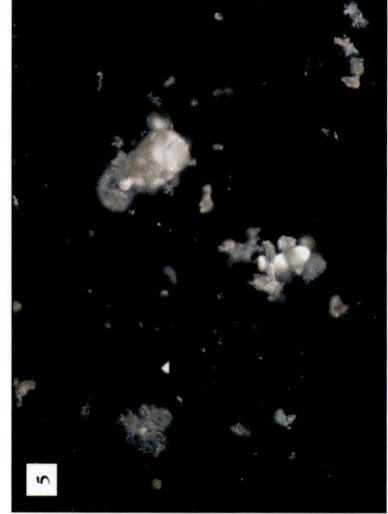

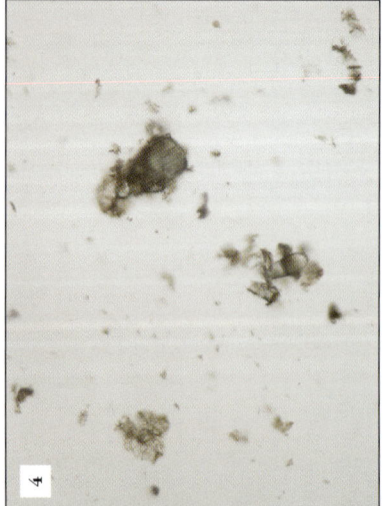

Key:

1. P1503: PPL/40×. Particles of sulfur. Grains have been rounded through melting upon contact with the warm mounting medium.

2. P1503: ~XPL/40×. The field of view as in Fig. 1 under crossed polars.

3. P1503: RPL/40×. The field of view as in Fig. 1 in reflected light.

4. P0917: PPL/40×. Sulfur plus other fine-grained impurities; again particle morphology has been modified through slight melting.

5. P0917: XPL/40×. The field of view as in Fig. 4 under crossed polars.

6. P0917: RPL/40×. The field of view as in Fig. 4 in reflected light.

Crystallinity	Optic sign	2V	n_α	n_β	n_γ	δ
S_8: Orthorhombic	+ve	69°–70°	1.9579	2.0377	2.2452	0.2873

(data from Webmineral, 2003)

In plane-polarised light orthorhombic sulfur is visible as translucent, colourless particles with very high relief and RI greater than that of the medium. Dispersion is relatively weak. When viewed microscopically in mounting media requiring temperatures of ~60°C to melt, sulfur is placed above its own melting point and particle margins will begin to convert to the amorphous form of sulfur. As a result, particle morphology will normally be seen as rounded droplets, regardless of its original habit. Particle surfaces are consequently also smooth and appear domed. Particle size ranges from medium to coarse.

In cross-polarised light sulfur has high birefringence, with high third order to fourth order interference colours in evidence; the order of colours increases with particle size. Extinction is complete or sweeping.

The Pigment Compendium

TIN(IV) SULFIDE

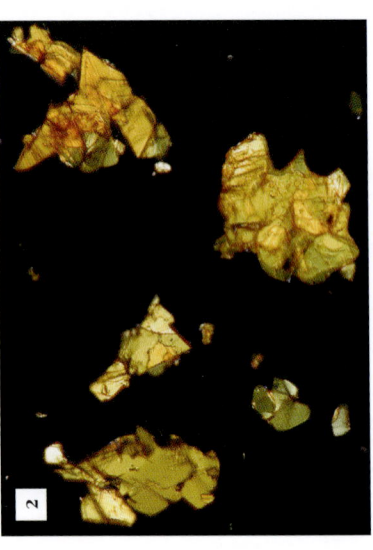

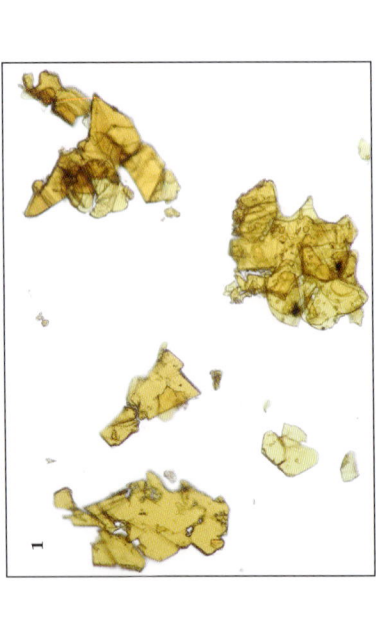

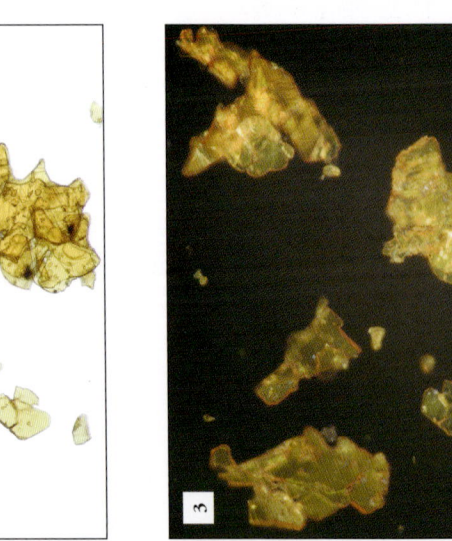

Crystallinity	Optic sign	n_ω	n_ε	δ
Trigonal	−ve	1.93	1.705	0.2250

(data for berndtite from Dana, 1997; Webmineral, 2003)

In plane-polarised light tin(IV) sulfide ('mosaic gold'), the synthetic analogue of the mineral berndtite, can be seen as forming transparent, strong lemon yellow particles. Relief is high and the RI is greater than that of the medium. Particles form anhedral irregular plates with a small proportion of euhedral hexagonal plates and bladed forms also present in the examples examined. Grain surfaces are fractured and scratched with a weak pitted texture. Some particles also exhibit weak concentric growth zones. Particle size is medium to very coarse, the size range therefore broad.

In crossed polars, tin(IV) sulfide has low birefringence, with some particles appearing to be isotropic. Low first order interference colours are masked by the body colour. Some particles show lamellar twins. Extinction in tin(IV) sulfide is sweeping crystals are length fast.

In reflected light tin(IV) sulfide is readily distinguished from gold as having a duller, more bronze-coloured reflection. The platy particles are also clear (Muñoz Viñas and Farrell, 1999).

Lit.: Muñoz Viñas & Farrell (1999)

The Pigment Compendium

AZO PIGMENTS GROUP [1]

[Variable]

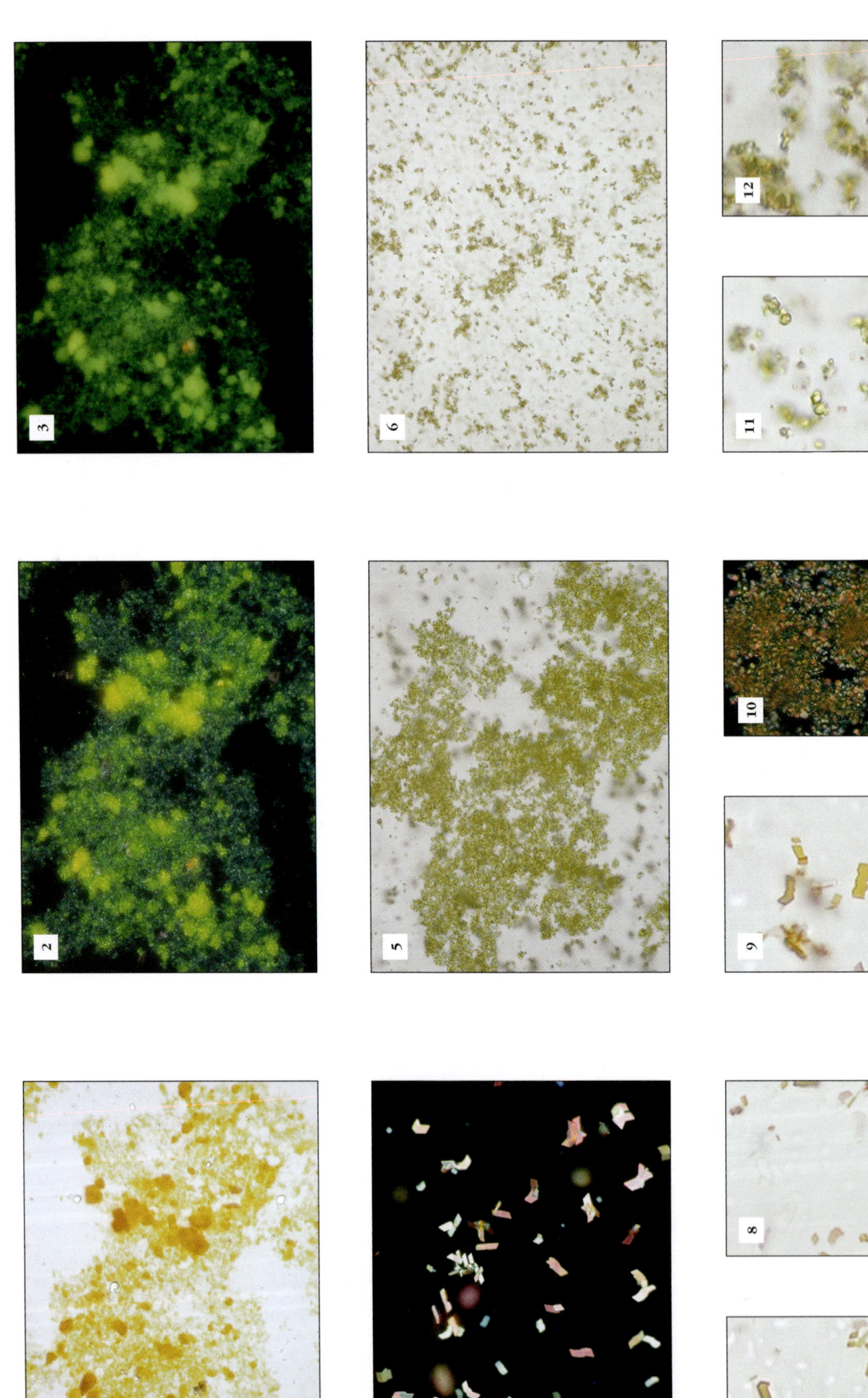

Key:

1. P0219: PPL/100×. A mixture of PY13 and PY1.

2. P0219: XPL/100×. The field of view as in Fig. 1 under crossed polars, note strong green internal reflections.

3. P0219: RPL/100×. The field of view as in Fig. 1 in reflected light.

4. P0220: PPL/100×. PY3; rounded particles.

5. P0220: XPL/100×. PY3; the same field of view as Fig. 4 under crossed polars.

6. P1396: PPL/100×. Particles of PY4.

7. P0039: PPL. PY1; chevron-shaped crystals.

8. P0039: PPL. PY1; the same field of view as Fig. 7 on rotation of polariser through 90°, note intense pleochroism from yellow to colourless.

9. P0039: PPL. PY1; chevron, bladed and zigzag-shaped crystals.

10. P0220: XPL. Particles from Fig. 5 under crossed polars.

11. P1395: PPL. Enlarged view of subrounded particles of PY3.

12. P1396: PPL. Enlarged view of bladed and elongate particles in PY4.

Crystallinity	n_α	n_β	n_γ	δ
CI Pigment Yellow 1: Monoclinic	1.449	1.700–1.800	>1.90	High
CI Pigment Yellow 3: Orthorhombic	1.449	1.700–1.800	>1.90	High
CI Pigment Yellow 4: Triclinic	1.439	1.700–1.800	>1.90	High
CI Pigment Yellow 13: [Unknown]		>1.662		Low

(data from Zona, 1996 and authors)

The sub-group of azo pigments known generically as 'Hansa' yellows form a large and diverse collection; the following selection therefore illustrates only a handful of typical examples.

In plane-polarised light, CI Pigment Yellow 1 (PY1) forms translucent yellow to orange-yellow pleochroic particles. Relief is moderate and RI is greater than that of the medium. Grain size is bimodal, with the majority of particles very fine grained with acicular and bacterioid habits. A medium grain fraction exists of distinctive euhedral crystals with single bladed grains, twinned X-shaped and chevron-shaped habits. Grain surfaces are smooth. Under crossed polars, PY1 has high birefringence and third order interference colours are typically observed. Interference colours are masked by the body colour. The fine-grained fraction appears as bright yellow and anomalous green particles. The medium crystals appear yellow, pink and blue-green. Extinction is complete and inclined.

In plane-polarised light, CI Pigment Yellow 3 (PY3) forms translucent yellow particles. Relief is moderate to high and RI is greater than that of the medium. Grain size is very fine with particle size close to that of the resolution of the microscope and therefore particles appear rounded and bacterioid. Zona (1996) has reported that crystals of this pigment are clumped together in medium-grained, rounded, polycrystalline aggregates. Under crossed polars, PY3 has high birefringence,

but third order interference colours are masked by the body colour and particles appear bright yellow, pink and green. Individual grains show complete extinction.

In plane-polarised light, CI Pigment Yellow 4 (PY4) appears as translucent, pale yellow particles. Relief is moderate to low and RI is greater than that of the medium. Grain size is extremely fine, with particles appearing rounded or bacterioid. In plane-polarised light, PY4 has high birefringence, but with interference colours masked by the body colour. Grains appear bright, anomalous pale green and yellow. Extinction is complete.

In plane-polarised light, CI Pigment Yellow 13 (PY13) forms translucent yellow particles. Relief is low and RI is just greater than that of the medium. Particle size is very fine individual grains or as medium-grained, polycrystalline aggregates. The grain surfaces of the latter are rough, and in colour they are brown-yellow. For the fine-grained fraction, crystal size is close to that resolvable with the optical microscope and therefore particles appear rounded or bacterioid. Aggregate particles are sub-rounded anhedral grains. In cross-polarised light, PY13 has low birefringence and weak, but strongly anomalous green, interference colours are observed. The poly-crystalline nature of the larger particles is distinct and complete extinction is not attained, but grains twinkle as the stage is rotated.

AZO PIGMENTS GROUP [2]

[Variable]

Key:

1. P1360: PPL/100×. PY73; fine-grained particles and aggregates.

2. P1361: PPL/100×. PY79; fine-grained particles.

3. P1361: XPL/100×. ● The field of view as in Fig. 2 under crossed polars, note high birefringence and strong green internal reflections.

4. P0042: PPL/100×. PY74; yellow particles.

5. P0042: XPL/100×. PY74; the view as in Fig. 4 under crossed polars.

6. P0042: RPL/100×. PY74; the view as in Fig. 4 in reflected light.

7. P1360: PPL. PY73; Enlarged view of particle aggregate from Fig. 1.

8. P1360: XPL ● The particle in Fig. 7 under crossed polars.

9. P1360: RPL. The particle in Fig. 7 in reflected light.

10. P1361: PPL. Acicular crystal forms of PY79.

Crystallinity

	n_α	n_β	n_γ	δ
CI Pigment Yellow 65: Orthorhombic	1.519	1.700–1.800	>1.90	High
CI Pigment Yellow 73: [Unknown]	1.448	1.700–1.800	>1.90	High
CI Pigment Yellow 74: [Unknown]		>1.662		Moderate
CI Pigment Yellow 79: [Unknown]		<1.662		High

(data from Zona, 1996 and authors)

In plane-polarised light, CI Pigment Yellow 65 has translucent, orange particles. Relief is moderate and RI is greater than that of the medium. Particle size is fine and particles appear rounded or bacterioid. Some particles clump together in crumb-like aggregates. Zona (1996) reports that recrystallised samples of PY65 are of rectangular habit and are pleochroic from pale yellow to yellow. Under crossed polars, PY65 has high birefringence. Individual particles appear bright anomalous green. Clumps of particles have moderate orange-brown internal reflections.

In plane-polarised light, CI Pigment Yellow 73 (PY73) forms translucent lemon yellow particles. Zona (1996) reports two polymorphs of PY73; a pleochroic yellow-colourless form and a non-pleochroic yellow form. Relief is low and RI is just greater than that of the medium. Individual particles are very fine grained and appear rounded or bacterioid in habit. Other particles are in rounded polycrystalline aggregates with medium grain size and rough, pitted surface textures. The overall particle size appearance is bimodal. Under crossed polars, PY73 has moderate to high birefringence with interference colours strongly masked by the body colour. Grains appear dull yellow. The polycrystalline nature of the larger particles is evident and complete extinction does not occur; grains twinkle as the stage is rotated.

In plane-polarised light, CI Pigment Yellow 74 (PY74) forms translucent yellow-orange particles. Relief is moderate and RI is greater than that of the medium. Grain size is

fine and close to that resolvable by the optical microscope and therefore appear rounded or bacterioid in shape. In plane-polarised light, PY74 has high birefringence, but with interference colours masked by the body colour. Grains appear bright, anomalous pale green, orange and yellow. Extinction is complete.

In plane-polarised light, CI Pigment Yellow 79 (PY79) forms translucent yellow particles. Relief is low and RI is just less than that of the medium. Grain size is very fine, and particles appear rounded or bacterioid in shape. A few fine-grained particles have euhedral acicular or bladed habits. Some grains are in rounded, polycrystalline aggregates. In plane-polarised light, PY79 has high birefringence, but with interference colours masked by the body colour. Grains appear bright, anomalous pale green and yellow. Extinction is complete. Bladed particles have straight extinction.

If sufficient material is available, sublimation recrystallisation techniques as described by Zona may be applied to this group of pigments. Numerous other azo and organic polycyclic pigments are also known; the complexities of the chemistry and the poor differentiation of these afforded by optical microscopy alone make their identification outside the scope of this text.

Lit.: Zona (1996)

The Pigment Compendium

GAMBOGE

[Complex]

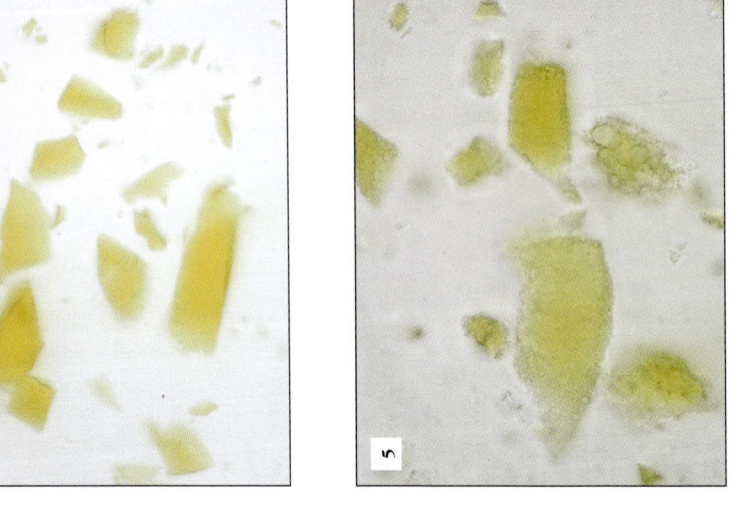

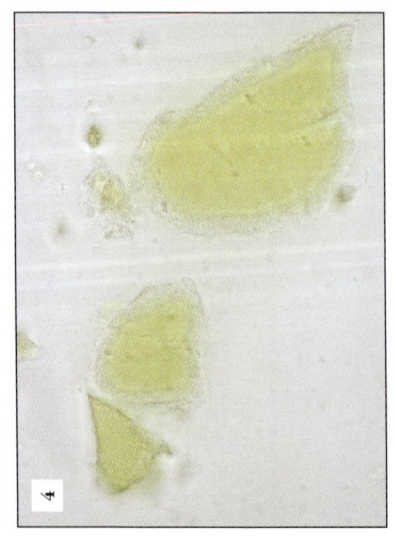

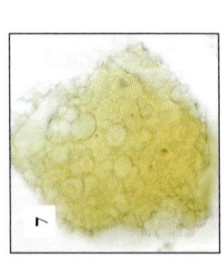

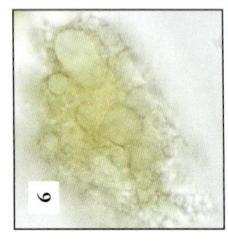

1. P0483: PPL/10×/H. Strongly coloured yellow angular flakes of gamboge.

2. P0483: PPL/40×/H. Yellow angular grains with moderate relief.

3. P0836: PPL/40× ● Angular yellow grains with pock marks at surface.

4. P0962: PPL/100×/ ●/H. Angular yellow grains developing reaction rims with the medium.

5. P1322: PPL/100×/H. Anhedral fragments of gamboge with sponge-like surfaces and bubbles.

6. P1322: PPL/H. Rounded and elongate bubbles in gamboge fragment.

7. P0837: PPL/ ● Angular yellow grains with pock marks at surface.

Crystallinity
Amorphous

n	δ
1.582–1.586	0.0

(data from McCrone et al., 1973–80)

When used as a watercolour, gamboge resin is dissolved and particles are not visible using the polarising light microscope. A yellow stain appears on white substrates such as shell white (see: Winter, 1997). The ground gum alone has not yet been identified as a pigment in its natural state; however, further identifications may be forthcoming.

Under plane-polarised light, gamboge resin appears as crushed, angular particles with a strong golden yellow colour, although it may be softened by heat such that mounting a sample may distort its morphology. It has moderate relief, with RI less than that of the medium. The surface is distinctive, with a spongy, pock-marked appearance. Similar bubbles are visible within the internal regions of the particles. The density and size of the bubbles is variable; bubble shape ranges from spherical to elongate. In reflected light the particles can be seen to be transparent, with little apparent colour.

Gamboge has been noted as showing a yellow-gold fluorescence under excitation from blue light (Townsend, 1993).

As a resin, gamboge is amorphous and therefore appears isotropic under cross-polarised light.

Lit.: McCrone et al. (1973–80); Townsend (1993); Winter (1997)

INDIAN YELLOW

Ca, Mg euxanthates

Key:

1. P1065: PPL/100×. Mats of acicular crystals.

2. P1065: XPL/100×. The same field of view as in Fig. 1 under crossed polars.

3. P1065: PPL/100×. Acicular and quatrefoil particles.

4. P1409: PPL/40×. Polycrystalline aggregates.

5. P1409: XPL/40×/○. The same field of view as in Fig. 4 under crossed polars.

6. P1409: UV/40×. The same field of view as in Fig. 4 with UV fluorescence.

7. P1065: PPL. Enlarged mat of acicular particles.

8. P1065: XPL/○ Same particle as in Fig. 7 under crossed polars showing anomalous blue and green interference colours.

9. P1065: UV. Same particle as in Fig. 7 with UV fluorescence.

10. P1409: XPL/○ Quatrefoil particle.

11. P1409: XPL. Quatrefoil particle.

12. P1409: XPL. Acicular crystals under crossed polars showing intense anomalous blues.

Crystallinity
Crystalline; unknown system

n	δ
~1.662	Anisotropic (data from authors)

In plane-polarised light, Indian yellow forms intensely coloured, golden yellow, translucent particles; Baer *et al.* (1986) have also noted the occurrence of some rare olive-green particles. Particle shapes are distinctive and adopt two forms. Common are acicular and lath-shaped crystals, which may exist as individual particles, as aggregates with rosette and variolitic habits and as mat-like aggregates with randomly orientated crystals. A second particle form is composed of finely fibrous spherulites, which appear as yellow circular or 'quatrefoil' forms in plane-polarised light. Samples may have both forms, or may be dominated by either form. Baer *et al.* (1986) also note the presence of 'colourless, highly birefringent particles'. Relief is low and RI is just less than that of the medium. Particle size is fine to coarse. Coarse particles are aggregates of needles or quatrefoils.

In crossed polars, all particles have high birefringence. However, the strong body colour causes particles to have bright yellow or anomalous blue-green interference colours, with this latter phenomenon primarily demonstrated by the acicular and lath-shaped forms. Those of coarse enough particle size often appear to have a dark line down their centre and some of these may represent

simple twin forms. Spherules generally have bright yellow interference colours and a well-developed standing extinction cross. All particles have straight extinction though in wider rectangular particles extinction is undulose (McCrone *et al.*, 1973–80). Elongation is length slow in acicular to lath forms, but the crystallites in the quatrefoils are distinct in being length fast.

With reflected light Indian yellow appears a brownish yellow colour. When viewed through the Chelsea filter the strong yellow colour of Indian yellow is not transmitted and the particles appear the same colour as the medium. Under ultraviolet light, Indian yellow has a strong yellow fluorescence; the peak excitation is at around 435 nm, with peak emission at 535 nm (Baer *et al.*).

Many substances have been substituted for Indian yellows, including yellow azo pigments, cobalt yellow, lead chromate and a variety of dyes. However, 'real' Indian yellow is so optically distinctive it is unmistakable. The optical, chemical and physical properties of Indian yellow have been reviewed by Baer *et al.* (1986).

Lit.: Baer *et al.* (1986); McCrone *et al.* (1973–80)

TURMERIC

1. P0408: PPL/40×/H. Showing starch granules and yellow vegetable matter.

2. P0408: XPL/40×/H. The same field of view as in Fig. 1 under crossed polars.

3. P0408: UV/40×/H. The same field of view as in Fig. 1 showing UV fluorescence.

4. P0786: PPL/10×. Yellow stained vegetable material.

5. P0786: XPL/10× ⊙. Same view as in Fig. 4 under crossed polars.

6. P0786: UV/10×. Same view as in Fig. 4 showing UV fluorescence.

7. P0839: PPL. Yellow stained starch granule.

8. P0839: XPL ⊙. The grain shown in Fig. 7 under crossed polars.

9. P0839: UV. The grain shown in Fig. 7 under UV illumination.

10. P0410: PPL/H. Yellow stained vegetable material.

11. P0410: RPL/H. Same view as in Fig. 10 with crossed polars.

12. P0410: UV/H. Same view as in Fig. 10 with UV fluorescence.

Crystallinity
Composite structure

n
≪1.662

δ
Anisotropic
(data from authors)

Turmeric, the root of *Curcuma* species, has been used as a pigment, both in the form of the ground root and as a dye. The ground root (more properly, the rhizome) is typically sliced and dried before grinding. The yellow coloration is due to the presence of curcumin, a compound pervasive to all parts of the rhizome. The rhizome is composed predominantly of cork cells with a distorted polygonal cell structure. Another cellular structure, the parenchyma, is filled with nodules of yellow-stained starch granules; these are commonly and predominantly observed in the powdered rhizome. Under plane-polarised light these appear as translucent, rounded, spherical particles or as aggregates of polyhedral particles with high relief and RI less than that of the medium. The surface textures are rough, and finely pitted. *Curcuma* starch forms particles in the range of 70–140 μm diameter.

In cross-polarised light, *Curcuma* starch is of characteristically low birefringence, leading to a poorly defined extinction cross ('Maltese cross'). The drying process used in the preparation of the pigment distorts the crystalline structure of the *Curcuma* starch. Particles do not go to complete extinction but typically appear finely mottled.

Other pigments either include starches or are based on starches as dye substrates; these are discussed under the entry for starch.

The Pigment Compendium

DRAGON'S BLOOD [Complex]

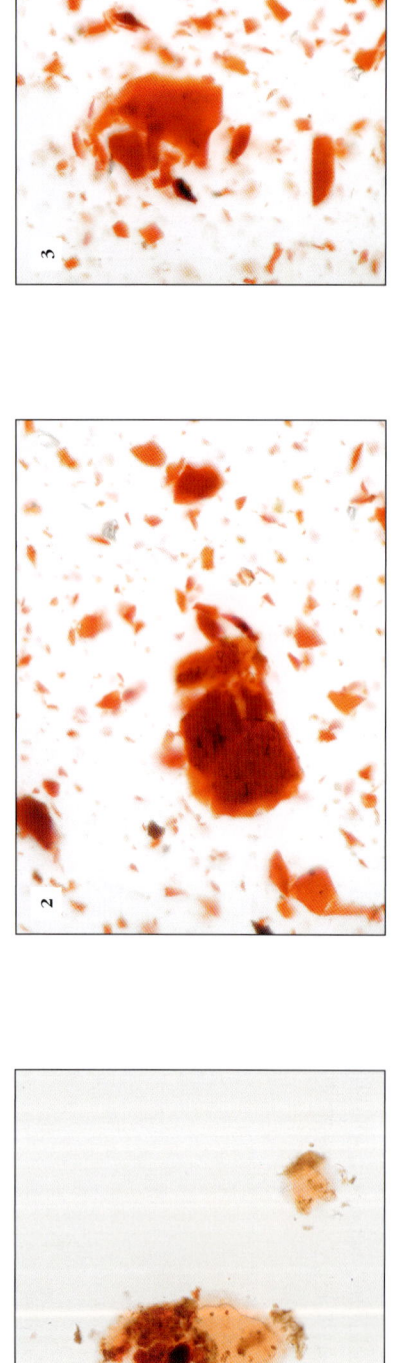

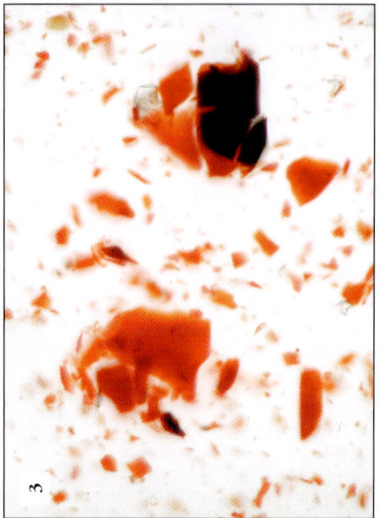

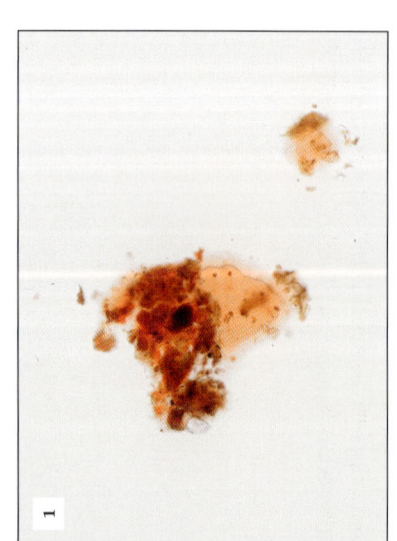

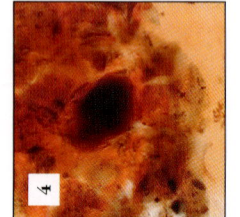

Key: **1.** P0833: PPL/40×. Varicoloured particle of dragon's blood.   **2.** P0482: PPL/40×/H. Red coloured flakes of dragon's blood.   **3.** P0482: PPL/40×/H. Red flakes of dragon's blood showing opaque portions.

4. P0833: PPL. Detail of particle in Fig. 1.

Crystallinity
Amorphous

n
<1.662

δ
Isotropic
(data from authors)

Dragon's blood is a red resinous material which forms as an exudate from a variety of plant species belonging primarily to the *Daemonorops* and *Dracaena* genuses, though other plant resins and materials have also been used and traded under the name *Dragon's blood*. The examples illustrated here are believed to be from *Daemonorops*.

In plane-polarised light, dragon's blood may be observed as flakes of the raw pigment. These appear as anhedral, thin flakes and plates, often cross cut by conchoidal fractures. In colour they are a strong orange-red, and the colour may be observed to bleed into the medium. Particles are translucent to opaque while particle surface texture ranges from smooth to rough; pitted and crenulated. Relief is low and RI is less than that of the medium. Particle size is variable but frequently medium to very coarse.

Since this material is an amorphous resin, under crossed polars particles of dragon's blood are isotropic. Areas having undergone deformation may be weakly birefringent, with the strong body colour masking the interference colours. In such areas, extinction is undulose.

Dragon's blood transmits a pink-red colour when viewed through the Chelsea filter.

BLOOD

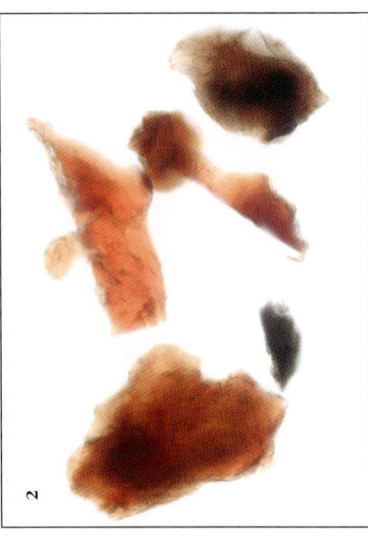

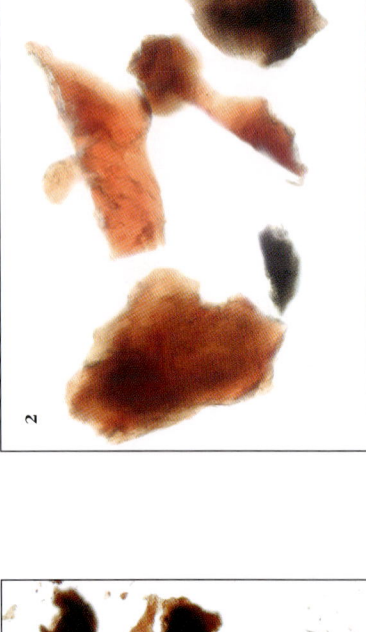

Key: 1. P1507: PPL/10×. Weakly translucent to opaque flakes of dried blood.

2. P1507: PPL/40×. Translucent of dried blood.

Crystallinity

Amorphous	*n*	δ
	<1.662	Isotropic
		(data from authors)

In plane-polarised light, dried blood forms translucent to almost opaque, brown-red particles. Relief is moderate and RI is less than that of the medium. Particle shape is of irregular, subrounded to subangular flakes. Grain surfaces are slightly uneven. Grain size distribution is broad with grains ranging from medium to very coarse.

Blood is isotropic under cross-polarised light.

BISMUTH

Bi

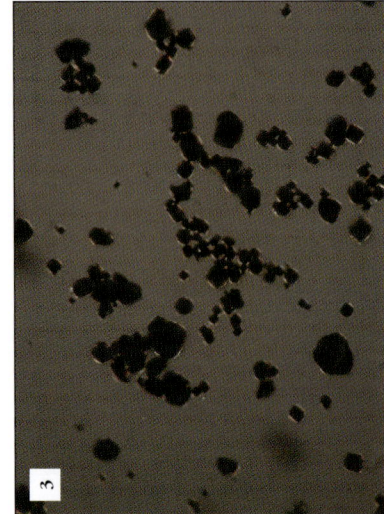

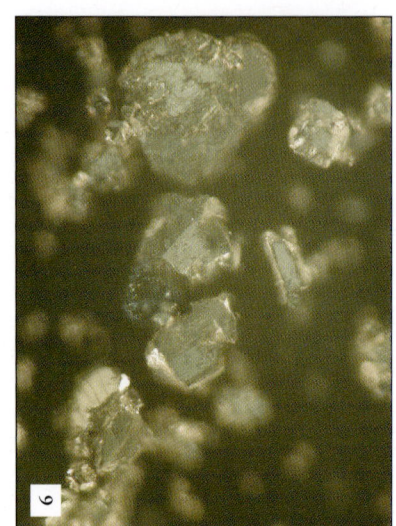

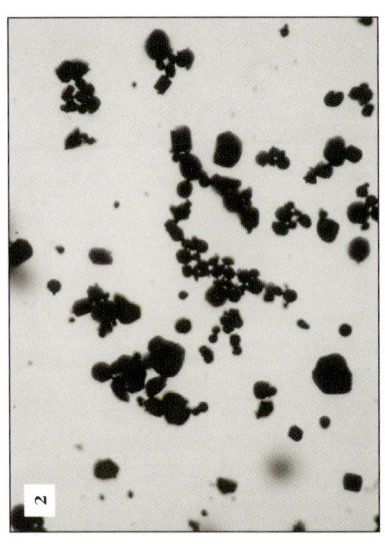

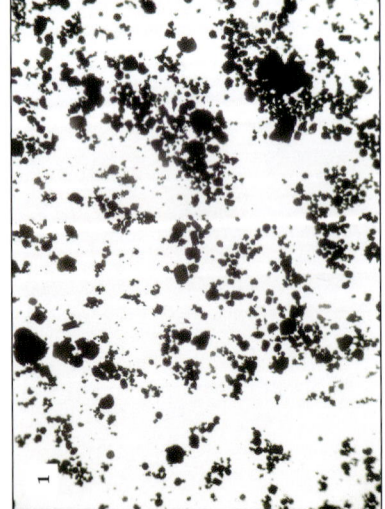

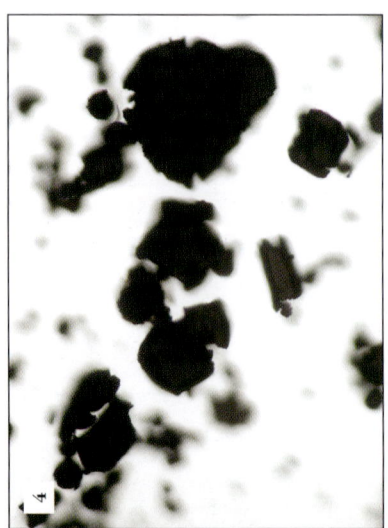

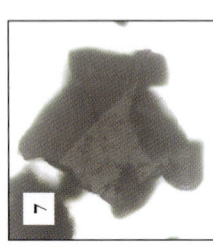

Key:

1. P0135: PPL/10×. Opaque particles of bismuth.

2. P0135: PPL/40×. Subrounded particles of bismuth.

3. P0135: ~XPL/40×. Particles of bismuth in partially crossed polars, showing weak anisotropy at grain boundaries. Same field of view as in Fig. 2.

4. P0135: PPL/40×. Coarse-grained particles of bismuth.

5. P0135: RPL/40×. The field of view as in Fig. 4 in reflected light.

6. P0135: RPL/40×. The field of view as in Fig. 4 in bireflected light.

7. P0135: PPL. Bismuth particle showing weak translucency with high illumination.

Crystallinity
Trigonal-Hexagonal

n
≫1.662

δ
Weak

(data from Webmineral, 2003 and authors)

Under plane-polarised light, bismuth appears as weakly translucent, grey-brown particles. Relief is very high with RI much greater than that of the medium, so that only the centres of large particles will transmit light and particle boundaries appear opaque. Particle shapes range from anhedral to euhedral. Many particles are rounded and equant, others elongate and many euhedral forms have a cylindrical to barrel-like appearance. Particle surfaces are relatively smooth, with fine striations. Some particles appear flattened, whereas others have domed surfaces.

Under crossed polars, bismuth is very weakly birefringent, but interference colours are strongly masked by the high body colour. The rounded, equant particles appear spherulitic with weak standing extinction crosses just visible. Other particles demonstrate sweeping or undulose extinction. There is notably very strong polarisation associated with particle boundaries.

COPPER HEXACYANOFERRATES

$CuK_2Fe(CN)_6$ and $Cu_2Fe(CN)_6 \cdot xH_2O$

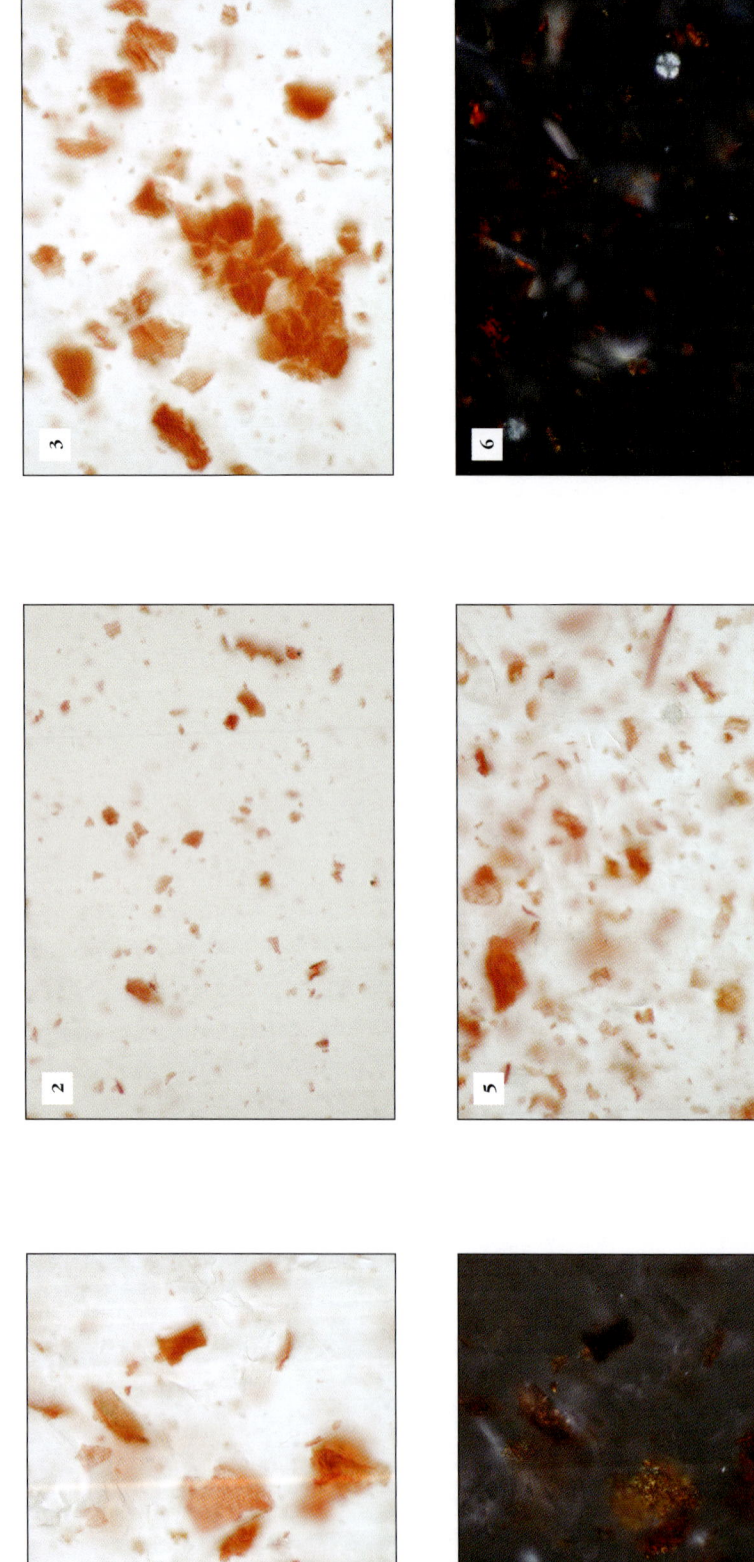

Key:
1. P0941: PPL/100×. Coarse-grained fibrous aggregates of copper hexacyanoferrates.
2. P0942: PPL/100×. Medium-grained plates of copper hexacyanoferrate.
3. P1527: PPL/40×. Coarse-grained plates of copper hexacyanoferrate.
4. P0941: ~XPL/100×. The field of view as in Fig. 1 under crossed polars.
5. P0941: PPL/100×. Plates and fibrous sheaves of copper hexacyanoferrate with colourless spherulites.
6. P0941: ~XPL/100×. The field of view in Fig. 5 under crossed polars.

Crystallinity	n	δ
Type A: Cubic	<1.662	Isotropic
Type B: Anisotropic	>1.662	Moderate
		(data from authors)

Two copper hexacyanoferrate compounds are described in the literature (*Colour Index*, 1971); copper dipotassium hexacyanoferrate γ(II) and dicopper hexacyanoferrate(II) hydrate. Preparation of samples by the authors did indeed show that two sorts appeared to be formed, one isotropic (type A'), the other anisotropic (type B').

In plane-polarised light copper hexacyanoferrate type A appears as rusty red-brown coloured particles. They have low relief with an RI just less than that of the medium. Particle shape varies from ragged-looking particles with an elongate, almost fibrous appearance to smoother flake-like particles. Particle shape is generally uneven and irregular, appearing crenulated. Particle size is moderate to coarse.

Under crossed polars the type A compound is isotropic, though some particles do show depolarisation at particle boundaries.

In plane-polarised light the appearance of copper hexacyanoferrate type B is very similar to that of copper hexacyanoferrate A. Particles have low relief, but

RI is just greater than that of the medium. In colour they are rust red-brown, although colour intensity varies within particles. Particles have a ragged fibrous habit, and surfaces are fractured and pitted. Characteristically present are colourless or dirty-brown spherical particles or fragments of these; these vary in diameter from fine to coarse.

Under crossed polars, the type B compound is anisotropic with moderate birefringence. However, interference colours are strongly masked by the body colour and red-brown colours are seen. The particles are finely polycrystalline and therefore do not show complete extinction, but appear to twinkle as the stage is rotated. The spherulitic particles are distinctive and are shown to be composed of fibrous or acicular spherulitic aggregates of crystals. These are birefringent and emit white interference colours and have distinct standing extinction crosses, from which it can be seen that individual crystals have straight extinction. These particles are length slow.

Lit.: *Colour Index* (1971)

COPPER IRON CHROMATE

$CuFe_{0.5}Cr_{1.5}O_4$

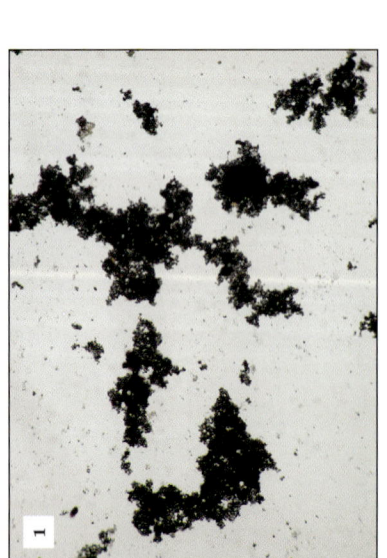

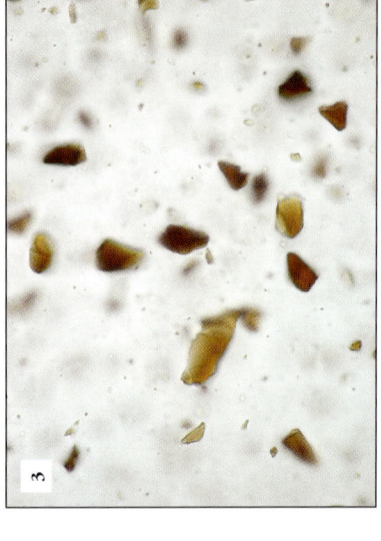

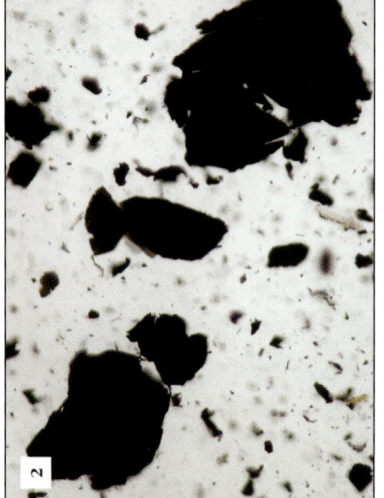

Key: **1.** P0125: PPL/100×. Fine-grained particles of opaque copper iron chromate.

2. P0323: PPL/40×/H. Coarse-grained particles of opaque copper iron chromate.

3. P1285: PPL/40×. Translucent brown particles of copper iron chromate.

Crystallinity
Cubic

n
>1.662

δ
Isotropic
(data from authors)

In plane-polarised light copper iron chromate ('spinel black') can be seen as being formed of opaque black particles. However, if particles are sufficiently coarse they can be seen to be weakly translucent and dark brown; in reflected light they typically appear as black to weakly brown-black. Particle size may range from very fine to very coarse and shape is typically as subhedral, equant, polyhedral crystals. Relief is high, and RI is greater than that of the

medium. Particle surfaces are crossed with straight, but randomly orientated, fractures.

Spinel blacks, like all spinel group compounds, are isotropic under crossed-polarised light. They appear weakly brown to black in reflected light.

The Pigment Compendium

GRAPHITE

C

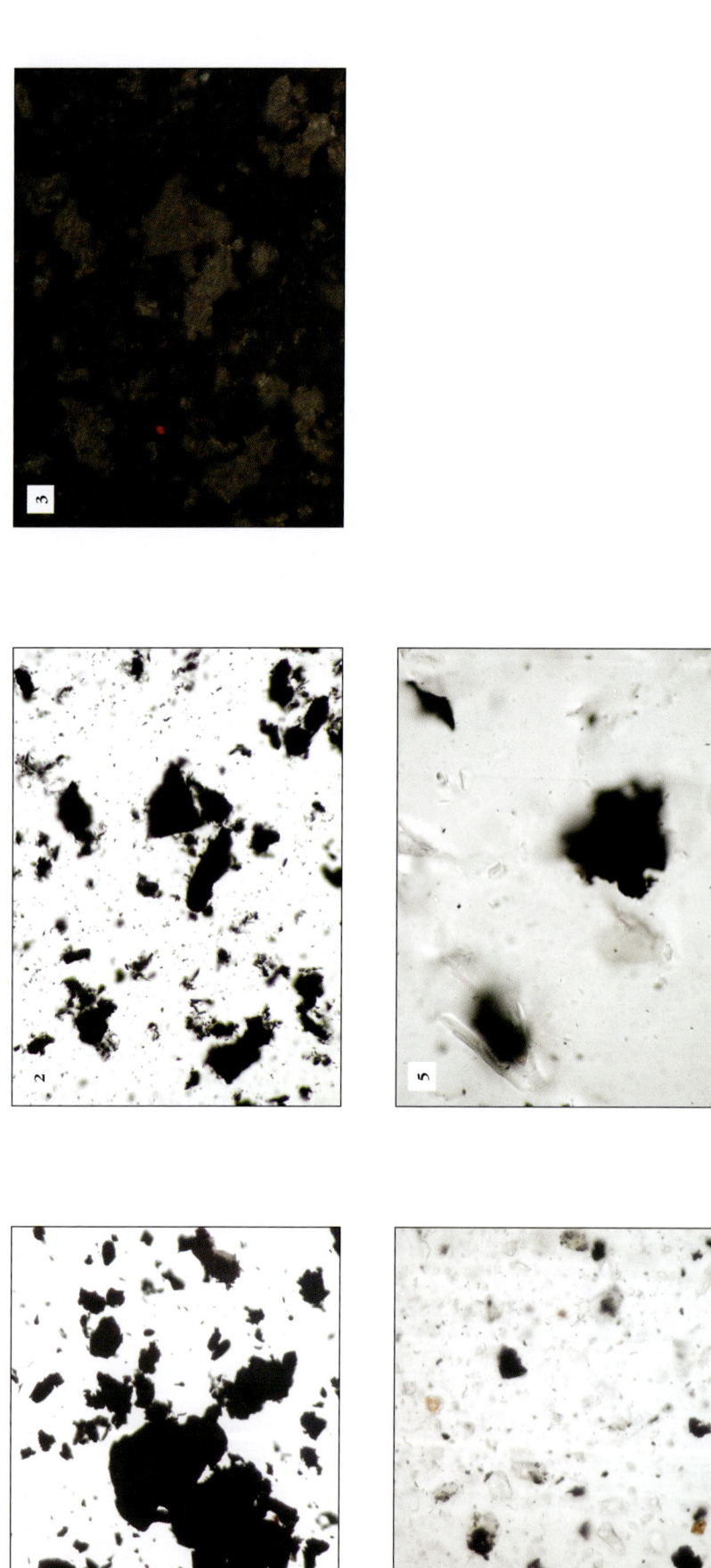

Key:

1. P0128: PPL/40×. Coarse-grained subhedral plates of graphite.

2. P0183: PPL/100×. Anhedral angular flakes of graphite.

3. P0183: RPL/100×. The same field of view as in Fig. 2 in reflected light showing weak yellow reflections.

4. P1201: PPL/40×. A graphite 'earth' containing opaque graphite, quartz and iron oxide particles.

5. P1201: PPL/100×. Enlargement of the graphite grain shown in Fig. 4 showing highly anhedral morphology.

Crystallinity	Optic sign	n_ω	n_ε	δ
Hexagonal		[Opaque] >1.662		

In plane-polarised light, graphite forms weakly translucent to (more commonly) opaque particles. Relief is moderate and RI is greater than that of the medium. Particle morphology ranges from irregular, crumb-like particles to flakes and subhedral, hexagonal plates. Particles may be distorted. Particle surfaces, when discernable, are fractured. Particle size ranges from fine to very coarse.

Under crossed polars Winter (1983) has pointed out that individual graphite particles are typically surrounded by a greyish halo, a consequence of the surface reflection of this compound; oblique illumination of the surface of graphite flakes and plates reveals that it has a characteristic silvery, metallic and highly anisotropic bireflectance (Ergun, 1968; cf. Winter). However, in reflected light the immediate impression is likely to be of weakly grey to brown-yellow particles. There is no fluorescence with UV excitation.

So-called 'black chalk' and 'black earth' is often graphite or defective graphite mixed with quartz, iron oxides and other minerals, though in practice the definition is broad and may therefore encompass a wide range of other compounds. Typically the graphite will be seen to coat the quartz particles to a greater or lesser extent, causing the masses to appear colourless through grey to opaque black; fine flakes of graphite may also be visible at high magnification (Winter).

Lit.: Ergun (1968); Winter (1983)

The Pigment Compendium

CARBON-BASED BLACK, CHARS

[Complex]

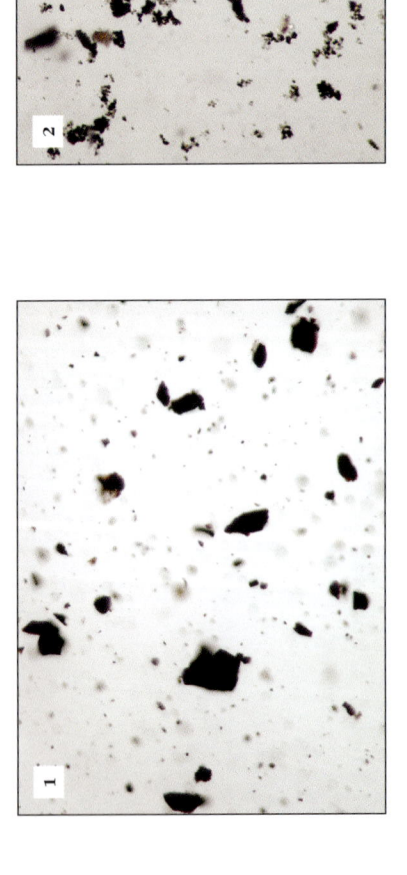

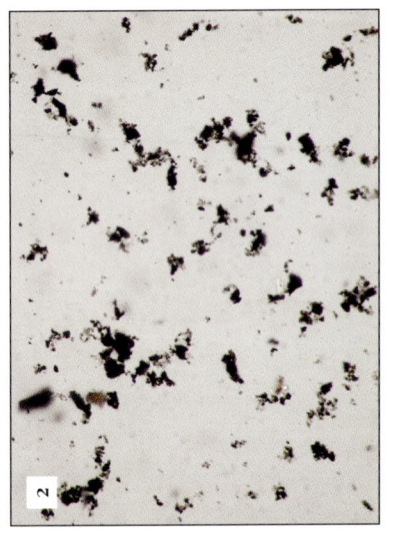

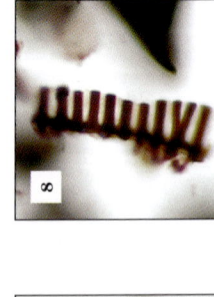

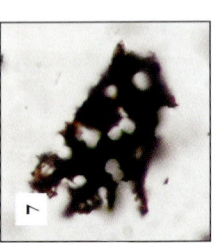

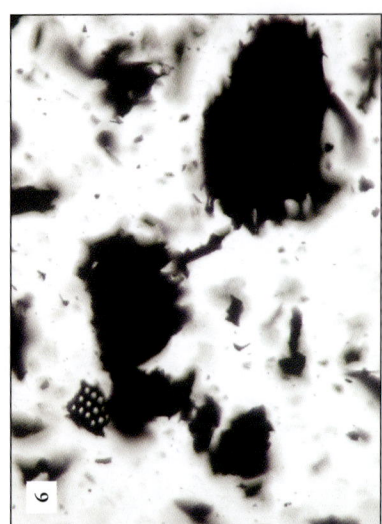

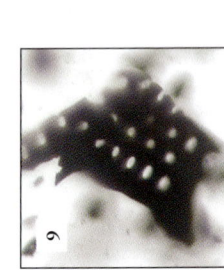

Key:

1. P0815: PPL/40×. Particles of vine black.
2. P1079: PPL/100×. Peachstone char.
3. P1078: PPL/40×. Angular particles of peachstone char.
4. P1080: PPL/100×. Peachstone char, note the occurrence of relict cellular material.
5. P1528: PPL/40×. Charcoal including splintery shards and relict cellular plant material.
6. P1528: PPL/40×. Coarse-grained particles of charcoal with flame-like grain boundaries.
7. P1078: PPL. Relict plant cellular material in peachstone char.
8. P1080: PPL. Detail of cellular fragment from Fig. 4.
9. P1528: PPL. Detail of cellular fragment from Fig. 5.

Crystallinity
Amorphous

n [Opaque]

δ (data from authors)

Chars are carbon-based black pigments formed from solid precursors that remain solid throughout the carbonisation process. As a result the end products retain much of the morphology of the starting material, such that cellular structure may be seen. Chars can be formed from a very wide range of materials, most notably cellulosic plant material. While no realistic limit can be placed on the sources that might have been employed historically and geographically, those consistently mentioned as starting material in the literature include barks, fruitstones, paper and wood. The following examples are illustrative of some morphologies that may be encountered.

Wood chars: Under plane-polarised light 'charcoal' can be observed as opaque particles with moderate to low relief. Particle shape varies from irregular ragged featureless flakes to particles that have a clearly organic origin with relics of cell structures still visible. Numerous particles occur as splinter-like shards, some of which may be seen to be branching. Larger particles often have splintered, flame-like grain boundaries. Irregular holes in the centres of particles are also not uncommon. Particle size is frequently coarse, but can range from fine to very coarse.

Examples of 'vine black' (a generally loose term, but one likely to denote a wood char) studied also typically form opaque particles, though partly combusted samples may appear dark brown and weakly translucent. Particle morphology is variable, but angular shards are typical. Other particles can be subrounded, or form elongate flakes. Organic structures were not visible in samples examined by the authors. Particle size distribution is broad with particles ranging from fine to coarse particle size.

Fruitstone chars: Under plane-polarised light, examples of peachstone chars examined were seen to form as opaque particles. Particle morphology is typically of anhedral angular particles and some crumb-like particles. Partly combusted examples may be weakly translucent and brown (this is particularly visible at particle edges) and some cellular organic material may be preserved. Particle size varies from fine to coarse.

The morphologies encountered in cork and birch-bark chars tend to be similar to those observed in wood chars, with preservation of cellular structure possible. The friability of cellulose fibre chars means that structural material is unlikely to survive preparation and use in this material.

Carbon-based blacks are isotropic under crossed polars. However, impurities or extenders may show up as being birefringent. Silica, calcium oxalate and calcite occur naturally in well-developed crystalline forms in certain plant structures; it is conceivable that these might remain following the charring process although they have not been explicitly observed by the authors in samples examined. The black colour of the particles is evident under reflected light.

Lit.: Winter (1983)

The Pigment Compendium

CARBON-BASED BLACK, BONE AND YEAST COKES

[Complex]

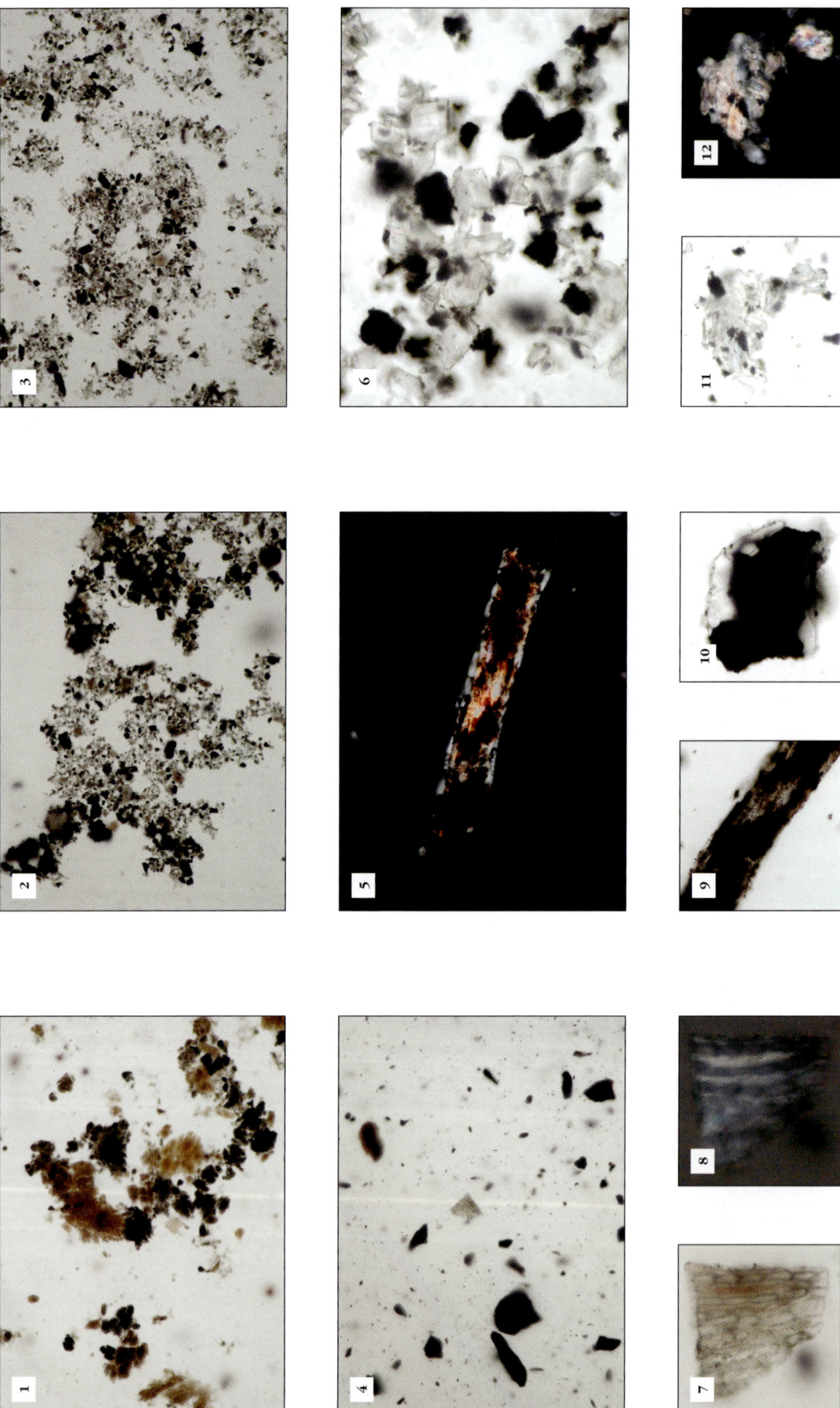

1. P0800: PPL/100×. Particles of carbon black produced by the Degussa process.

2. P1067: PPL/100×. Rounded particles of carbon black.

3. P0601: PPL/10×/H. Very coarse-grained crumb-like irregular particles of carbon black.

4. P0754: PPL/100×/H. Particles of carbon black produced by the Degussa process.

5. P1068: PPL/100×. Rounded particles of carbon black.

6. P0601: XPL/10×/●/H. The field of view as in Fig. 3 under partially crossed polars, note birefringent impurities.

n

Crystallinity
Amorphous

δ

[Opaque]

(data from authors)

Flame carbons may include materials prepared as diversely as soot collected from lamp flames or chimneys, to that derived from highly controlled processes (such as so-called 'gas' blacks), or even with surface modification of functional groups. Consequently the composition and morphologies of the group as a whole are diverse with shape and particle size strongly dependent on the production method used for their manufacture.

In plane-polarised light particles of flame carbon black produced by historical methods appear as opaque, but finely porous, irregular flakes. Relief is low and RI is just greater than the medium. Surfaces are irregular and finely pitted. Grain size may be very coarse – early methods produced blacks with broad particle size distributions and particle morphologies ranging from rounded to subrounded, to irregular crumb-like particles. In crossed polars the particles are predominantly isotropic although inclusions can be birefringent.

On the other hand 'modern' flame carbons, such as those from the Degussa process (manufactured from 1935 onwards) have very fine particle sizes and a narrow distribution of particle size. Such flame carbon-based blacks are also notable by their high level of purity (Buxbaum, 1998). Particles are likely to be observed mostly in small clumps or chains, individual particles being difficult to disperse (Winter, 1983). The black colour of the particles is evident under reflected light.

Bistre (*q.v.*) is a special case of flame carbon.

Lit.: Buxbaum (1998); Winter (1983)

BISTRE AND SEPIA

[Complex]

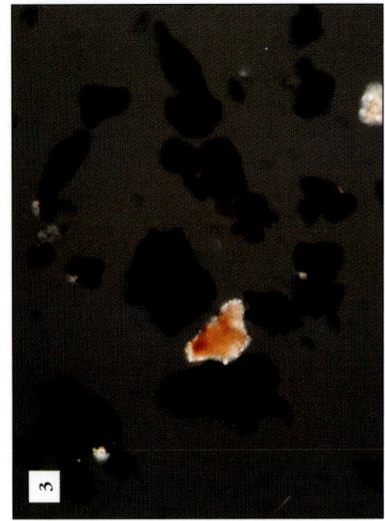

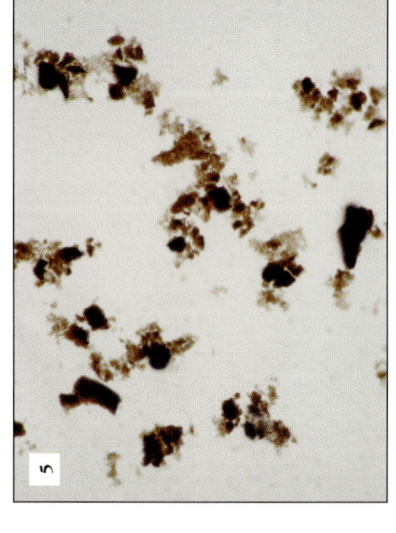

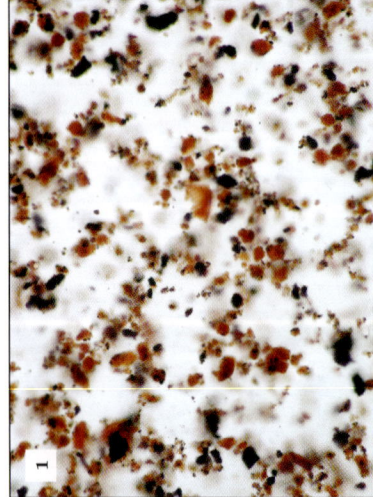

Key:

1. P1088: PPL/100×. Anhedral platy particles of bistre, note range of body colour from opaque to rust red.

2. P0028: PPL/40×. Predominantly opaque particles of bistre with a few translucent grains.

3. P0028: ~XPL/40×. The field of view as in Fig. 2 under partially crossed polars, note the high birefringence of the translucent particles.

4. P0029: PPL/40×. Flakes of sepia.

5. P0029: PPL/100×. Grains from the sample in Fig. 4 at higher magnification.

Crystallinity

	n	δ
Bistre: Amorphous	<1.662	Isotropic
Sepia: Amorphous	>1.662	Isotropic
		(data from authors)

Bistre is a special form of flame carbon that was prepared from wood soot, preferably beechwood (*Fagus* species). After the soot was collected close to the flames, it was treated with hot or boiling water. When the particles had settled out, the supernatant liquors were decanted and the sediment taken to dryness. It is a mixture of flame carbon, char and possibly some coke along with uncarbonised tarry material. According to Winter (1983), bistre shows very similar characteristics to wood soot taken from domestic chimneys.

In plane-polarised light, bistre forms translucent to opaque, red-brown to black particles. Relief is low and RI is less than that of the medium. Particle habits are of angular to subangular, anhedral flakes. Particle surfaces appear smooth. Particle size may be extremely variable from sample to sample, ranging from fine to very coarse. Since bistre is largely amorphous, under crossed polars it is generally seen to be isotropic. However, some particles and areas of particles may be weakly anisotropic and appear orange-brown,

with sweeping or undulose extinction. Under the Chelsea filter, bistre appears deep red.

Sepia is a brown pigment derived from the ink sac of various species of cephalopoda, principally the cuttlefish *Sepia officinalis* L. The pigment from *Sepia* species contains the complex compound *melanin*.

Under plane-polarised light, sepia forms flakes of translucent dark olive brown to (more commonly) opaque particles. These have irregular subrounded to subangular morphologies and sometimes form crumb-like particles. Particle surfaces are smooth. Relief is low and RI is just greater than that of the medium. Particle size varies from fine to coarse. Particle opacity increases with particle size. Sepia is amorphous and therefore isotropic under crossed polars.

Lit.: Winter (1983)

HUMIC EARTH

[Complex]

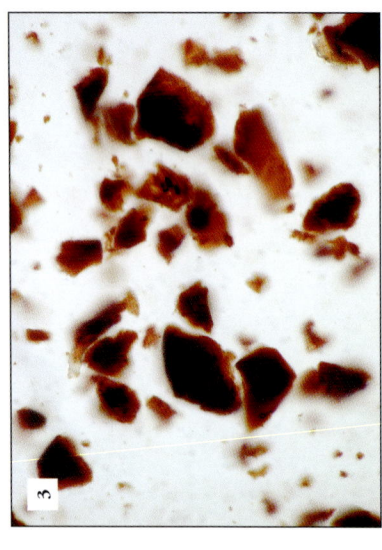

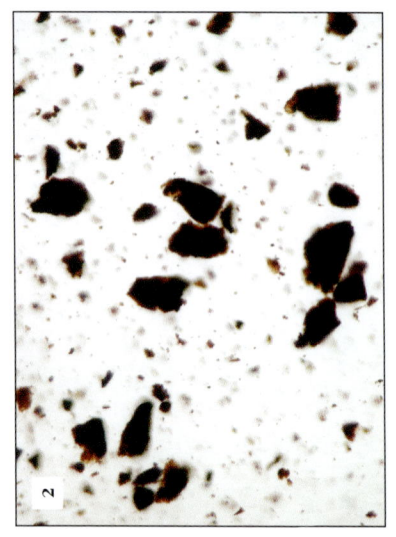

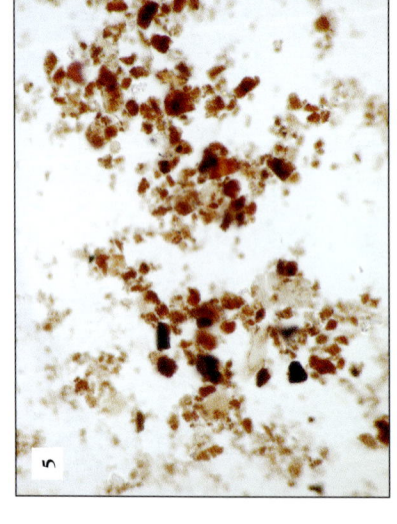

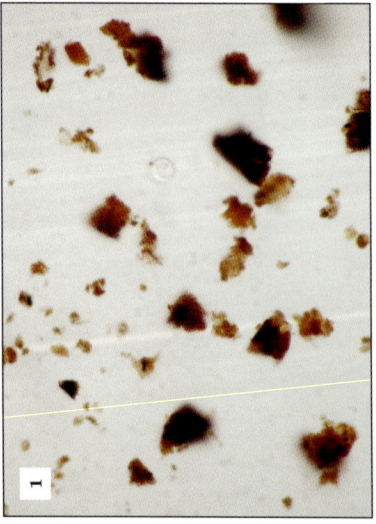

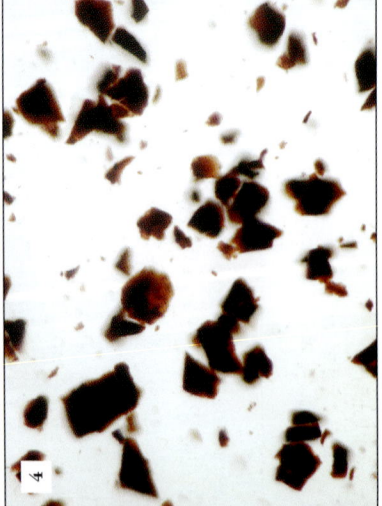

Key:

1. P0070: PPL/100×. Weakly transparent flake-like particles of humic earth.

2. P0360: PPL/40×/H. Predominantly opaque flakes of humic earth.

3. P0361: PPL/100×/H. Translucent to opaque coarse-grained flakes of humic earth.

4. P0756: PPL/40×/H. Translucent to opaque coarse-grained flakes of humic earth.

5. P0264: PPL/100×. Medium- to fine-grained predominantly translucent rounded flakes of humic earth.

Crystallinity
Amorphous

n
1.62–1.69

δ
Isotropic
(data from Gettens & Stout, 1966)

In plane-polarised light, the organic matter of humic earths ('Vandyke brown') forms weakly transparent to opaque dark red-brown particles, with low relief and RI just less than that of the medium; Gettens and Stout (1966) do, however, give the RI as lying in the range 1.62–1.69, such that this may appear to be either lower or higher than that of the mounting medium. Particles form plate-like shards with anhedral angular habits or irregular crumb-like particles. Conchoidal fracture surfaces are common. Particle surfaces vary in appearance from smooth to grooved and pitted; Feller and Johnston-Feller (1997) remark that they 'resemble the convolutions of the brain'. Particle sizes range from medium to coarse.

Under crossed polars, the majority of particles are amorphous and isotropic. Varying amounts of iron oxides can affect the crystallinity of the particles and some grains or parts of grains can appear weakly anisotropic, exhibiting masked orange interference colours.

The optical and chemical properties of Vandyke brown pigments have been discussed by Feller and Johnston-Feller (1997).

Lit.: Feller & Johnston-Feller (1997); Gettens & Stout (1966)

ASPHALT

[Complex]

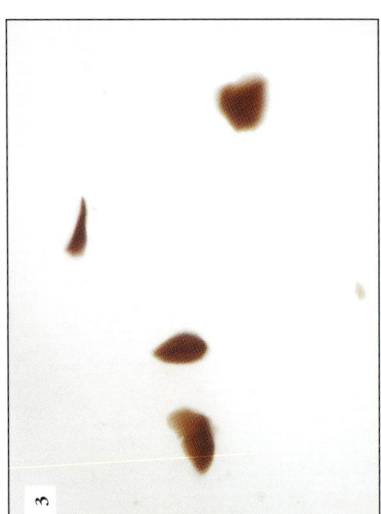

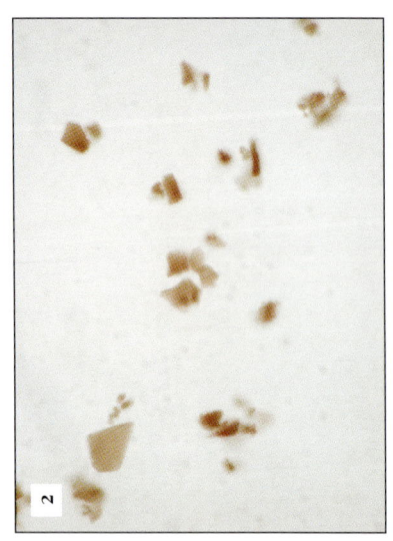

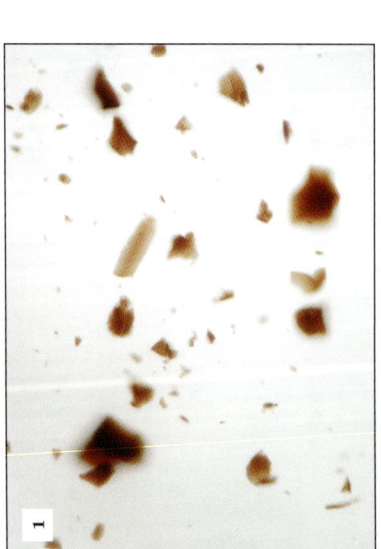

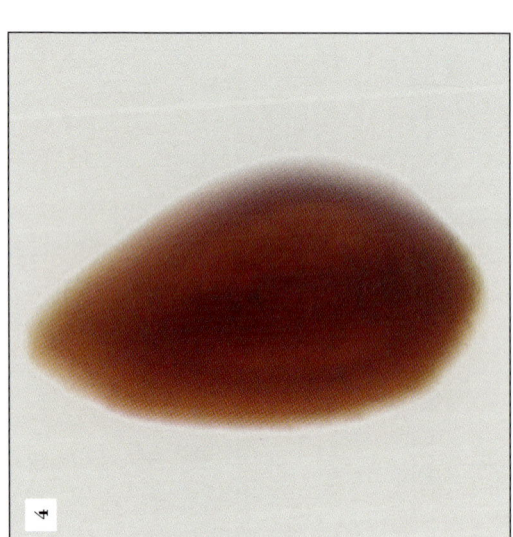

Key: **1.** P0127: PPL/40×. Brown platy grains often with conchoidal fracture and translucent to opaque diaphaneity.

2. P0186: PPL/100×. Brown tabular particles of asphalt.

3. P0809: PPL/40×. Brown rounded flakes of asphalt.

4. P0809: PPL. Enlarged particle from Fig. 3 with concentric ridges.

Crystallinity
Amorphous

n

<1.662

δ

Isotropic

(data from authors)

Asphalt forms plate-like particles with low relief. RI is less than that of the medium. The plates are coloured a rich brown, and range from translucent to opaque, depending largely on particle size/thickness. Particle surfaces are smooth, although some may show irregular, fine, concentric ridges. Particle shapes are developed through crushing the pigment, and are angular shards. Fracture is conchoidal and therefore gently curved particle boundaries are the norm. Other particles are rounded.

Asphalt is an amorphous substance and therefore isotropic under cross-polarised light.

There is no apparent difference by optical microscopy between asphalts from different sources.

COAL, ANTHRACITE AND CANNEL TYPES

[Complex]

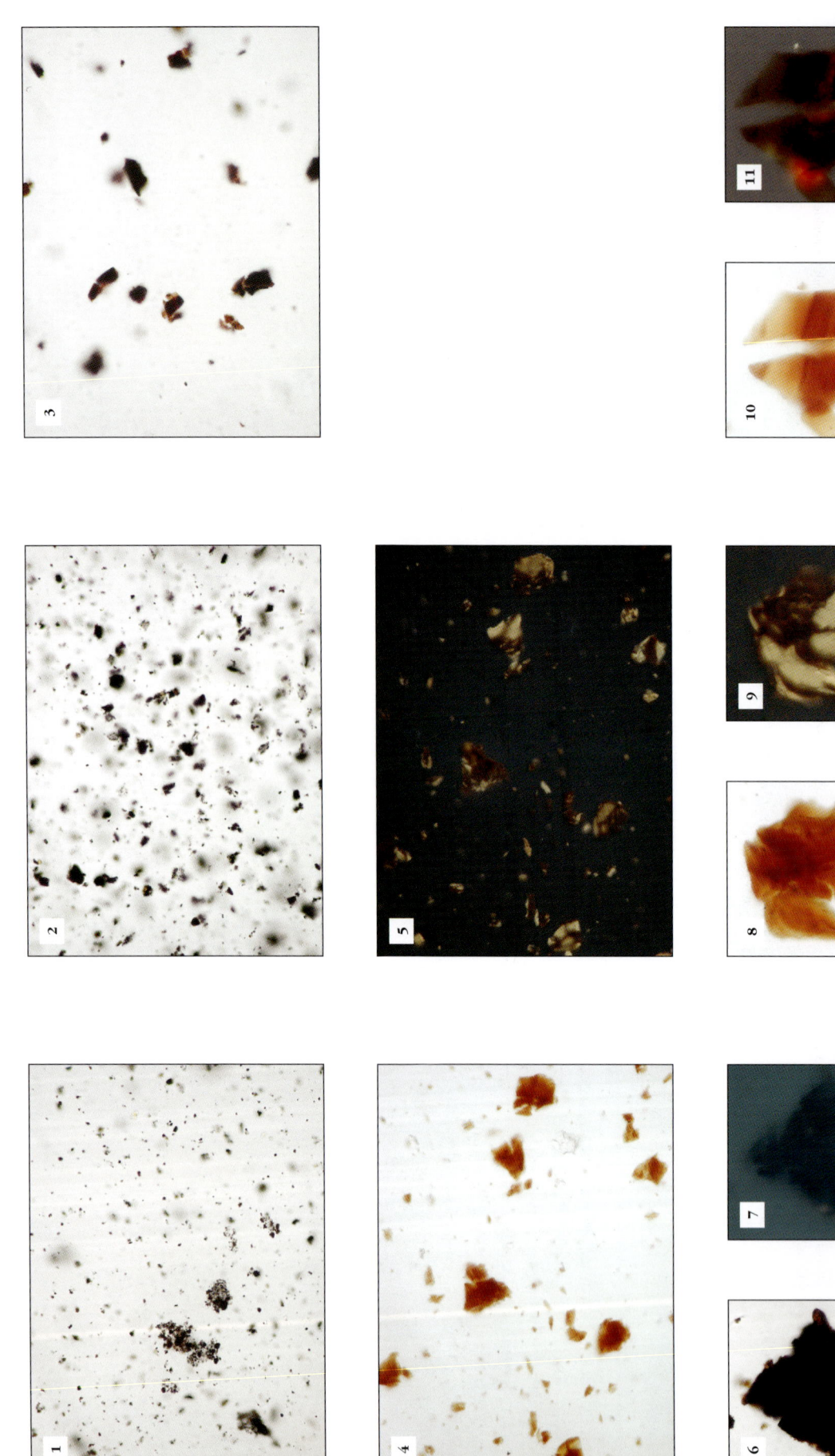

Key:

1. P1105: PPL/100×. 'Bideford black', with particles showing broad range of grain size.

2. P1105: PPL/100×. Particles of Bideford black.

3. P1176: PPL/100×. Weakly translucent to opaque particles of cannel coal.

4. P1177: PPL/40×. Flakes of translucent brown jet.

5. P1177: UV/40×. The field of view in Fig. 4 showing UV fluorescence.

6. P1176: PPL. Particle of cannel coal showing weakly translucent grain margins.

7. P1176: UV. The particle in Fig. 6 with areas of UV fluorescence corresponding to translucent margins.

8. P1177: PPL. Enlarged view of jet particle.

9. P1177: UV. The particle in Fig. 8 showing UV fluorescence.

10. P1177: PPL. Particles of jet, note banding.

11. P1177: ~XPL. The particle in Fig. 10 under crossed polars, showing weak anisotropy.

Crystallinity

Amorphous

n [Opaque] **δ**

(data from authors)

Under plane-polarised light anthracitic coal such as that known as 'Bideford black' presents as opaque, irregular particles with particle margins of a feathery appearance. As a ground material, particle size ranges from very fine to coarse and the particle size distribution is broad. Winter (1983) has described a sample of anthracite coal he examined by electron microscopy; he remarks upon the fact that it shows features one would anticipate from a pulverised brittle solid. Under crossed polars, anthracite is isotropic. Impurities may be anisotropic and include quartz, calcite, clay minerals and micas.

Under plane-polarised light, examples of cannel coal examined by the authors appear as opaque, irregular particles with feathery margins; it is black in reflected light. Particle size ranges from very fine to coarse and the particle size distribution is broad. Colourless, euhedral cubic impurities are present. These have high relief, with RI less than that of the medium. Under crossed polars, cannel coal is isotropic, and so are the cubic impurities, which are probably

halite. Other impurities may be anisotropic and include quartz, calcite, clay minerals and micas. An example of a bituminous coal studied by electron microscopy by Winter showed a marked lamellar cleavage.

The appearance of certain coals is notable under UV fluorescence in that the hydrocarbon phase can fluoresce. Rost (1995) notes that the fluorescence colour observable in coals can vary between greenish yellow, yellow, orange and dull orange to brown. This phenomenon is especially visible in paint cross-sections, but is also generally discernible in dispersions.

Coals generally appear brown to black in reflected light with grey-white reflective surfaces.

The coals form a complex group of materials and properly require specialist knowledge to differentiate.

Lit.: Rost (1995); Winter (1983)

The Pigment Compendium

COPPER(II) OXIDE, TENORITE TYPE AND TENORITE

CuO

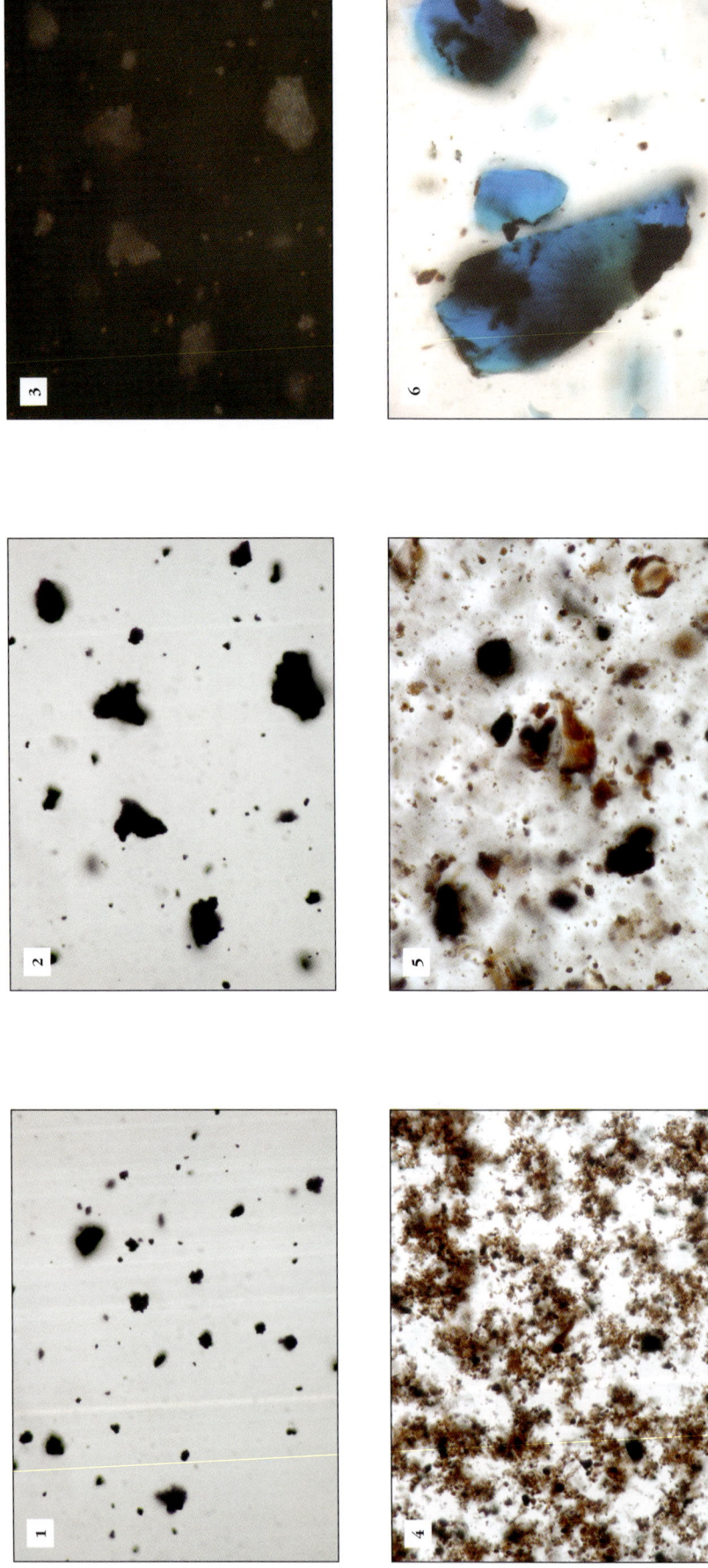

Key:

1. P0891: PPL/100×. Crumb-like aggregates of fine-grained copper(II) oxide.

2. P0891: PPL/100×. Irregular anhedral particles of copper(II) oxide.

3. P0891: RPL/100×. The field of view in Fig. 2 in reflected light.

4. P1263: PPL/40×. Translucent brown to opaque particles of copper(II) oxide.

5. P1263: PPL/100×. Enlarged view of particles from Fig. 4.

6. P1149: PPL/40×. Tenorite forming opaque patches of alteration in coarse-grained azurite.

Crystallinity	Optic sign	2V	n_α	n_β	n_γ	δ
Triclinic			[Functionally opaque]			

(data from Webmineral, 2003)

The black copper oxide mineral tenorite and its synthetic analogue copper(II) oxide, tenorite type are of similar appearance. Under plane-polarised light both forms are almost opaque; however, at high power with a strong transmitted light source small particles and the margins of larger particles are seen to be weakly translucent and brown. Typically, the particle morphology is of crumb-like masses, which, in synthetic precipitates, appear to be composed of radiating fibrous and acicular particles. Translucent particle margins may be observed to have a feathered texture. Natural tenorite may show similar morphologies, although more common are structureless single particles.

The natural mineral would be expected to contain impurities. Common impurities associated with tenorite mineral are other secondary copper minerals including azurite, malachite and cuprite, plus quartz, baryte, calcite, chalcopyrite and suchlike. Tenorite itself occurs as an impurity and alteration product of azurite, visible in the form of opaque inclusions and patches within the mineral.

Copper(II) oxide is isotropic. It can appear weakly yellow-orange (natural) to metallic grey (synthetic) in reflected light.

Lit.: Scott (2002)

The Pigment Compendium

LEAD(IV) OXIDE AND PLATTNERITE

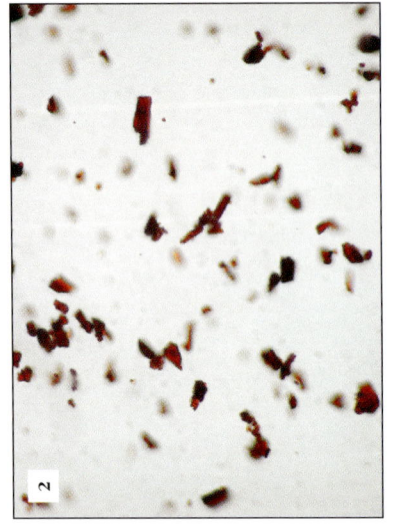

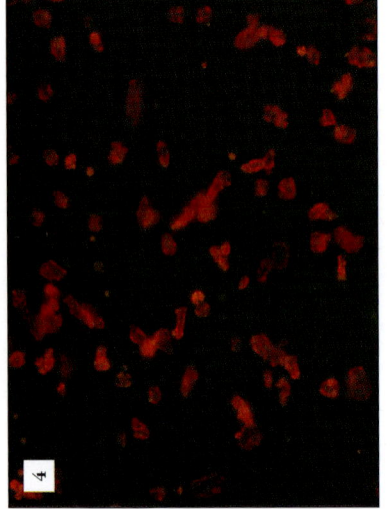

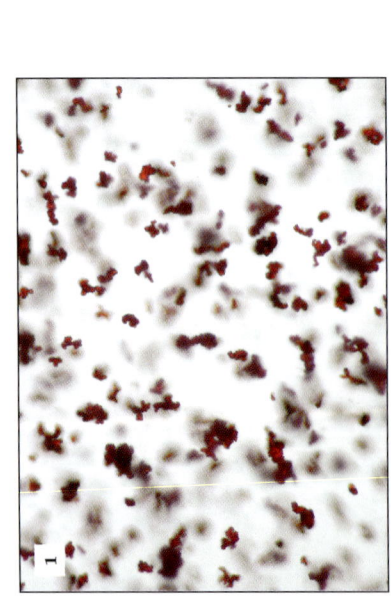

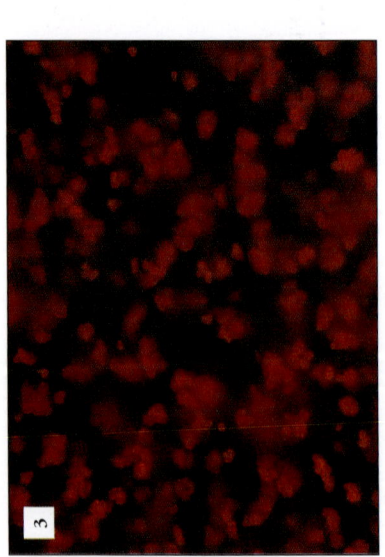

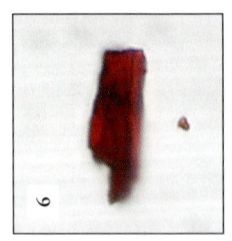

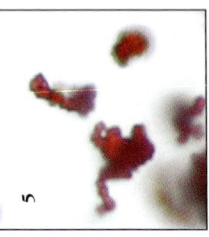

Key:

1. P0902: PPL/100×. Fine-grained, anhedral particles of red lead(IV) oxide.

2. P1430: PPL/100×. Bladed subhedral particles of lead(IV) oxide.

3. P0902: RPL/100×. The field of view as in Fig. 1 under crossed polars.

4. P1430: RPL/100×. The field of view as in Fig. 2 under crossed polars.

5. P0902: PPL. Detail of particle morphology from Fig. 1.

6. P1430: PPL. Detail of particle morphology from Fig. 2.

Crystallinity	Optic sign	n_ω	n_ε	δ
Tetragonal	−ve	2.35	2.25	0.1000

(data for mineral from Webmineral, 2003)

Under plane-polarised light lead(IV) oxide, the synthetic analogue of the mineral plattnerite, forms dark red-brown translucent to almost opaque particles, with high relief and RI greater than that of the medium. Particles in modern precipitates are typically very fine and appear as rounded, often slightly elongate particles and framboidal aggregates of these. Particle surfaces appear smooth and domed.

Under crossed polars lead(IV) oxide particles appear weakly birefringent. However, the low first order interference colours typically observed when fine particles are present are heavily masked by the strong body colour of the particles; consequently the birefringence may be barely visible. Lead(IV) oxide has low, deep red internal reflections. In reflected light a weak red-orange colour is visible at particle boundaries.

Lead(IV) oxide appears deep red when viewed through the Chelsea filter.

The Pigment Compendium

GALENA AND LEAD(II) SULFIDE

PbS

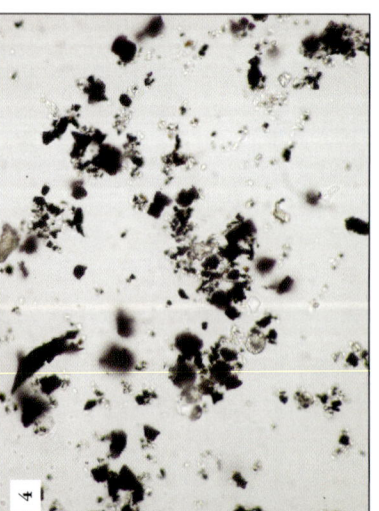

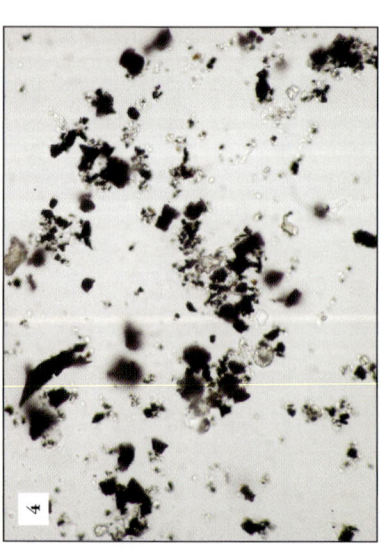

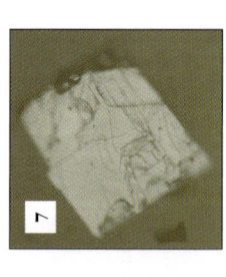

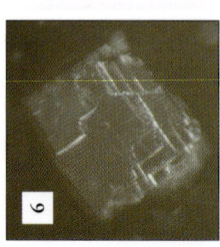

Key:

1. P1527: RPL/40×. Cubic particle of galena.

2. P1527: PPL/40×. Crushed particles of galena defined by the strong perfect cleavages.

3. P1527: RPL/40×. The crystals as in Fig. 2 in reflected light.

4. P0012: PPL/100×. Particles of synthetic lead(II) sulfide.

5. P0900: PPL/40×. Crumb-like aggregates and fine-grained particles of synthetic lead(II) sulfide.

6. P1527: RPL. A crystal of galena in bi-reflected light, note the step-like grain surface defined by the cleavage.

7. P1527: RPL. The particle in Fig. 6 in reflected light.

Crystallinity
Cubic

n	δ
3.91	Isotropic

(data for mineral from Deer *et al.*, 1993)

In plane-polarised light lead(II) sulfide and its mineral analogue galena are very weakly translucent and almost opaque, being charcoal grey or black in colour. The relief is accordingly very high and the RI much greater than that of the medium. Particle shape ranges from euhedral square or rectangular crystals to earthy, crumb-like masses. Particle surfaces are striated and often show triangular etch pits. Particle size ranges from fine to very coarse and size distributions are frequently broad.

Under crossed polars, since galena is cubic, it appears as isotropic. However, in reflected light it has a grey colour.

Galena is often found in association with minerals including fluorite, calcite, dolomite and baryte.

Galena may be differentiated from the otherwise similar-appearing mineral stibnite (antimony(III) sulfide, *q.v.*) by the differences in cleavage and its lack of bireflectance.

249

The Pigment Compendium

GALENA AND LEAD(II) SULFIDE

The Pigment Compendium

STIBNITE AND ANTIMONY(III) SULFIDE, STIBNITE TYPE

Sb$_2$S$_3$

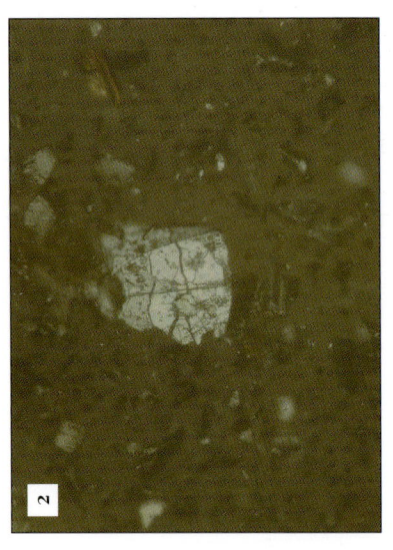

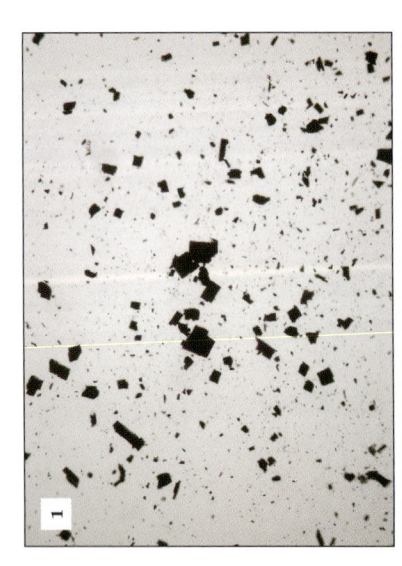

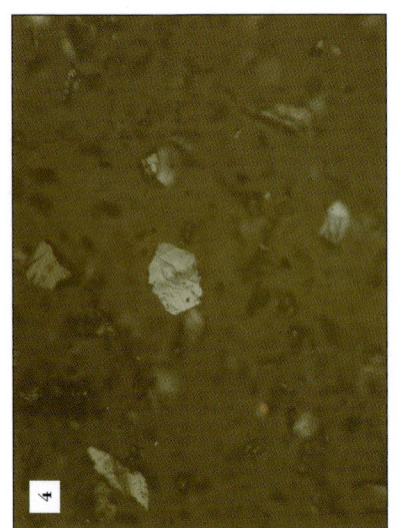

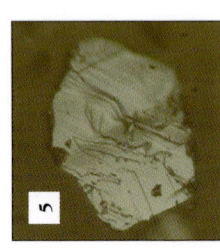

Key:
1. P0322: PPL/10×/H. Opaque particles of stibnite, note cubic grain morphology.

2. P0014: RPL/40×. Coarse-grained particle of stibnite in reflected light.

3. P1260: ~XPL/100×. Stibnite under partially crossed polars; note red weakly birefringent particles of kermesite.

4. P0014: RPL/40×. Particles of stibnite in reflected light.

5. P0014: RPL. Detail of grain from Fig. 4 showing stepped surface as a result of cleavage.

Crystallinity Orthorhombic	Optic sign	2V	n_α [Functionally opaque]	n_β	n_γ	δ

(data from Webmineral, 2003)

At first glance under plane-polarised light stibnite and its synthetic analogue antimony(III) sulfide appear opaque and black. However, with a strong light source the crystals can be observed to be a dark, charcoal grey and weakly translucent. Stibnite has very high relief. The compound has one perfect cleavage and this, coupled with the prismatic habit, typically influences the particle shape, forming bladed and tabular particles. Cleavage planes are often knife-sharp and clearly visible. Additionally, stibnite has a weaker platy cleavage, which appears as laminae, peeling off the crystal faces; this phenomenon is also frequently visible on well-formed crystals. Angular and subangular anhedral and earthy forms are also common. Particle size distribution is broad, ranging from fine to very coarse.

Under cross-polarised light, because of the low translucency and strong body colour, the mineral appears almost opaque. It is in fact weakly birefringent

with masked interference colours; bright, white fringes are observed around the margins of the crystal, which fade and reappear as the stage is rotated. Stibnite has straight extinction.

Stibnite may be differentiated from the otherwise similar-appearing mineral galena (lead(II) sulfide, q.v.) by the differences in cleavage and also that it is bireflectant, unlike the cubic galena.

Alteration to the red antimony oxide and oxide sulfide valentinite and kermesite may be present in the naturally occurring mineral, visible as bright red or rust red particles, as mottling within a particle, or as fringes around the edges of particles. Natural stibnite may also include impurities of valentinite and kermesite, plus other associated phases including (most commonly) quartz, baryte and calcite.

BISMUTHINITE

Bi_2S_3

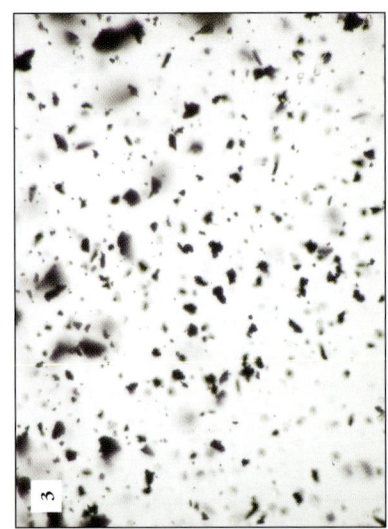

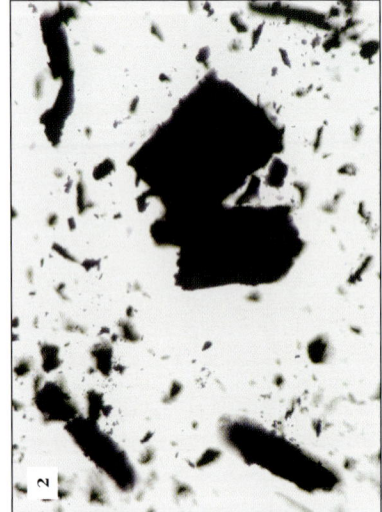

Key: 1. P1511: PPL/100×. Anhedral shards of bismuthinite with broad grain size distribution.

2. P1512: PPL/40×. Coarse-grained bladed and tabular particles of bismuthinite.

3. P1512: PPL/100×. Fine-grained crumb-like particles of bismuthinite.

Crystallinity	Optic sign	2V	n_α	n_β	n_γ	δ
Orthorhombic				[Opaque]		

(data from Webmineral, 2003)

In plane-polarised light, bismuthinite forms weakly translucent to opaque, iron grey to black particles. Relief is high and RI is greater than that of the medium. Particle shape of crushed samples varies from anhedral, angular shards, to finely divided crumb-like grains. Particle shape may be strongly influenced by the perfect prismatic cleavage that is visible on grain surfaces as knife-sharp striations. Particle size distribution may be broad, with grain size ranging from fine to very coarse.

Particles of bismuthinite are predominantly isotropic under crossed polars. However, a few grains have very low birefringence and show very weak, low first order interference colours that are only visible at high illumination levels.

The Pigment Compendium

IRON(II,III) OXIDE, MAGNETITE TYPE, MAGNETITE AND PYRITE

Fe$_3$O$_4$ and FeS$_2$

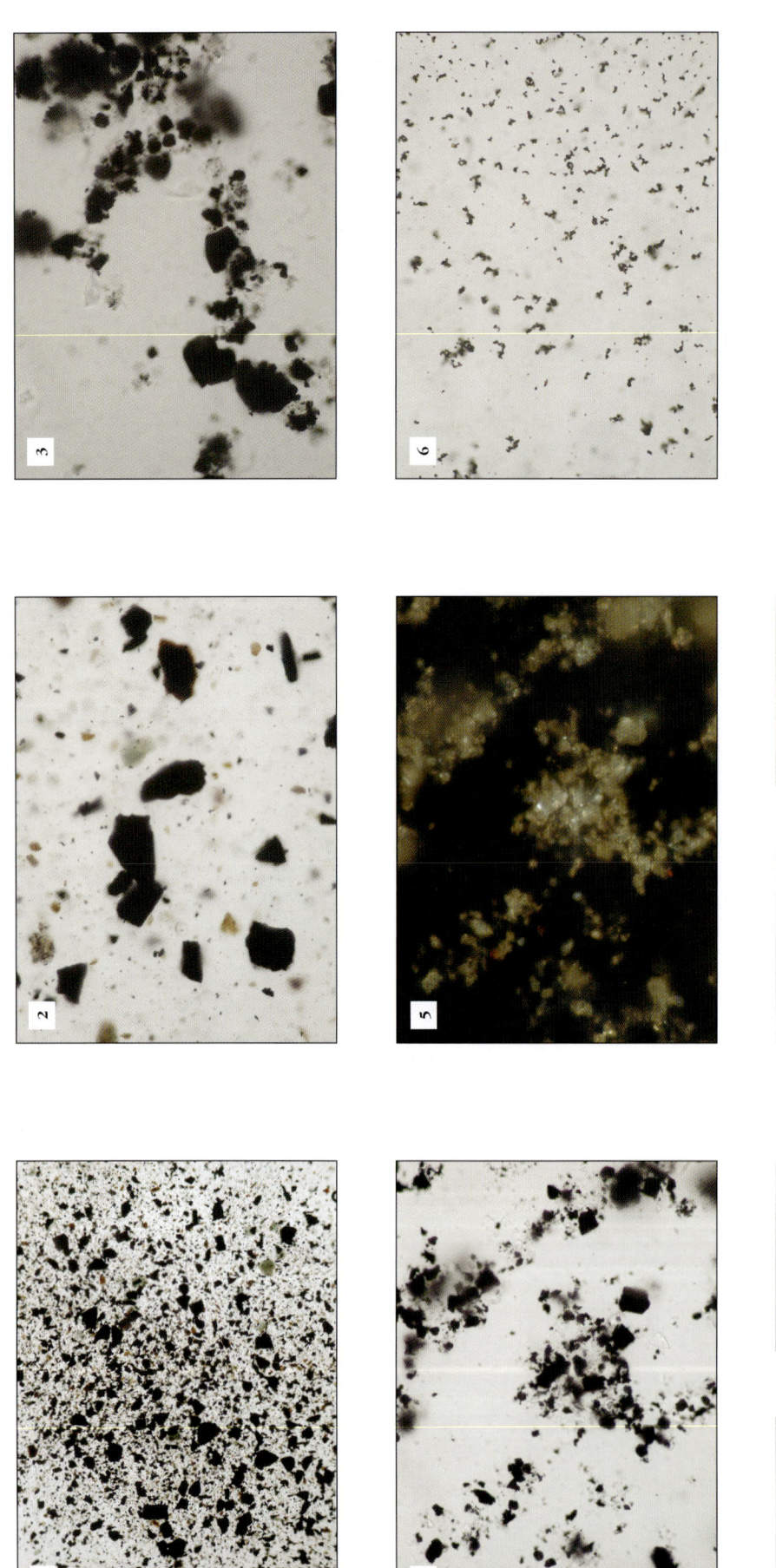

Key:

1. P1225: PPL/10×. Crushed magnetite with green and brown and chlorite.

2. P1225: PPL/40×. Subhedral to anhedral particles of magnetite plus chlorite impurities.

3. P0013: PPL/100×. Particles of pyrite.

4. P0013: PPL/100×. Anhedral particles of pyrite.

5. P0013: RPL/100×. The field of view as in Fig. 4 in reflected light.

6. P1072: PPL/100×. Synthetic magnetite.

7. P1225: PPL. Anhedral and subhedral grains of magnetite.

8. P1072: PPL. Enlargement of Fig. 6 showing particle morphologies.

9. P1247: PPL. Anhedral particle of pyrite.

10. P1247: RPL. Pyrite particle in Fig. 9 in reflected light.

Crystallinity

Cubic

n

[Opaque]

δ

(data from Webmineral, 2003)

Synthetic black iron(II,III) oxide is crystallographically identical to the mineral magnetite. Under plane-polarised light, the compound forms very fine-grained dark brown translucent to black opaque particles. Because the particles are at the resolution of the microscope they appear to have round to bacterioid habits. However, according to Cornell and Schwertmann (1996) the principal habit for magnetite is octahedral, with intergrown octahedral twins, rhombic dodecahedra, cubes, spheres and bullet-shaped particles also possible. It also seems to form agglomerated chains due to its magnetic properties. For translucent particles relief is high and RI is greater than that of the medium. The iron oxides have strong dispersion, and therefore refractive index will vary depending on the light source. Iron(II,III) oxide black is isotropic, though where the pigment has been produced by thermal conversion of another iron oxide trace amounts of partially converted particles may be seen to be birefringent. Such samples typically also contain goethite and hematite forms, the bulk pigment appearing brown.

Since magnetite and pyrite both belong to the cubic system, they are therefore isotropic. They are, however, additionally opaque in plane-polarised light, while the particle shape in both minerals ranges from anhedral, angular shards to euhedral cubic or polyhedral forms. Particle size can range from fine to very coarse, with particle size distributions that are frequently broad. Both phases occur as impurities in other natural pigments and both may be present in earth pigments. Pyrite is commonly also found with lazurite.

The compounds may be distinguished with reflected light, pyrite being characterised by its gold-coloured metallic appearance (pyrite is also known as 'fool's gold'), magnetite by its black colour.

Lit.: Cornell & Schwertmann (1996)

MANGANESE OXIDES

[Various]

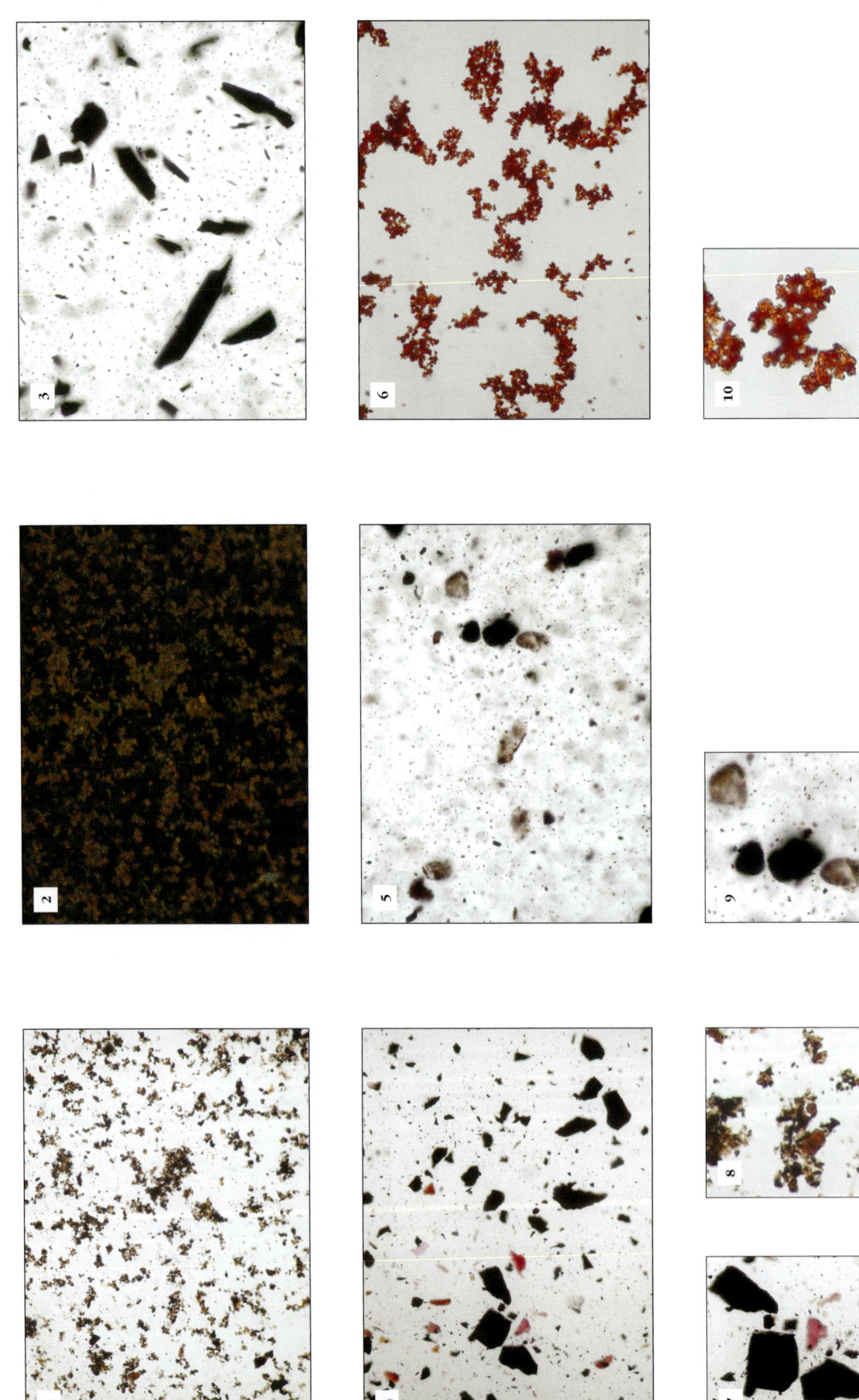

Key:

1. P0126: PPL/100×. Fine-grained mixture of opaque and dark brown particles of pyrolusite.

2. P0126: RPL/100×. Same view as in Fig. 1 in reflected light.

3. P1248: PPL/100×. Prismatic opaque laths of pyrolusite.

4. P1157: PPL/100×. Anhedral and subhedral opaque grains of bixbyite and impurities of purpurite.

5. P1207: PPL/100×. Finely disseminated particles of hausmannite coating quartz grains.

6. P0062: PPL/100×. Synthetic hausmannite, 'manganese brown'.

7. P1157: PPL. Enlargement of group of grains from Fig. 4.

8. P0126: PPL. Enlarged view of particles from Fig. 1.

9. P1207: PPL. Enlarged group of grains from Fig. 5.

10. P0062: PPL. Enlarged view of particle morphologies from Fig. 6.

Crystallinity

	n	δ
Hausmannite: Tetragonal	$n_\omega = 2.46$, $n_\varepsilon = 2.15$	0.3100
Braunite: Tetragonal	[Unknown]	[Unknown]
Pyrolusite: Tetragonal	~2.86–3.06?	[Unknown]
Manganite: Monoclinic	$n_\alpha = 2.25$, $n_\beta = 2.25$, $n_\gamma = 2.53$	0.2800

(data from Webmineral, 2003)

There is a wide range of naturally occurring manganese oxides that are, along with their synthetic analogues, used as black pigments. All are black and opaque or weakly translucent, brown and isotropic or very weakly anisotropic. Consequently, they are difficult to distinguish using the optical microscope. The main phases are hausmannite (Mn_3O_4), braunite (Mn_2O_3), pyrolusite (MnO_2) and the oxide hydroxide manganite ($MnO[OH]$). When naturally occurring, in earth pigments such as umbers and wads (q.v.), they are likely to be found with each other and other manganese-bearing minerals including purpurite (q.v.), lithiophyllite and psilomelane (romanechite), and associated ore and gangue minerals such as quartz, feldspar, calcite and so forth.

All phases have either opaque black or very dark brown and weakly translucent particles. Typically when crushed they produce subangular anhedral particles. Surface textures may be observed and have irregular fracture surfaces.

Bixbyite may demonstrate a conchoidal fracture. Pyrolusite is frequently fibrous and elongate particles are typical; striated fibrous surfaces may also be observed. Hausmannite has a brown-black cast and forms subangular granular particles.

Synthetic manganese oxide blacks are distinctive by their high purity.

The pigment known as wad is an earth pigment composed of one or more of the above manganese oxide minerals, plus other phases such as iron oxides (goethite, hematite and magnetite).

The Pigment Compendium

ALUMINIUM OXIDE, CORUNDUM TYPE

Al$_2$O$_3$

Crystallinity	**Optic sign**	**n_ω**	**n_ε**	**δ**
Trigonal – Hexagonal	−ve	1.768	1.760	0.0080
			(data for corundum from Sinkankas, 1966; cf. Webmineral, 2003)	

Under plane-polarised light, aluminium oxide, the synthetic analogue of the mineral corundum, forms transparent, colourless particles with low to moderate relief and RI just greater than that of the medium. Grain shape is of rounded, polycrystalline plates and very fine particles with a grain size close to that resolvable by the optical microscope. These therefore appear to have rounded habits.

Under crossed polars, aluminium oxide has very low, first order interference colours. Some grains appear almost isotropic, but the presence of anisotropy may be proven by use of the sensitive tint plate. Extinction is undulose.

Corundum itself may be found in certain contexts as a secondary phase in mineral aggregates. Under plane-polarised light it is seen as colourless to pale pinkish tan coloured particles. Particle morphology exhibits as sharp-edged fragments and chips, the form of the particles being influenced by the good to perfect basal parting. The relief is moderate with RI greater than that of the medium; other values of RI are cited by McCrone *et al.* (1973–80) as $n_\omega = 1.7594–1.7598$ and $n_\varepsilon = 1.7676–1.7682$.

As stated above, the birefringence is low; consequently, low first order colours (greys) are likely to be visible.

Lit.: McCrone *et al.* (1973–80)

The Pigment Compendium

ALUMINIUM HYDROXIDE, BAYERITE AND GIBBSITE TYPES

Al(OH)$_3$

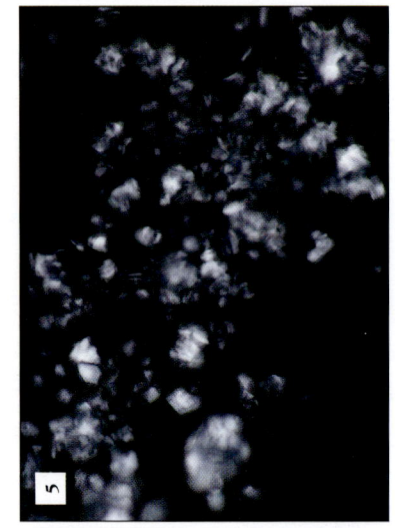

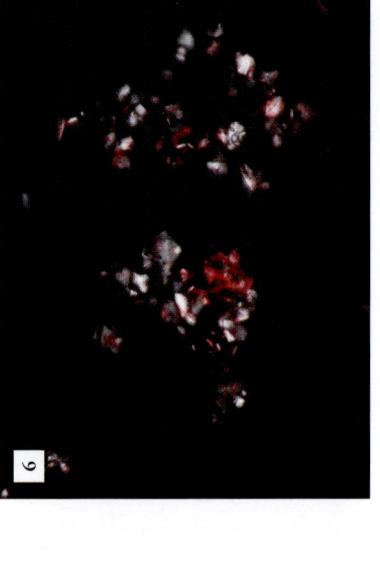

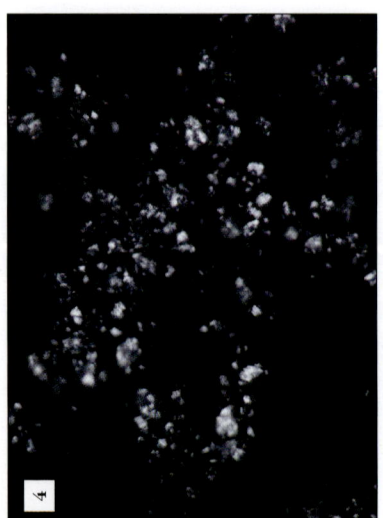

1. P0161: PPL/40×. Plates of aluminium hydroxide, gibbsite type.

2. P0161: PPL/100×. The particles in Fig. 1 further magnified.

3. P0048: PPL/40×. Aluminium hydroxide, gibbsite type as a substrate for carmine.

4. P0161: XPL/40×/ ● ● The field of view as in Fig. 1 in crossed polars.

5. P0161: XPL/100×/ ● ● The field of view as in Fig. 2 in crossed polars.

6. P0048: ~XPL/40×. The field of view as in Fig. 3 in crossed polars.

Crystallinity	Optic sign	2V	n_α	n_β	n_γ	δ
Bayerite: Monoclinic				~1.583		
Gibbsite: Monoclinic	+ve	0°–5°	1.568–1.570	1.568–1.570	1.586–1.587	0.0170–0.0180
						(data for mineral from Heinrich, 1965; cf. Webmineral, 2003)

Bayerite is generally only encountered as an artificial compound, being named from its formation as part of the Bayer process of purifying the rock bauxite, a common commercial source of aluminium. According to Winchell (1931), this compound is produced by precipitating aluminium hydroxide from solution; it is therefore a form likely to be produced during preparation of lake pigments where a dyestuff is co-precipitated. In the context of pigments it therefore typically occurs as a lake substrate or as a filler.

Recent research by Lee *et al.* (1997) has shown that formation of bayerite over the closely related gibbsite during preparation of aluminium hydroxide by the Bayer process, as well as the particle morphologies obtained, is dependent on a number of factors. For example bayerite is typically obtained from room temperature reactions, whereas gibbsite is favoured at higher temperatures. Carbonation of sodium aluminate during preparation formed 'web-like' and frond-shaped crystals whereas lowering pH could form elongate corrugated lamellar particles.

Under plane-polarised light particles of the gibbsite type of aluminium hydroxide were found in samples examined to form transparent, colourless crystals with low relief and a RI just less than that of the medium. Particle shapes vary from euhedral and subhedral hexagonal plates to angular anhedral plates. Overall, the particles have a pearlescent quality, with smooth particle surfaces. Particle size is medium and particle size distribution is relatively narrow. The research by Lee *et al.* has also shown that the nature of the alkaline metal present may also have an impact on morphology; in gibbsite, hexagonal plates arise with precipitation from sodium aluminate, elongated hexagonal prisms from potassium aluminate.

Under crossed polars, the gibbsite form of aluminium hydroxide illustrated here is weakly birefringent and shows low first order grey interference colours. Many particles are revealed to be finely polycrystalline, and have sweeping extinction.

Lit.: Lee *et al.* (1997); Winchell (1931)

ANTIMONY(III) OXIDE, SENARMONTITE TYPE

Sb_2O_3

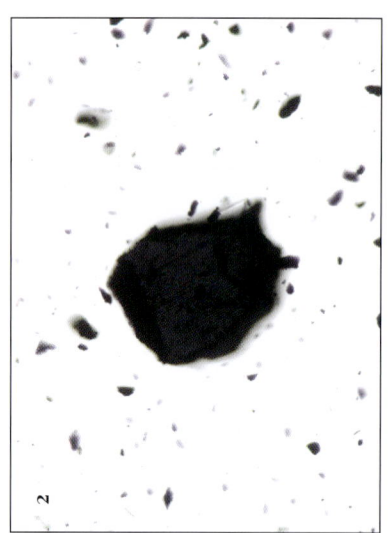

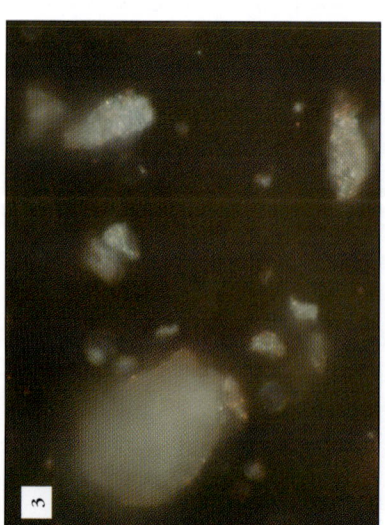

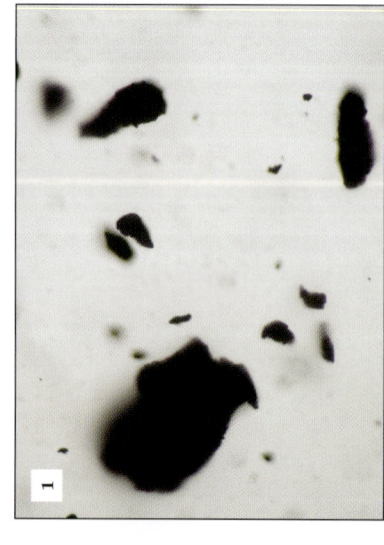

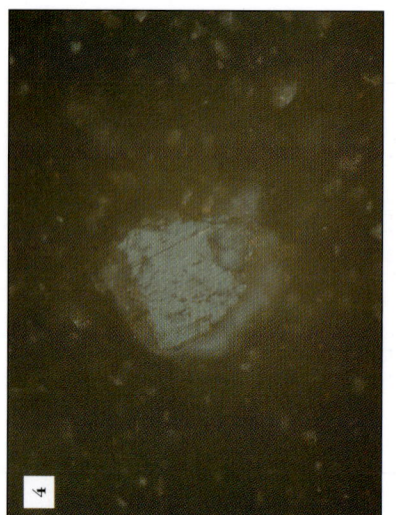

Key:

1. P0874: PPL/100×. Subrounded particles of antimony(III) oxide.

2. P0874: PPL/40×. Coarse-grained particles of antimony(III) oxide, note at high illumination the crystals are weakly translucent.

3. P0874: RPL/100×. The field of view as in Fig. 1 in reflected light.

4. P0874: RPL/40×. The grain in Fig. 2 in bireflected light.

Crystallinity
Cubic

n
2.087

δ
Isotropic

(data for mineral from Ford, 1932; cf. Webmineral, 2003)

Synthetic antimony oxide, senarmontite type can be seen in plane-polarised light to form colourless transparent particles with moderate relief and RI greater than that of the medium. Grain size ranges from fine to very fine and the smallest particles appear rounded and bacterioid. Some fine particles are clearly observed as forming euhedral rods. Particle surfaces are smooth and grains are relatively inclusion free. According to Gloger and Hurley (1973), pigmentary grades of this compound typically show at high magnification simply as finely divided and irregular in shape, though at electron microscopy scales the morphology resolves into cubic, tabular, triangular and colummar-shaped particles. The commercial pigment is generally produced by a fume process, this resulting in an average particle size of about 1 μm.

Under crossed polars particles can be seen to have very low birefringence, such that first order dark greys predominate and grains appear almost isotropic. However, use of the sensitive tint plate proves anisotropy. The rod-shaped particles are length slow, with straight, complete extinction. Particles also have weak grey internal reflections. With UV excitation samples examined appeared weakly grey; in reflected light they were a grey-white colour.

Lit.: Gloger & Hurley (1973)

The Pigment Compendium

ARSENIC(III) OXIDE, ARSENOLITE TYPE

As_2O_3

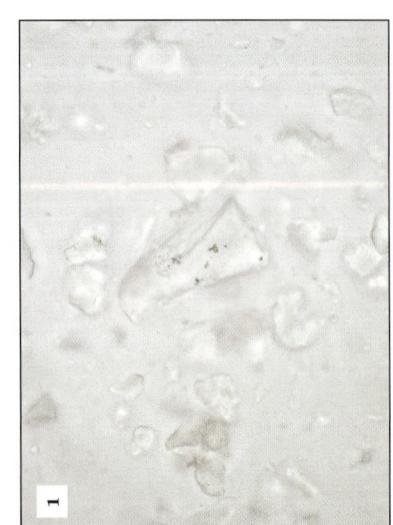

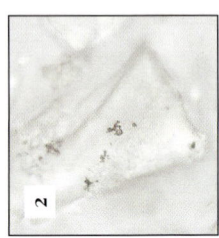

1. P0493: PPL/100×/H. Angular particles of arsenic(III) oxide.

2. P0493: PPL/H. Detail of
   crystal morphology
   from Fig. 1.

Crystallinity
Cubic

n	δ
1.755	Isotropic
	(data for mineral from Ford, 1932; cf. Webmineral, 2003)

In plane-polarised light the particles of arsenic(III) oxide, the synthetic analogue of the mineral arsenolite, are colourless and translucent. However, they have a pearlescent appearance, while the surfaces appear slightly roughened and etched. Some of the particles contain strings of inclusions that are weakly translucent and brown. They are formed of very fine crystals with acicular habits and arranged in rosettes and spherulitic aggregates. Inclusion-bearing particles have deeply etched and striated surfaces. Other inclusions present within the particles are colourless and have a rhombic form. The particles have moderate relief, with RI greater than that of the medium.

Under crossed polars the compound is predominantly isotropic, although a few particles are weakly anisotropic with very low first order interference colours. Distinctive white internal reflections are associated with the inclusion trails. The particles are length slow.

Arsenic(III) oxide appears transparent in reflected light.

Lit.: FitzHugh (1997)

The Pigment Compendium

BARIUM CARBONATE, WITHERITE TYPE AND WITHERITE

BaCO₃

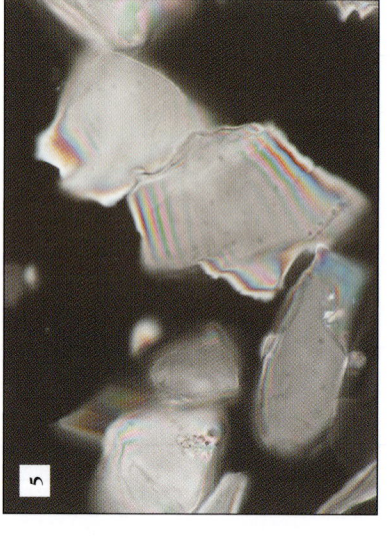

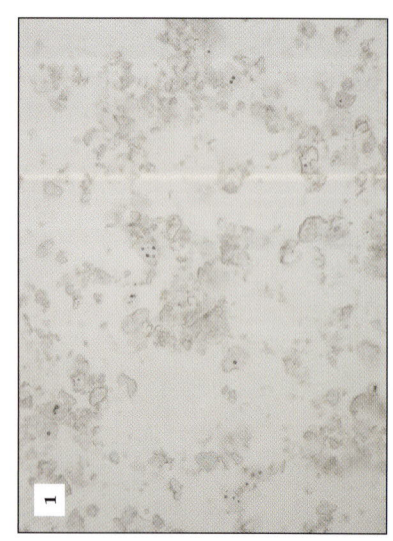

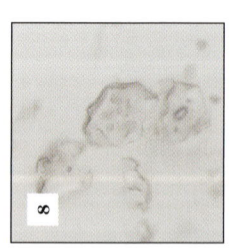

Key:

1. P0164: PPL/40×. Synthetic barium carbonate.

2. P1270: PPL/100×/○ Coarse-grained particles of natural witherite.

3. P1270: PPL/100×/○. The particles in Fig. 2 rotated through 90° to show change in relief.

4. P0164: XPL/40×. The field of view in Fig. 1 under crossed polars.

5. P1270: XPL/100×/○. The particles in Fig. 2 rotated through 90° under crossed polars.

6. P1270: RPL/40×. Particles of natural witherite in reflected light.

7. P0164: PPL. Enlargement of particles from Fig. 1 showing inclusion-rich grains.

8. P0164: PPL. Enlargement of particles from Fig. 1 showing subhedral rhombic grains.

Crystallinity	Optic sign	2V	n_α	n_β	n_γ	δ
Orthorhombic	−ve	8°–16°	1.529	1.676	1.677	0.1480

(data for mineral from Mason, 1968; cf. Webmineral, 2003)

Under plane-polarised light, witherite is visible as transparent, colourless grains, with strongly variable relief. RI varies from just greater than that of the medium to just less than that of the medium. Particle shape encompasses irregular, angular shards and plates, while particle surfaces are smooth. For crushed natural samples particle size will vary from fine to very coarse with broad size distribution.

Under cross-polarised light, witherite has very high birefringence and fourth order and greater interference colours are typical. Concentric fringes of interference colours are commonly seen at particle margins. Witherite has straight and complete extinction and is length fast. McCrone et al. (1973–80) have noted that twinning is almost always present in samples of the mineral;

however, this has distinctly not been noticed by the authors and Mactaggart (2002).

It is worth noting that witherite often forms in geological environments as an alteration product of baryte (q.v.). Relicts of this mineral may be present in witherite crystals, noticeable from their higher and non-variable relief, and lower birefringence. It occurs chiefly in association with the minerals galena, anglesite, baryte and calcite (Deer et al., 1992).

Witherite also appears very similar to aragonite (q.v.).

Lit.: Deer et al. (1992); Mactaggart (2002); McCrone et al. (1973–80).

The Pigment Compendium

BARIUM SULFATE

BaSO$_4$

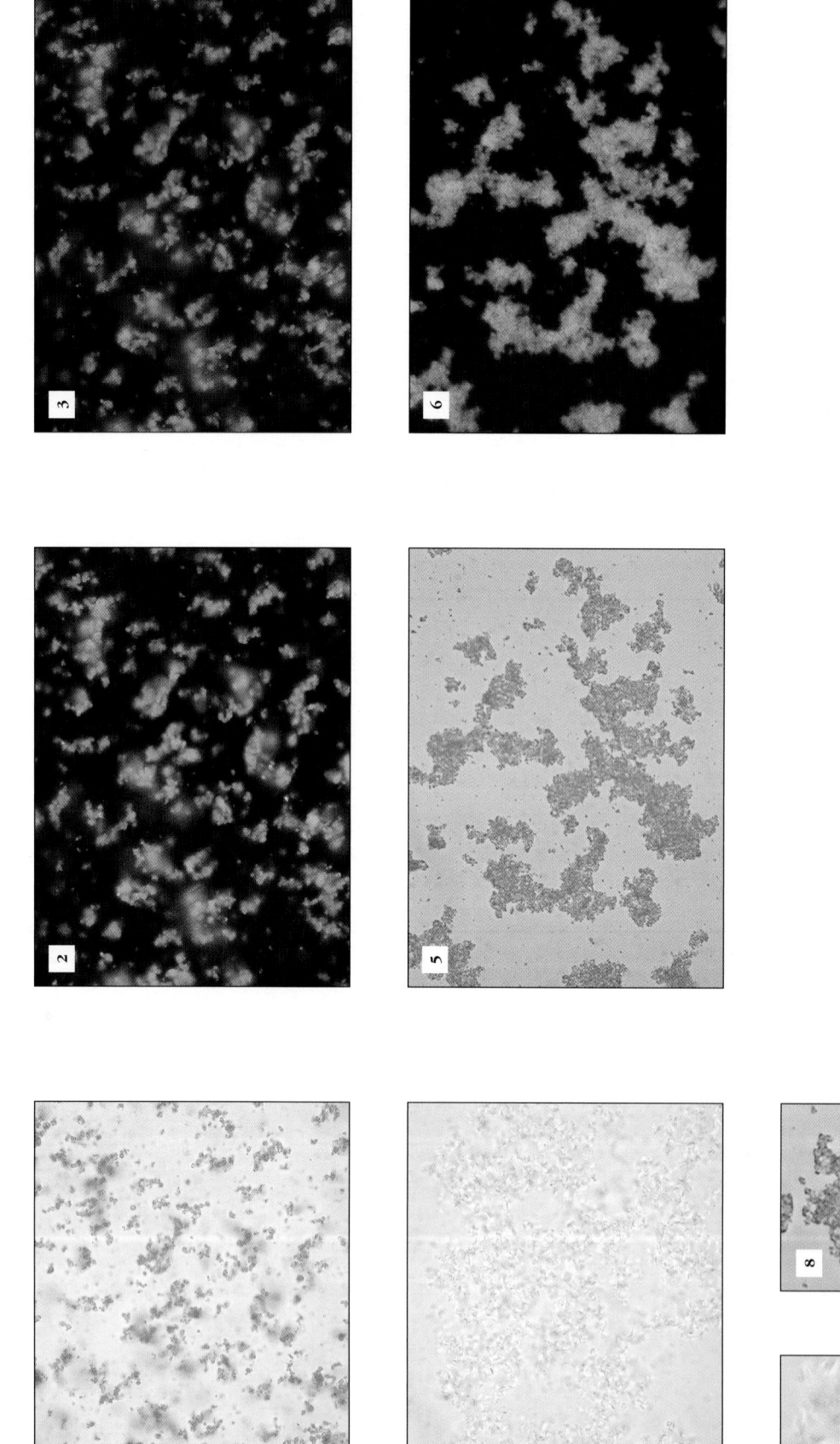

Crystallinity	Optic sign	2V	n_α	n_β	n_γ	δ
Orthorhombic	+ve	36°–42°	1.634–1.637	1.636–1.638	1.646–1.648	0.010
					(data for mineral from Heinrich, 1965; cf. Feller, 1986)	

In plane-polarised light synthetic barium sulfate commonly appears as extremely fine-grained, colourless particles. The particle size is often close to that resolvable with the optical microscope so that particles appear to have round or bacterioid habits. However, Kittel (1960; cf. Feller, 1986) illustrates two other forms: one that occurs as large lump-shaped masses not dissimilar to the mineral and another, said to be a rarer habit, where needles occur. Under scanning electron microscopy Moser (1973) has stated that some elongation can sometimes be seen. Feller also documents a special German product where the particles form rhombs that are discernible by optical microscopy. Relief of barium sulfate is low and just less than that of 1.662RI mounting media.

Under crossed polars barium sulfate can be seen to have low birefringence; this, and the very fine particle size of barium sulfate, causes it to appear almost isotropic. However, very low first order greys can be observed, the particles seeming to twinkle as the stage is rotated.

Barium sulfate is a common extender pigment and may therefore be encountered in numerous contexts, including use as a lake base. Of special interest among these is lithopone, where barium sulfate is co-precipitated with zinc sulfide. This forms an intimate mixture that is extremely difficult to differentiate from barium sulfate itself without recourse to other analytical techniques (Feller, 1986). An electron micrograph of lithopone published by Buxbaum (1998) illustrates the intimate mixture achieved in lithopones as well as a difference in particles size – the barium sulfate phase has a mean particle diameter of 1.0 μm and the zinc sulfide 0.3 μm.

The chemistry and identification of both mineral baryte and synthetic barium sulfate in the context of pigments have been reviewed by Feller (1986).

Lit.: Buxbaum (1998) 75; Feller (1986); Kittel (1960); Moser (1973)

The Pigment Compendium

BARYTE

BaSO$_4$

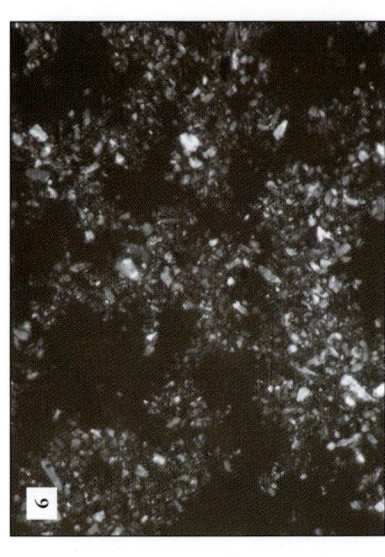

Key:

1. P1151: PPL/40×/○ Colourless angular particles of baryte.

2. P0158: PPL/100×. Colourless flakes of baryte.

3. P0158: PPL/40×. Fine-grained particles of baryte.

4. P1151: XPL/40×/○ The field of view as in Fig. 1 under crossed polars.

5. P0158: XPL/100×. The field of view as in Fig. 2 under crossed polars.

6. P0158: XPL/40×. The field of view as in Fig. 3 under crossed polars.

Crystallinity	Optic sign	2V	n_α	n_β	n_γ	δ
Orthorhombic	+ve	36°–42°	1.634–1.637	1.636–1.638	1.646–1.648	0.010

(data for mineral from Heinrich, 1965; cf. Feller, 1986)

Under plane-polarised light the mineral baryte appears as transparent, colourless particles. Relief is low and RI is just less than that of the medium. Dispersion is weak with $r < v$.

Particle shape is typically of crushed angular, anhedral, shards. Baryte has three cleavages and these will influence particle shape, with straight particle boundaries common. Particle surfaces are smooth, and cleavages are rarely observed within particles because of the low relief. Particle size ranges from fine to very coarse and size distribution is typically broad.

Under crossed polars, baryte has low birefringence and low first order greys are typically observed. Coarser particles may show up to first order yellows. Extinction is straight and complete.

Baryte may be found in association with quartz, fluorite, pyrite and a wide range of ore minerals, typically with lead and zinc mineralisation. Particle morphology should permit differentiation from the otherwise optically identical synthetic analogue.

The chemistry and identification of both mineral baryte and synthetic barium sulfate in the context of pigments has been reviewed by Feller (1986).

Lit.: Feller (1986)

ARAGONITE

CaCO$_3$

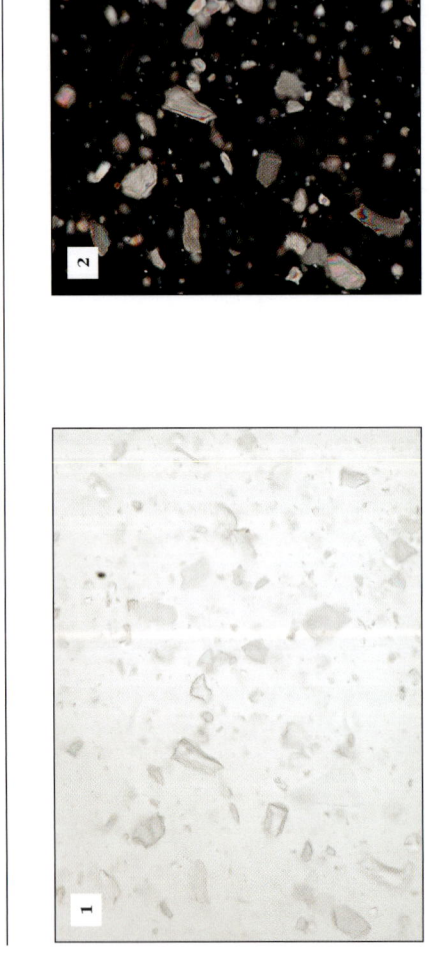

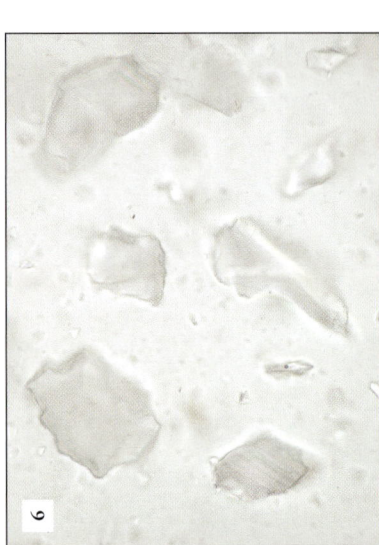

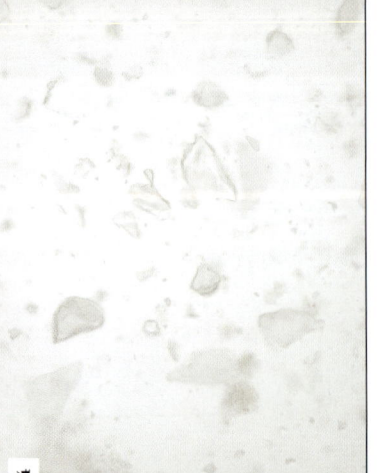

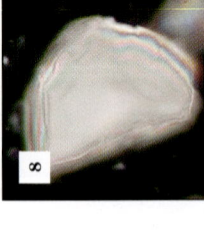

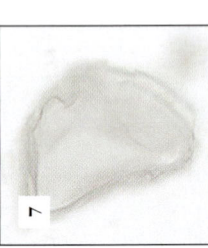

Crystallinity	Optic sign	2V	n_α	n_β	n_γ	δ
Orthorhombic	−ve	16°–19°	1.529–1.530	1.680–1.682	1.685–1.686	0.1560

(data for mineral from Heinrich, 1965; cf. Webmineral. 2003)

Aragonite is the prime constituent of most unfossilised seashells (apart from oyster shells, for which see: Calcite, oyster shell). However, it also occurs in non-biogenic 'mineral' settings, where it appears microscopically similar to calcite.

Under plane-polarised light mineral aragonite can be seen to form as colourless, anhedral, platy particles. Euhedral crystals form prisms, but these will rarely be visible in samples crushed and ground for use in pigments. McCrone et al. (1973–80) document a number of other particle morphologies including elongated prisms terminating in slightly oblique faces; subcircular plates with saw-toothed edges due to cleavage; triangular plates with rough to sharp angular edges; striated rhombs that closely resemble calcite except that the striations bisect the obtuse rhomb angle. Aragonite has one cleavage,

which makes it distinctive from calcite, both in general appearance and in the particle shapes formed on crushing. However, like calcite, aragonite has strongly variable relief, though this is not as extreme as it is in calcite. Dispersion is weak, $r < v$.

Under cross-polarised light mineral aragonite is seen to be strongly birefringent, showing high (typically third order and greater) interference colours. Concentric rims of birefringent fringes are often observed on the crystals. Aragonite has straight extinction and lamellar twinning may be present in some samples. The striations as noted above by McCrone et al. are due to this twinning; they are thus thin crystal bands rather than grooves.

Lit.: McCrone et al. (1973–80)

CALCIUM CARBONATE, CALCITE TYPE

CaCO$_3$

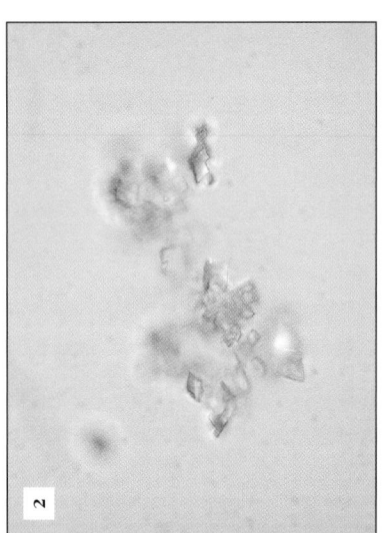

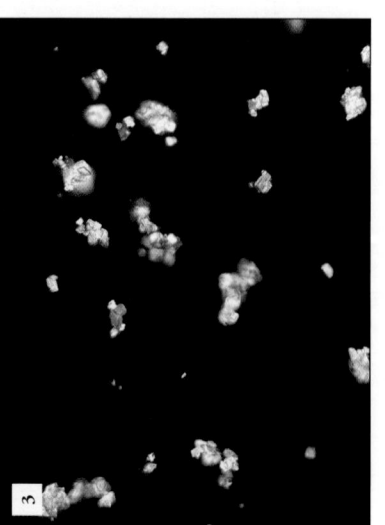

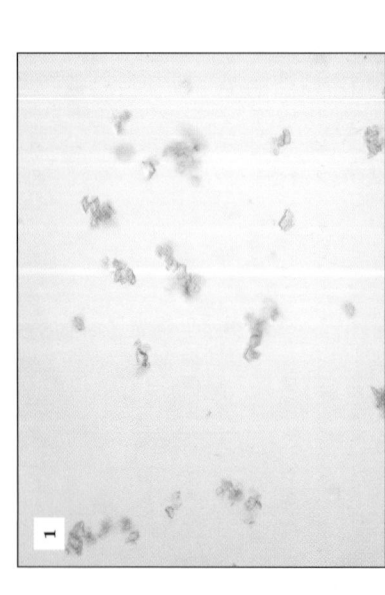

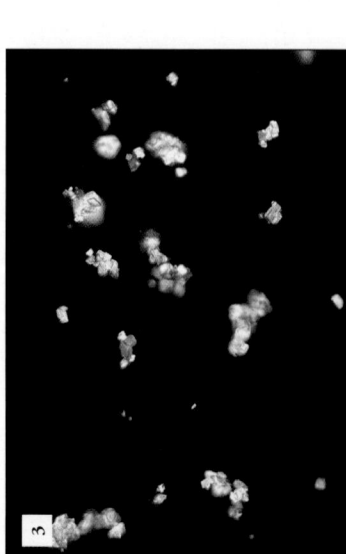

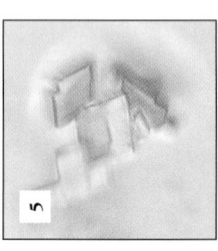

Key:

1. P0880: PPL/40×. Fine-grained euhedral particles of synthetic calcite.

2. P0880: PPL/100×/ **O**. Calcite rhombs.

3. P0880: XPL/40×/ **O**. The field of view as in Fig. 1 under crossed polars.

4. P0880: XPL/100×/ **O**. The field of view as in Fig. 2 under crossed polars.

5. P0880: XPL/ **O**. Euhedral rhombs of synthetic calcite.

Crystallinity	**Optic sign**	n_ω	n_ε	δ
Trigonal	−ve	1.64–1.66	1.486	0.1540–0.1740

(data for calcite from Webmineral, 2003)

In plane-polarised light, the most distinctive feature of all forms of calcite is the highly variable relief, which ranges from low ('RI equals that of the medium) to high (RI ≪ medium); effectively the particle 'disappears' as the stage is rotated. Synthetic calcium carbonate additionally shows distinctive, well-formed euhedral rhombic crystals that typically range in size from fine to medium. Woerner (1973) describes a range of commercial products with particle sizes that could be as small as 0.03 μm, to as large as 10 μm; moreover, despite the use of controlled precipitation, size ranges within a sample might also be broad. Calcite has three sets of perfect cleavage (at 60° to each other, parallel to the crystal faces), but these are rarely visible in the fine synthetic crystals. The particles are colourless.

Under cross-polarised light, calcite has high birefringence and therefore strong interference colours, typically third order or greater (increasing with crystal size). Calcite has inclined extinction.

Calcium carbonate-based pigments have been reviewed by Gettens et al. (1993a).

Lit.: Gettens et al. (1993a); Woerner (1973)

The Pigment Compendium

CALCITE

CaCO$_3$

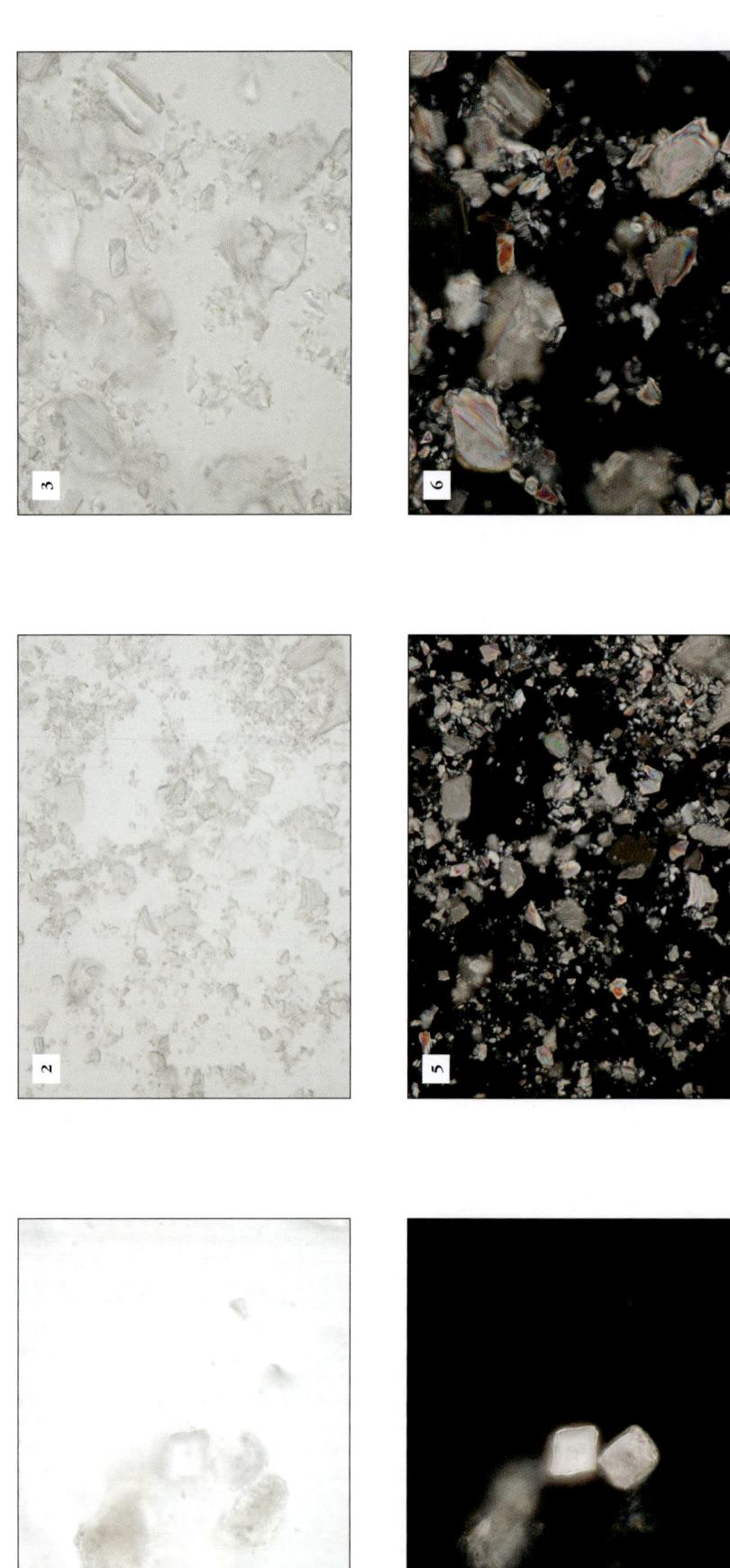

1. P0860: PPL/100×. Euhedral rhombic and inclusion-rich cloudy anhedral particles of calcite.

4. P0860: XPL/100×/●. The field of view as in Fig. 1 under crossed polars.

2. P0162: PPL/40×. Crushed fragments of calcite marble.

5. P0162: XPL/40×/●. The field of view as in Fig. 2 under crossed polars.

3. P0162: PPL/100×. Crushed fragments of calcite marble.

6. P0162: XPL/100×/●. The field of view as in Fig. 3 under crossed polars, note the lamellar twins in the crystal at the top left.

Crystallinity	Optic sign	n_ω	n_ε	δ
Trigonal	−ve	1.64–1.66	1.486	0.1540–0.1740
				(data from Webmineral, 2003)

Calcite occurs naturally in two main forms: as a fine cryptocrystalline micrite (as in chalks; see calcite, chalk) and as crystalline sparite. Individual particles are indistinguishable using optical microscopy in micrites; crushed examples may appear as being finely polycrystalline. The features of sparite are described below.

Under plane-polarised light, the most distinctive feature of all forms of calcite is the highly variable relief, which ranges from low (RI much less than the medium) to high (RI much less than the medium); effectively the particle disappears as the stage is rotated. Calcite has three sets of perfect cleavage (at 60° to each other, parallel to the crystal faces), and this strongly influences the particle habit on crushing, producing rhombic particles. As the particle margins are formed by cleavage faces, they are technically angular and *anhedral* (that is, they are straight edges but they are not crystal faces). Uneven anhedral particles also occur. Cleavage planes are frequently visible in larger particles. The crystals are colourless. According to Hall (1973), ground calcium carbonate was supplied commercially at that time in grades

from that for fine paint (particle size ~2.5 μm) through to that for putty (~12 μm), though finer and coarser material may also be encountered. Since this is a ground mineral, the particle size distribution is likely to be broad.

Under cross-polarised light, calcite has high birefringence and therefore strong interference colours, typically third order or greater (increasing with crystal size). Calcite has inclined extinction. Multiple twins are common in well-crystallised particles.

Calcite derived from crushed marble is always sparite and generally shows well-developed multiple twins that are clearly visible under cross-polarised light. Frequently this source of calcite will form good rhombic particles on crushing. There will be no biological material in such rocks.

Calcium carbonate-based pigments have been reviewed by Gettens *et al.* (1993a).

Lit.: Gettens *et al.* (1993a); Hall (1973)

CALCITE, CHALK

CaCO₃

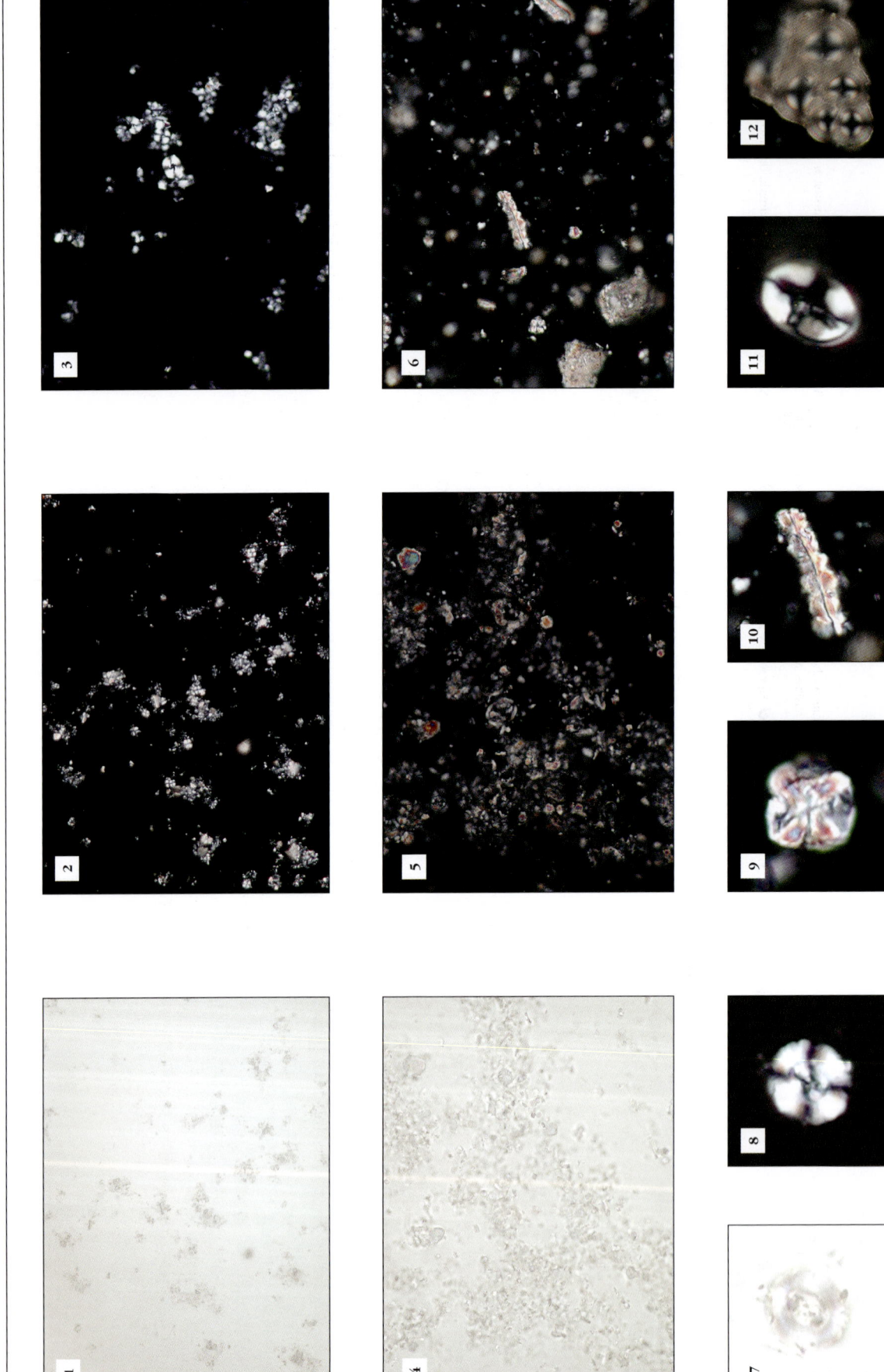

Key:

1. P0146: PPL/40×. Chalk with grains of calcite and microfossils.

2. P0146: XPL/40×. Same field of view as in Fig. 1, the coccoliths are clearly visible with standing extinction crosses.

3. P0146: XPL/100× ● A cluster of *Watznuaeria barnesiae* coccoliths.

4. P0147: PPL/100× ● Chalk with grains of calcite and microfossils.

5. P0147: ~XPL/100×/● Same view as in Fig. 4 under crossed polars.

6. P1004: XPL/100×/●/H. Sample of chalk with microfossils.

7. P0146: PPL ● *Watznuaeria barnesiae*.

8. P0146: ~XPL ● *Watznuaeria barnesiae* as in Fig. 7 under crossed polars.

9. P1004: XPL/●/H. Micula species nannofossil.

10. P1004: XPL/●/H. *Lucianorhabdus cayeuxii*.

11. P1004: XPL/●/H. *Eiffellitbus eximius*.

12. P1004: XPL/●/H. A planktonic foraminfera.

Crystallinity	Optic sign	n_ω	n_ε	δ
Trigonal	–ve	1.64–1.66	1.486	0.1540–0.1740

(data for calcite from Webmineral, 2003)

Geologically speaking, chalk refers to rocks composed of fine-grained calcite. The term has, in other fields including art historical ones, been applied to synthetic calcium carbonate and also to talc. This section deals with 'true' chalks – those derived from naturally occurring sources.

Chalks are fine-grained sediments, so much so that this particle size is preserved at the scale at which the pigment is ground. Under plane-polarised light, the variable relief is visible, even with very fine samples. Under cross-polarised light, the particles show the typical high birefringence and strong, bright interference colours associated with calcite. However, the most diagnostic feature of such materials is the occurrence of microfossils, predominantly the marine phytoplankton fragments known as coccoliths. Their presence confirms natural rather than synthetic carbonate. These are also within the size range that is preserved by grinding, and they are typically larger than the micritic groundmass. Other fossil fragments may be present, especially other varieties of nannofossils, including *Micula* species and

elongate holococcoliths. Identification down to species level should only be undertaken by an experienced micropalaeontologist. Nevertheless, coccoliths are of distinctive shape and optical properties and thus are easily observed. These fossils have the same relief as calcite (of which they are composed). In overall shape, they resemble buttons. They are constructed from radially organised fibrous microcrystals or plates. The former variety, and the most common, are readily identified under cross-polarised light by the appearance of a standing extinction cross. Accurate identification of these floras can yield a geological age for the chalk and thus narrow down the source regions.

Calcium carbonate-based pigments have been reviewed by Gettens *et al.* (1993a). Further information concerning the description of nannofossils may be found in Young *et al.* (1997) and Bown (1998).

Lit.: Bown (1998); Gettens *et al.* (1993a); Young *et al.* (1997)

CALCITE, CORAL

CaCO$_3$

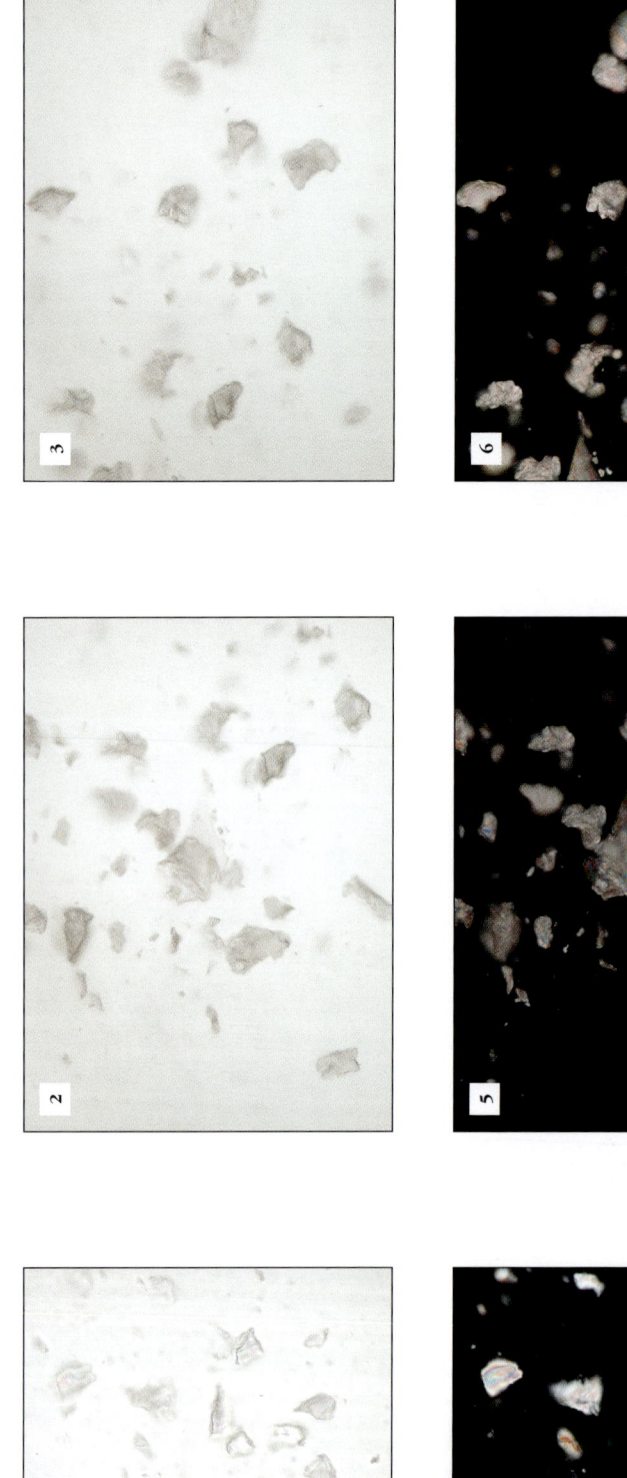

Key:

1. P1162: PPL/100×. Transparent anhedral particles of coral.
2. P1508: PPL/40×. Translucent fibrous particles of coral.
3. P1508: PPL/40×. Translucent fibrous particles of coral.
4. P1162: XPL/100×/ ❂. Same field of view as in Fig. 1 under crossed polars.
5. P1508: XPL/40×/ ❂. Same field of view as in Fig. 2 under crossed polars.
6. P1508: XPL/40×/ ❂. Same field of view as in Fig. 3 under crossed polars.

Crystallinity	Optic sign	n_ω	n_ε	δ
Trigonal	–ve	1.64–1.66	1.486	0.1540–0.1740

(data for calcite from Webmineral, 2003)

Under plane-polarised light biogenic calcium carbonate in the form of coral forms particles with a translucent, colourless or cloudy appearance. In common with other forms of calcite, relief is strongly variable from very low, with an RI almost equal to that of the medium, to high, with an RI much less than that of the medium. For crushed samples, particle shape is typically angular shards; evidence of biogenic structure is not apparent. Particle size is likely to be medium to coarse, though this depends upon processing.

Under crossed polars, coral shows the typical calcium carbonate high birefringence, with fourth order interference colours typical. Concentric fringes of colours are observed on many grains. Extinction is never complete. Some particles are finely polycrystalline and twinkle as the stage is rotated. Others show sweeping or undulose extinction.

Coral appears white or weakly yellow-white in reflected light. Variable amounts of fluorescence were observed in samples with UV excitation, from none to a weak white.

Calcium carbonate-based pigments have been reviewed by Gettens et al. (1993a).

Lit.: Gettens et al. (1993a)

The Pigment Compendium

CALCITE, EGGSHELL

CaCO$_3$

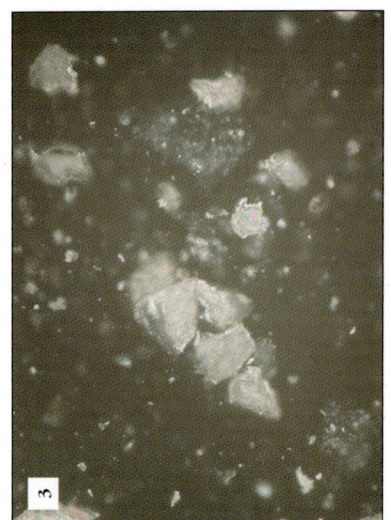

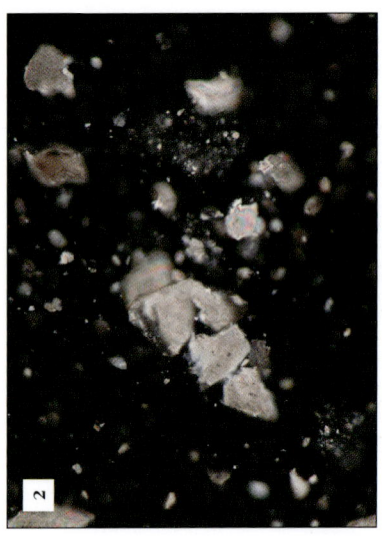

Key:

1. P1488: PPL/40×. Angular particles of eggshell and fibrous mats of membrane.

2. P1488: XPL/40×/●. The field of view as in Fig. 1 under crossed polars, note membrane is isotropic.

3. P1488: RPL/40×. The field of view as in Fig. 1 in reflected light.

4. P1488: PPL/40×. Cloudy inclusion-rich, finely banded fragments of eggshell.

5. P1488: XPL/40×/●. The field of view as in Fig. 4 under crossed polars.

Crystallinity	Optic sign	n_ω	n_ε	δ
Trigonal	−ve	1.64–1.66	1.486	0.1540–0.1740
				(data for calcite from Webmineral, 2003)

Avian eggshells are general composed of calcite, although a few species may have vaterite shells. The structure of avian eggshells, from internal to external surfaces, is composed of a thin 'membrane', a thin 'mamillary' layer, a thicker 'palisade' layer and a thin outer cuticle. These various components may be observed in pigments based on eggshells. The membrane is generally removed before pigment preparation because it impairs grinding. However, fragments are not uncommon and are distinctive under the microscope appearing as a fine mesh of randomly orientated and branching fibres; this material is isotropic under crossed polars.

Under plane-polarised light, the particles of the shell proper are colourless, but very rich in inclusions which give an overall cloudy, dirty appearance. Particle surfaces appear rough and finely striated; many fragments show evidence of laminations, which lay concentric to the surface of the egg. Relief is highly variable, ranging from very low to moderate, RI varying from almost equal that of the medium to lower than that of the medium. Particle size distribution is broad, ranging from very fine- to very coarse-grained particles. Particle shape is typically of angular shards.

Under cross-polarised light the typically high birefringence of calcite is observed, with fourth order and greater interference colours present. In larger particles, the polycrystalline nature of the shell may be observed, especially in particles of the palisade layer where broadly columnar crystals in parallel orientations are easily recognised by their different crystallographic orientations. Extinction of the individual crystals is complete, though extinction in the polycrystalline fragments as a whole is mottled.

The Pigment Compendium

CALCITE, OYSTER SHELL

$CaCO_3$

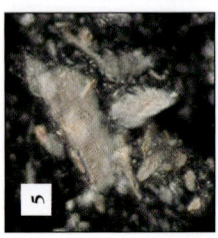

1. P0121: PPL/40×. Colourless particles of crushed oyster shell.

2. P0121: PPL/40×. Colourless particles of crushed oyster shell.

3. P0121: XPL/40× ⦿. The field of view as in Fig. 1 showing fibrous and laminated crystal structures.

4. P0121: XPL/40× ⦿. The field of view in Fig. 2 in crossed polars.

5. P0121: XPL ⦿. Detail of crystal morphology from Fig. 3.

6. P0121: XPL ⦿. Detail of fibrous crystal structure from Fig. 4.

Crystallinity	**Optic sign**	n_ω	n_ε	δ
Trigonal	–ve	1.64–1.66	1.486	0.1540–0.1740

(data for calcite from Webmineral, 2003)

Shell white pigments, also known as *gofun*, are normally derived from oysters. These are unusual shells in that they are composed of calcite rather than aragonite.

Under plane-polarised light particles of oyster shells are colourless and have strongly variable relief, from equal to that of the medium to moderate. The particles are colourless. Oyster shell is composed of finely laminated sheets of calcite, and this is reflected in the particle shape. Typically, lath-like particles and plates with an internal laminated structure are present. Particle sizes of the unmodified pigment are uneven, ranging from very fine to coarse; however, elaborate sedimentation techniques have been used in Japan that may lead to more selective, narrower size ranges.

Under cross-polarised light, particles show the usual high birefringence and corresponding interference colours of calcite, typically fourth order, except in the case of very small particles. Extinction is sweeping, reflecting the laminated structure of the shells.

Calcium carbonate-based pigments have been reviewed by Gettens *et al.* (1993a).

Lit.: Gettens *et al.* (1993a).

DOLOMITE AND ANKERITE

CaMg(CO$_3$)$_2$ and Ca(Mg$_{0.67}$Fe$_{0.33}$)(CO$_3$)$_2$

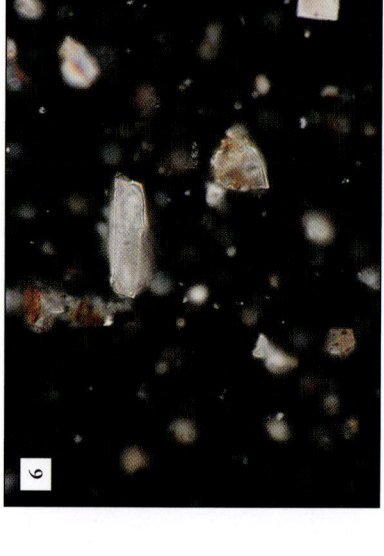

Key:

1. P1187: PPL/40×. Rhombic particles of dolomite.

2. P1212: PPL/40×. Inclusion-rich pale brown particles of dolomite.

3. P1141: PPL/100×. Particles of colourless ankerite and orange iron oxides.

4. P1187: XPL/40×/ ● . The field of view as in Fig. 1 under crossed polars.

5. P1212: XPL/40×/ ● . The field of view as in Fig. 2 under crossed polars.

6. P1141: XPL/100×/ ● . Particles in Fig. 3 under crossed polars.

Crystallinity	Optic sign	n_ω	n_ε	δ
Dolomite: Trigonal	–ve	1.679–1.681	1.500	0.1790–0.1810
Ankerite: Trigonal	–ve	1.690–1.750	1.510–1.548	0.1820–0.2020
				(data from Deer *et al.*, 1992)

Dolomite and ankerite are indistinguishable from each other and from calcite when viewed through the polarising microscope.

Under plane-polarised light the most distinctive feature of all forms of dolomite and ankerite is the highly variable relief, which ranges from low (RI equals that of the medium) to high (RI much less than the medium); effectively the particle disappears as the stage is rotated. Ankerite has variation in relief that is slightly less extreme (from low to moderate) than that in dolomite or calcite. Dolomite and ankerite have three sets of perfect cleavage (at 60° to each other, parallel to the crystal faces), and this strongly influences the particle habit on crushing, producing rhombic particles. As the particle margins are formed by cleavage faces they are technically angular and anhedral (that is, they have straight edges but these are not crystal faces). Uneven anhedral particles also occur. Cleavage planes are frequently visible in larger particles. Some texts describe a crystal morphology that appears as having curved or saddle-shaped faces; this is rarely observed in microscopic samples. The crystals are colourless. Both minerals have strong dispersion.

Under cross-polarised light, dolomite and ankerite have high birefringence and therefore strong interference colours, typically third order or greater (increasing with crystal size). Both minerals have inclined extinction. Multiple twins are common in well-crystallised particles.

The magnesian minerals may be distinguished from calcite by using the dye Alizarin Red S; calcite will take up the dye and be stained pink, whereas a mineral such as dolomite will remain colourless (see Adams *et al.*, 1984). This is not normally a practicable technique for polarised light microscopy of pigments, although it might be used for thick sections. Calcite and dolomite may occur together in the same rock.

Ankerite occurs in hydrothermal mineral veins in association with galena, sphalerite, fluorite and baryte.

Lit.: Adams *et al.* (1984); Deer *et al.* (1992)

The Pigment Compendium

HUNTITE

$Mg_3Ca(CO_3)_4$

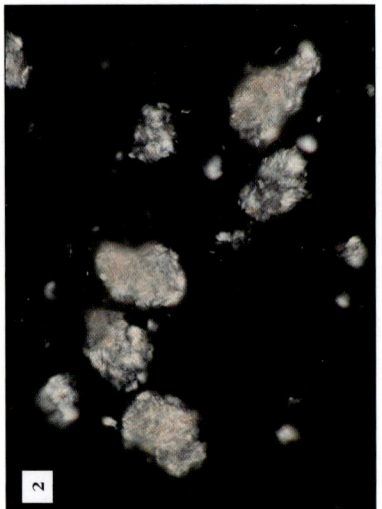

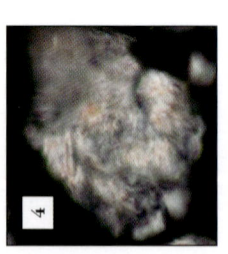

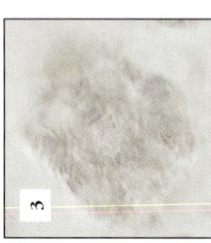

Key:

1. P1494: PPL/100×. Fibrous aggregates of huntite.

2. P1494: XPL/100×/○. The same view as in Fig. 1 under crossed polars, showing high birefringence.

3. P1494: PPL. An enlargement of particle from Fig. 1 showing crystallites.

4. P1494: XPL ○. The particle in Fig. 3 under crossed polars.

Crystallinity	Optic sign	n_ω/n_ε	δ
Trigonal	–ve	~1.625	High

(data from Deer et al., 1992)

In plane-polarised light huntite (a mineral that can resemble chalk when viewed as a large specimen) is seen as translucent, colourless crystals. It has low relief and RI just less than that of the medium. Particle size is fine. Particle morphology is seen as fibrous or feathery aggregates that resemble pom-poms. Aggregates may be medium to coarse grained. Huntite is additionally distinguishable from chalk in that it does not contain microfossils.

Under crossed polars, huntite has moderate birefringence and typically high first order and second order interference colours are observed. Extinction is sweeping and mottled.

The optical properties of huntite have been briefly described by Deer et al. (1992), though it should be noted that the fine particle size of huntite precludes accurate determination of refractive indices.

Lit.: Deer et al. (1992)

The Pigment Compendium

CALCIUM SULFATE, ANHYDRITE TYPE AND ANHYDRITE

CaSO$_4$

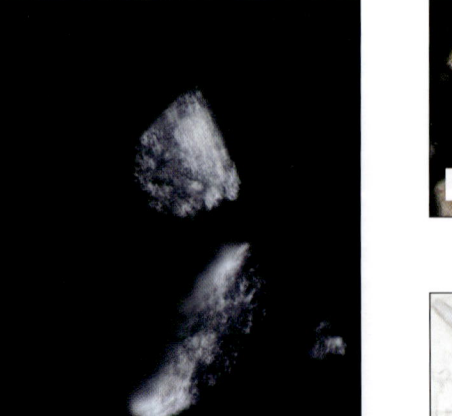

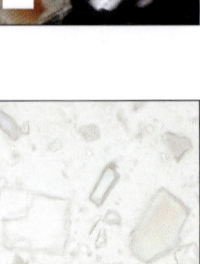

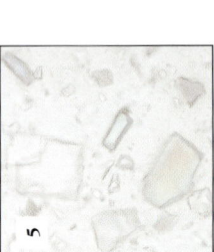

Key:

1. P0882: PPL/100X. Colourless fibrous anhydrite crystals.

2. P1140: PPL/40X/ ● . Colourless bladed anhydrite crystals; coloration observed in some grains is due to high reflectance from the crystal surface.

3. P0882: XPL/100X/ ● . The field of view as in Fig. 1 under crossed polars.

4. P1140: XPL/40X/ ● . The field of view as in Fig. 2 under crossed polars.

5. P1140: PPL/ ● . An enlarged view of a group of crystals from Fig. 2.

6. P1140: XPL/ ● . Anhydrite grains in inclined positions exhibiting up to second order interference colours.

Crystallinity	Optic sign	2V	n_α	n_β	n_γ	δ
Orthorhombic	+ve	42°–44°	1.569–1.574	1.574–1.579	1.609–1.618	0.0400–0.0450

(data for mineral from Deer et al., 1993)

Under plane-polarised light anhydrous calcium sulfate and the mineral analogue anhydrite form relatively inclusion-free crystals that therefore have 'clean' appearing particles. It has moderate and variable relief, while the dispersion is stated in the literature to be $r < v$.

There are three sets of cleavage at high angles to each other. As a result it characteristically forms rhombic, almost cubic to rectangular, tabular or platy fragments on crushing. Euhedral crystals are prismatic or sometimes fibrous and radiating. In anhydrite converted from gypsum ('dead burned'), particles may also retain morphological features of the origi-nating material. Love (1973) describes commercial products formed from gypsum where the pig-ment is ground and classified to a fine powder of less than 20 μm.

Under crossed polars, this pigment shows bright, second order interference colours, yellows, reds, blues and greens. Smaller particles show first order greys and yellows. Extinction is straight and fibrous particles are length slow.

In most contexts anhydrite is likely to be found with calcium sulfate in other hydration states as a consequence of either moisture take-up or dehydration processes acting on the hemihydrate/bassanite or gypsum forms.

Lit.: Love (1973)

CALCIUM SULFATE, BASSANITE TYPE AND BASSANITE

CaSO$_4$.xH$_2$O, where x ~ 0.5–0.8

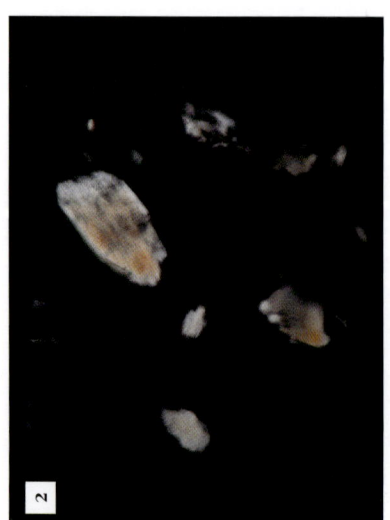

1. P0884: PPL/40×. Mottled brown bassanite particles.

2. P0884: ~XPL/40×. The field of view as in Fig. 1 under crossed polars.

Crystallinity [Uncertain]	Optic sign	n_ω	n_ε	δ
	+ve	1.55	1.57	0.02
				(data uncertain)

The partially hydrated forms of calcium sulfate, which include the naturally occurring mineral bassanite, synthetic hemihydrate and other closely similar hydration states, are optically indistinguishable. They are also very similar in appearance to the calcium sulfate dihydrate, gypsum. Birefringence is slightly higher, so that first order yellows typically occur rather than first order greys.

Significant confusion is evident in the mineral literature, with various sources documenting different crystal systems including orthorhombic, monoclinic

and pseudo-hexagonal. The only seemingly reliable optical data notes it as a uniaxial mineral.

In most contexts hemihydrate/bassanite is likely to be found with calcium sulfate in other hydration states as a consequence of moisture absorption or dehydration processes acting on the anhydrite or gypsum forms.

The Pigment Compendium

CALCIUM SULFATE, GYPSUM TYPE AND GYPSUM

$CaSO_4.2H_2O$

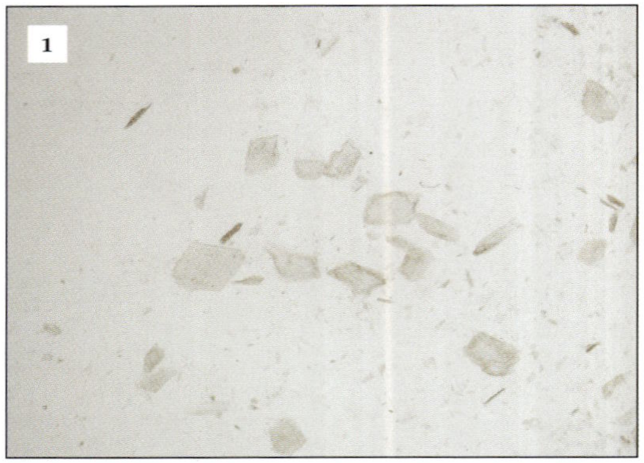

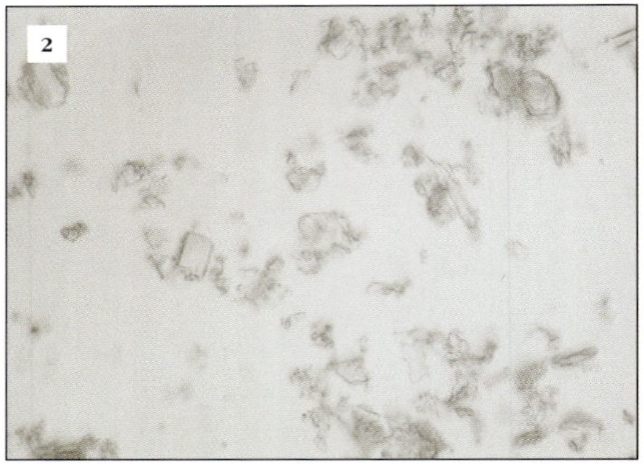

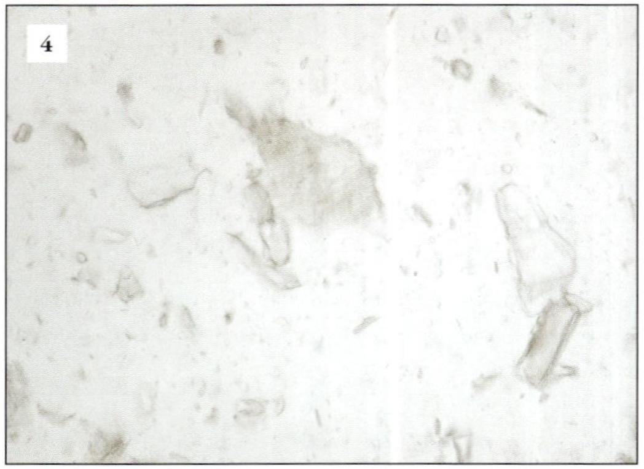

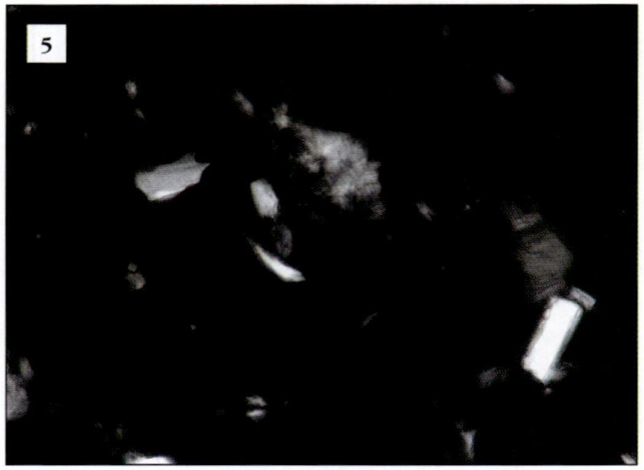

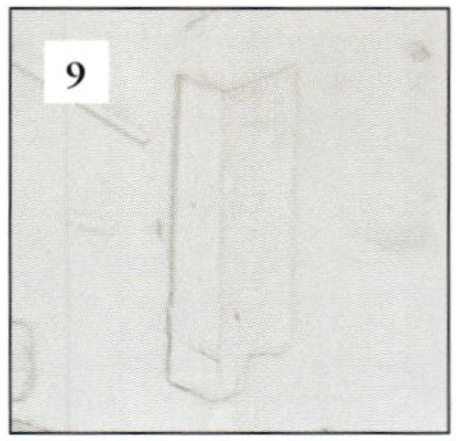

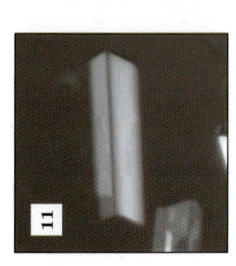

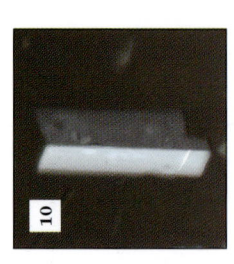

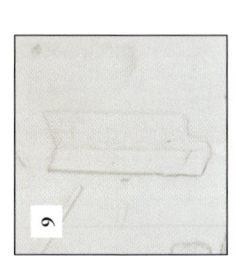

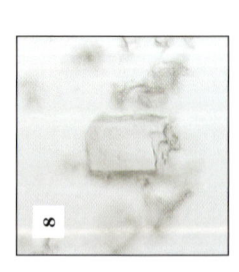

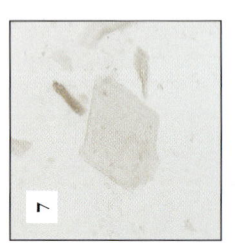

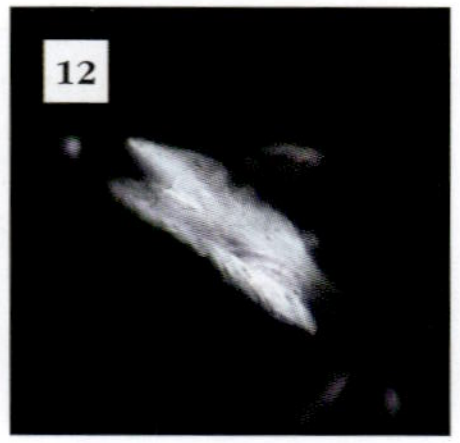

Key:

1. P0026: PPL/10×. Elongate fibrous and rhombic selenite grains.

2. P0764: PPL/40×/H. Euhedral and anhedral crystals of gypsum.

3. P0153: XPL/40×/ ●. Anhedral, fibrous gypsum crystals showing low first order interference colours.

4. P1202: PPL/40×. Colourless and mottled anhedral gypsum crystals.

5. P1202: XPL/40×/ ●. The same view as in Fig. 4 showing low first order interference colours.

6. P0883: XPL/40×. Shows low first order interference colours and euhedral swallowtail-twinned crystals.

7. P0026: PPL. Euhedral selenite rhomb.

8. P0764: PPL/H. Colourless gypsum crystal with rhombic habit.

9. P0883: PPL. Shows swallowtail-twinned crystal in plane polarised light.

10. P0883: XPL. The same crystal as in Fig. 9 under crossed polars.

11. P0883: XPL. Shows the crystal in Fig. 10 rotated through 45°.

12. P0153: XPL/ ●. Fibrous 'satin spar' crystal with low first order colours.

Crystallinity	Optic sign	2V	n_α	n_β	n_γ	δ
Monoclinic	+ve	58°–68°	1.519–1.521	1.522–1.523	1.529–1.530	0.0090–0.0100

(data from Heinrich, 1965; cf. Webmineral, 2003)

Calcium sulfate dihydrate is generally known as 'gypsum', although technically this name should only be applied to the naturally occurring form. When colourless and transparent, gypsum is also known as selenite or *marienglass*. The two varieties, however, are not clearly distinguished when viewed under the microscope. Synthetic calcium sulfate dihydrate may be prepared chemically and will also form in the hydration of calcium sulfate-based plasters ('plaster of Paris').

Under plane-polarised light, this compound has moderate relief (RI < medium) and is colourless. It has strong dispersion with $r > v$. Euhedral particles of gypsum adopt euhedral rhombic to elongate-rhombic habit, which in naturally occurring samples is generally lost during crushing. However, gypsum possesses three perfect cleavages that allow the crystals to form rhombic particles when ground. The particle morphology may also closely resemble that of the calcium carbonate minerals, calcite and dolomite (*qq.v.*); however, gypsum does not exhibit variable relief, a readily observed diagnostic feature. Alternatively, gypsum can occur in fibrous or acicular habits, when it is known as 'satin spar', or it may also adopt anhedral forms with ragged particle boundaries.

Precipitated synthetic gypsum forms well-developed euhedral crystals. Simple twinning is common in all forms of gypsum, whereby two mirror images of the crystals are effectively joined together to produce a characteristic 'swallowtail' (sometimes called 'arrowhead') habit.

All forms of gypsum contain inclusions; naturally occurring varieties are noticeably richer in inclusions than prepared varieties such that they appear 'dirty' in comparison. When these inclusions are clearly visible they are often seen to mimic the habit of the enclosing crystal, forming rhombic voids.

Under crossed polars gypsum can be seen to exhibit low birefringence, showing first order grey interference colours. It has inclined extinction. Twinning is optically apparent when one half of the crystal is in extinction while the other is illuminated. For naturally occurring examples, there is an overall mottled, 'brindled' appearance to the particles. In the fibrous and acicular habits crystals are length slow.

Gypsum may be differentiated from quartz by its lower birefringence and distinct crystal morphology.

CARBONATE HYDROXYLAPATITE

$Ca_5(PO_4,CO_3)_3(OH)$

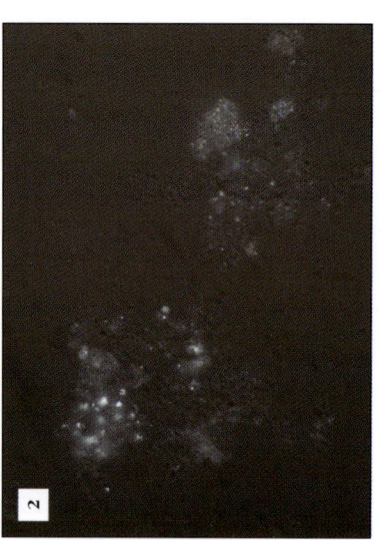

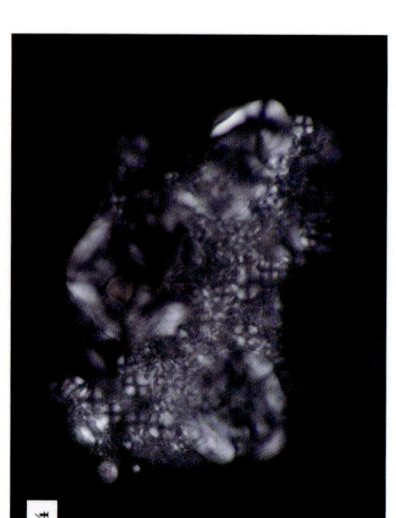

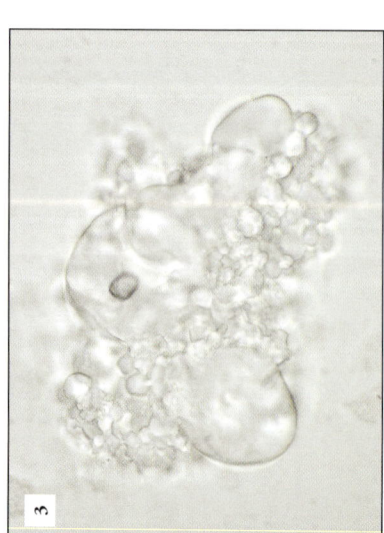

Key:

1. P0160: PPL/100×/ ⊙. Colourless and fine-grained bone ash particles.

2. P0160: ~XPL/100×. Field of view as in Fig. 1 under crossed polars, note very low birefringence.

3. P0160: PPL/100×. Spherulitic particles.

4. P0160: XPL/100×. The field of view as in Fig. 3 under crossed polars, note standing extinction crosses.

Crystallinity	**Optic sign**	n_ω	n_ε	δ
Hexagonal-monoclinic	–ve	1.629–1.667	1.624–1.666	0.001–0.007
				(data for mineral from Deer *et al.*, 1993)

This compound is normally encountered in the conversion of bone to ash; other authors refer to this under the term 'tricalcium phosphate' (McCrone *et al.*, 1973–80) and similar chemical names. Samples of bone ash examined by the authors matched X-ray diffraction patterns for hydroxylapatites.

Under plane-polarised light, bone ash forms translucent, colourless particles with low relief and RI just less than that of the medium; RI values reported in the literature specifically for bone ash were $n_\varepsilon = 1.626$ and $n_\omega = 1.629$ (McCrone *et al.*), thus falling within the range of known values for carbonate hydroxylapatite. Particle habits are of irregular to rounded, anhedral polycrystalline flakes, with clusters of moderate relief inclusions. Some grains contain circular voids; this phenomenon has also been noted by McCrone *et al.* Particle surfaces are rough. Particle size ranges from fine to coarse, though McCrone *et al.* add that the particle size is typically coarse, ranging from 15 to 75 µm.

Under crossed polars, bone ash has very low birefringence with low first order interference colours and some particles that can appear almost isotropic. Characteristically, the birefringence of bone ash is lower than that of non-calcined bone. The polycrystalline nature of the particles is clear and they have undulose or sweeping extinction.

Lit.: McCrone *et al.* (1973–80)

The Pigment Compendium

LEAD CARBONATE

PbCO$_3$

Key:

1. P0509: PPL/100×/H. Pale brown particles of lead carbonate.

2. P0509: XPL/100×/●/H. The field of view as in Fig. 1 under crossed polars.

3. P0509: RPL/100×/H. The field of view as in Fig. 1 in reflected light.

4. P0510: PPL/40×/H. Crumb-like, finely polycrystalline particles of lead carbonate.

5. P0510: XPL/40×/●/H. The field of view as in Fig. 4 under crossed polars.

6. P0510: RPL/40×/H. The field of view as in Fig. 4 in reflected light.

7. P0507: PPL/H. Detail of particle morphology of lead carbonate.

8. P0509: PPL/H. Detail of angular crystal morphology from Fig. 1.

9. P0510: PPL/H. Detail of crystal morphology from Fig. 4.

10. P0897: PPL. Euhedral hexagonal crystals.

Crystallinity	Optic sign	2V	n_α	n_β	n_γ	δ
Orthorhombic	−ve	8.5°	1.803	2.074	2.076	0.273
					(data for mineral from Deer *et al.*, 1992)	

Under plane-polarised light examples of lead carbonate, the synthetic analogue of the mineral cerussite, form translucent, colourless particles of very fine particle size, close to the resolution of the optical microscope. Consequently, they appear as rounded plates and framboidal aggregates of these. According to Dunn (1973), lead carbonate particles usually grow to a larger particle size than those of lead carbonate hydroxide, exhibiting a prismatic morphology. Relief is very high, and RI is greater than that of the medium.

Under crossed polars, lead carbonate has high birefringence and therefore third order and greater interference colours are typical. Birefringence is nearly twice that of the hydrocerussite form of lead carbonate hydroxide, though this may be difficult to judge. Extinction is complete.

Lit.: Dunn (1973)

CERUSSITE

PbCO₃

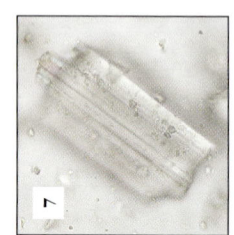

1. P1120: PPL/40×. Angular shards of cerussite.

2. P1120: XPL/40×/ ○. The field of view as in Fig. 1 under crossed polars.

3. P1120: RPL/40×. The field of view as in Fig. 1 in reflected light.

4. P1120: PPL/100×. Bladed crystal of cerussite, showing cleavage.

5. P0656: PPL/10×/H. Nineteenth century sample labelled 'lead earth' containing cerussite and opaque galena.

6. P1376: PPL/40×/H. Sample containing a mixture of 'Egyptian blue' and cerussite.

7. P1120: PPL. Bladed crystal of cerussite, showing cleavage.

8. P1120: XPL. The crystal in Fig. 4 under crossed polars.

Crystallinity	Optic sign	2V	n_α	n_β	n_γ	δ
Orthorhombic	−ve	8.5°	1.803	2.074	2.076	0.273
					(data from Deer *et al.*, 1993)	

In plane-polarised light, cerussite forms transparent colourless crystals with moderate to high, variable relief. RI varies from greater than to much greater than that of the medium. Particle surfaces appear smooth and glassy, cleavage may be visible, and many particles are fluid inclusion rich. On crushing, particles typically adopt angular, shard-like habits. Particle size varies from fine to coarse.

Under crossed polars, cerussite has very high birefringence with fourth order and above interference colours typical. Frequently, fringes of Newton's scale of colours are seen on the crystals of uneven thickness. Extinction is straight and complete.

Lit.: Deer *et al.* (1992)

LEAD CARBONATE HYDROXIDE AND HYDROCERUSSITE

$2PbCO_3.Pb(OH)_2$

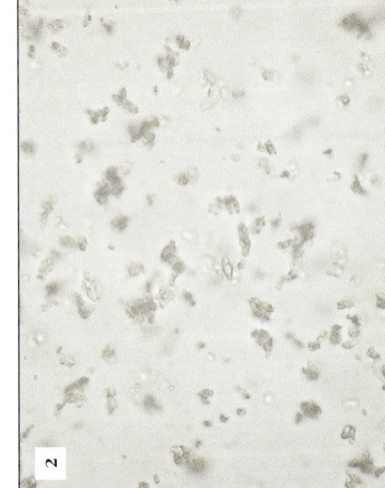

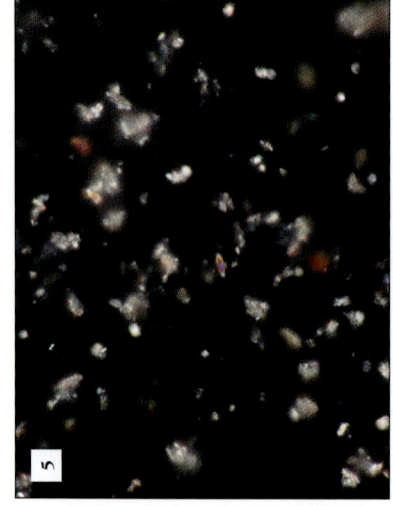

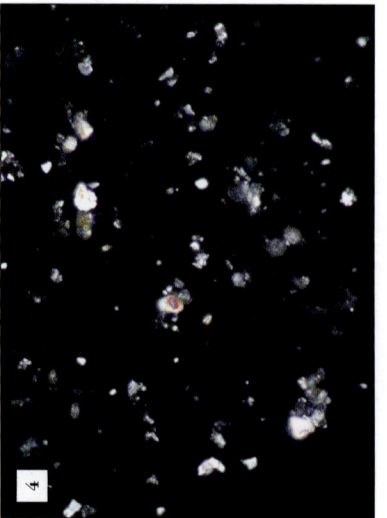

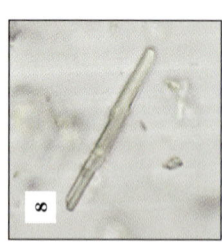

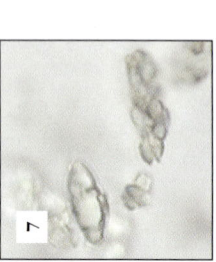

Key:

1. P0191: PPL/100×. Particles of lead carbonate hydroxide, hydrocerussite type.

2. P0813: PPL/100×. Particles of lead carbonate hydroxide, hydrocerussite type.

3. P1434: PPL/100×. Particles of lead carbonate hydroxide, hydrocerussite type.

4. P0191: XPL/100×/ ⚬. The field of view as in Fig. 1 under crossed polars.

5. P0813: XPL/100×/ ⚬. The field of view as in Fig. 2 under crossed polars.

6. P1434: XPL/100×/ ⚬. The field of view as in Fig. 3 under crossed polars.

7. P0813: PPL. Detail of typical cigar-shaped crystals.

8. P0814: PPL. Well-formed, acicular crystal.

Crystallinity	Optic sign	n_ω	n_ε	δ
Trigonal	−ve	2.09	1.94	0.15

(data from Gettens et al., 1993b)

In plane-polarised light, the synthetic lead carbonate hydroxide analogue of the mineral hydrocerussite forms colourless transparent crystals. Relief is moderate and variable and RI is higher than that of the medium. Particle shape varies according to the manufacturing process. Early preparations made by the Dutch or stack processes produced fine-to medium-grained samples, with euhedral and subhedral hexagonal plates. Other samples with finer particle size have developed acicular euhedral crystals and anhedral, polycrystalline aggregates. For pigments made by the Cremnitz process and more modern processes, particle size is very fine and close to the resolution of the optical microscope and therefore difficult to define, particles appearing rounded or bacterioid.

Under crossed polars, lead carbonate hydroxide has high birefringence, with third and fourth order interference colours typical. Many particles are shown to be finely crystalline aggregates or polycrystalline and therefore complete extinction is not achieved and particles twinkle as the stage is rotated. Plates and acicular forms demonstrate complete and straight extinction; elongated particles are length slow.

The natural mineral hydrocerussite is extremely rare and unlikely to be encountered.

Lit.: Dunn (1973); Gettens *et al.* (1993b)

LEAD CARBONATE HYDROXIDE, PLUMBONACRITE TYPE

$Pb_{10}(CO_3)_6O(OH)_6$

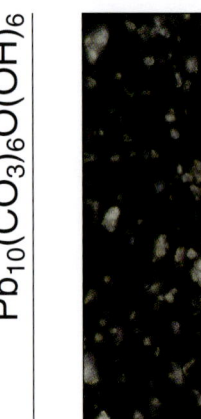

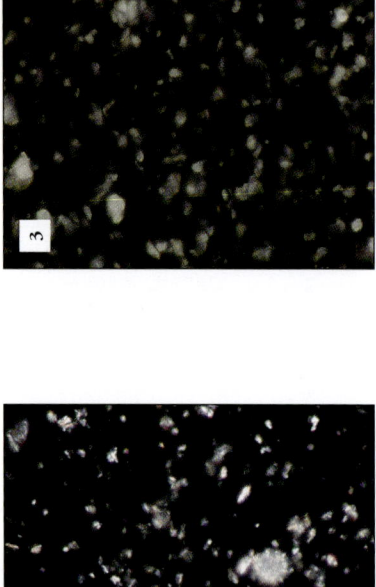

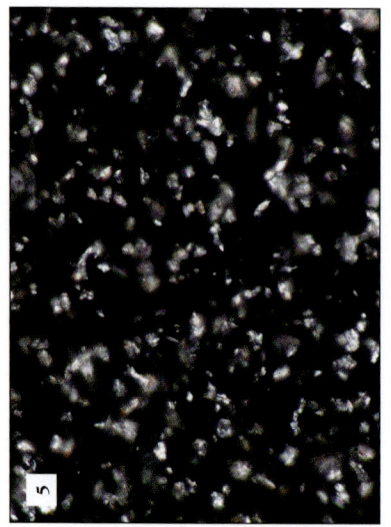

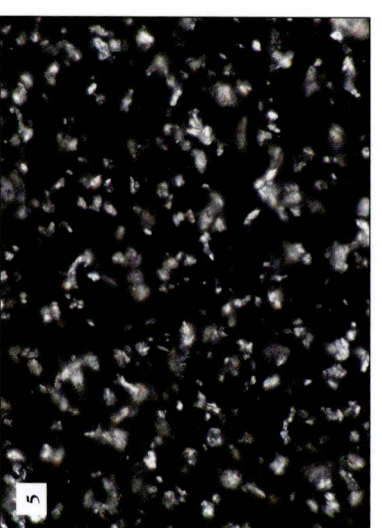

Key:

1. P0115: PPL/100×. Particles of lead carbonate hydroxide, plumbonacrite type.

2. P0115: XPL/100×. The field of view as in Fig. 1 under crossed polars.

3. P0115: RPL/100×. The field of view as in Fig. 1 in reflected light.

4. P1435: PPL/100×. Particles of lead carbonate hydroxide, plumbonacrite type.

5. P1435: XPL/100×. The field of view as in Fig. 4 under crossed polars.

Crystallinity	Optic sign	n_ω/n_ε	δ
Hexagonal	[−ve]	~2.03–2.04	High

(data for mineral from Webmineral, 2003)

In plane-polarised light, the lead carbonate hydroxide mineral plumbonacrite forms translucent, colourless crystals with high relief and an RI greater than that of the medium. Particles form euhedral and subhedral hexagonal plates and acicular crystals with pyramidal terminations. Particle size is fine.

Under cross-polarised light, plumbonacrite has high birefringence with fourth order interference colours typical. Extinction is straight and complete.

LEAD CHLORIDE HYDROXIDE

$Pb(OH)_2.PbCl_2$, $Pb_2Cl(O,OH)_{2-x}$ (where $x \sim 0.325$) and others

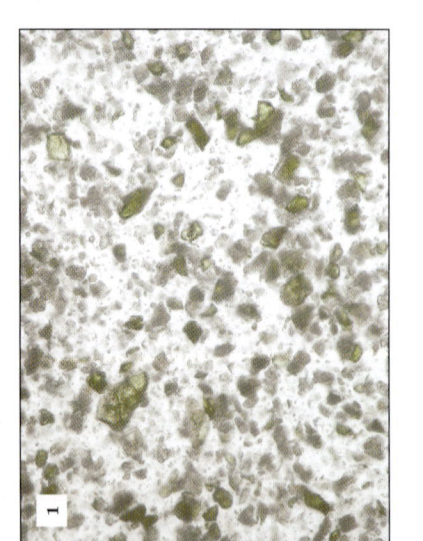

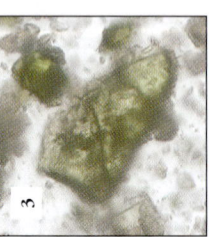

Key:

1. P0503: PPL/40×/H. Particles include deep yellow laurionite crystals.

2. P0503: XPL/40×/●/H. The field of view as in Fig. 1 under crossed polars.

3. P0503: PPL/H. Detail of laurionite particle morphology from Fig. 1.

4. P0503: PPL/H. Detail of laurionite particle morphology from Fig. 1.

Crystallinity	Optic sign	2V [-]	n_α	n_β [-]	n_γ	δ
Blixite: Orthorhombic	+ve	80	2.05	2.05	2.20	0.1500
Fiedlerite: Triclinic	–ve	[-]	1.98	2.04	2.10	0.1200
Laurionite: Orthorhombic	–ve	70–90	2.077	2.116	2.158	0.0810
Paralaurionite: Monoclinic	–ve	[-]	2.05	2.15	2.20	0.1500

(data for minerals from Webmineral, 2003)

The synthetic analogues of the lead chloride hydroxide minerals blixite ($Pb_2Cl(O,OH)_{2-x}$, where $x \sim 0.3$), fiedlerite ($Pb_3Cl_4F(OH)_2$), and laurionite ($Pb(OH)_2.PbCl_2$) have been occasionally found in a variety of pigment contexts; a polymorph of laurionite, paralaurionite, also exists but has not been identified thus far on paintings. No systematic studies of morphology of pigmentary examples by light microscopy appear to have been undertaken; however, some general observations can none-the-less be made.

Under plane-polarised light, the sample of laurionite illustrated above consists of anhedral (angular to subangular) and subprismatic crystals with a broad grain size distribution (5–70 µm). The particles are weakly pleochroic particles (yellow-green to pale green or yellow) and show variable surface character (rough and smooth). They have high relief and RI > 1.662. Under crossed-polars, the particles are anisotropic, with subprismatic grains showing straight extinction. Interference colours are masked by the crystal body colour, such that they appear green-yellow. On this basis, laurionite may be distinguished from lead chloride oxide, which exhibits anomalous interference colours in crossed-polars. In UV light, the laurionite particles show a weak green-yellow fluorescence. In the sample analysed, the laurionite

grains are associated with anhedral (subangular to subrounded) colourless cerussite and hydrocerussite (*qq.v.*) particles, which also show a broad grain size distribution (<2–50 µm). This sample was an unnamed specimen from the nineteenth century Dutch Hafkenscheid collection of pigments.

Examples of laurionite and a compound very similar to blixite have been described by Winter (1981) in the context of identifications in twelfth to fourteenth century AD Japanese paintings. While these are both noted as white in colour, the mineral blixite is yellow and a synthesis of the laurionite form carried out by Winter passed through a yellow stage during the formative process. It would seem that small variations in composition can lead to yellow to white colouration in these compounds.

Lead chloride hydroxide, fiedlerite type, has been found in studies of characteristic translucent inclusions that are known to form in lead-based paint layers. Noble and Wadum (1998) observed this compound in Rembrandt's *The Anatomy Lesson of Dr Nicolaes Tulp*, suggesting that it was a result of purification of lead white using salt water.

Lit.: Noble & Wadum (1998); Winter (1981)

The Pigment Compendium

LEAD(II) SULFATE, ANGLESITE TYPE

PbSO$_4$

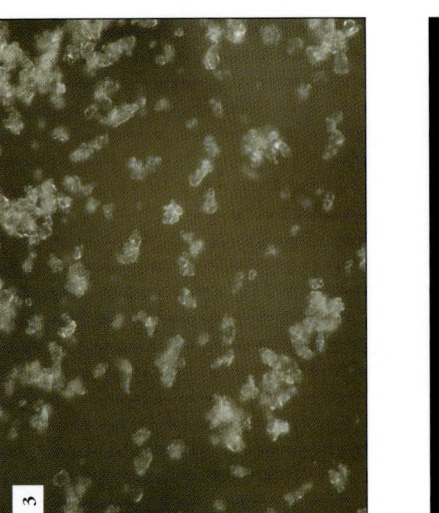

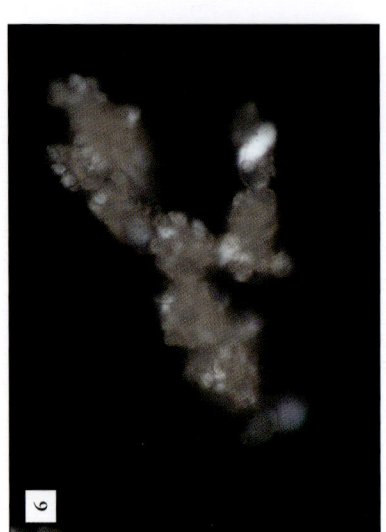

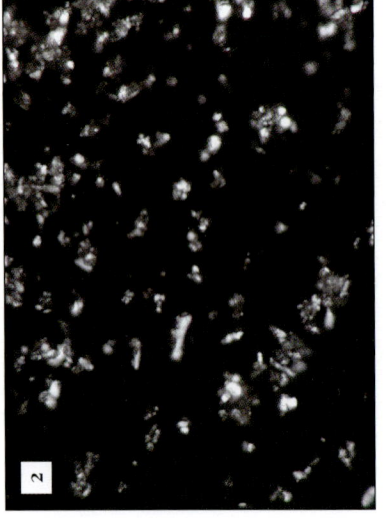

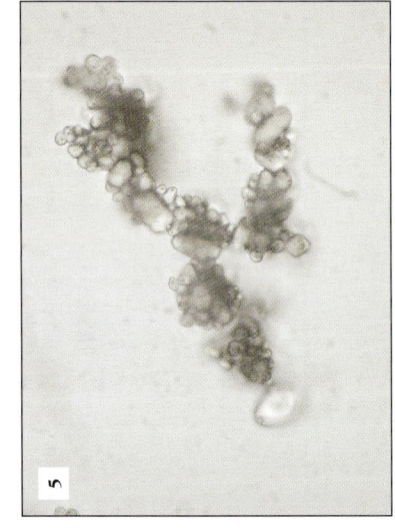

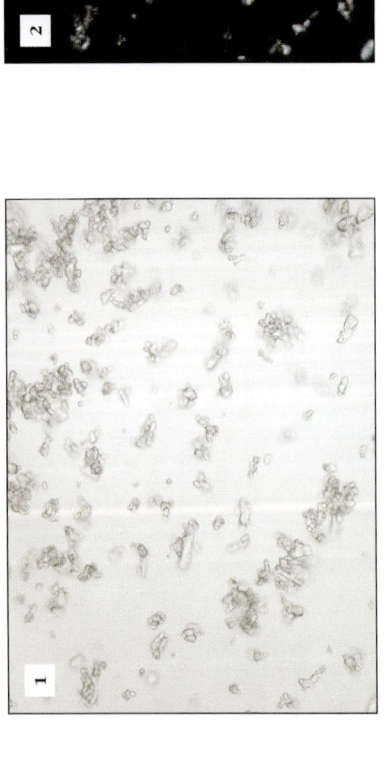

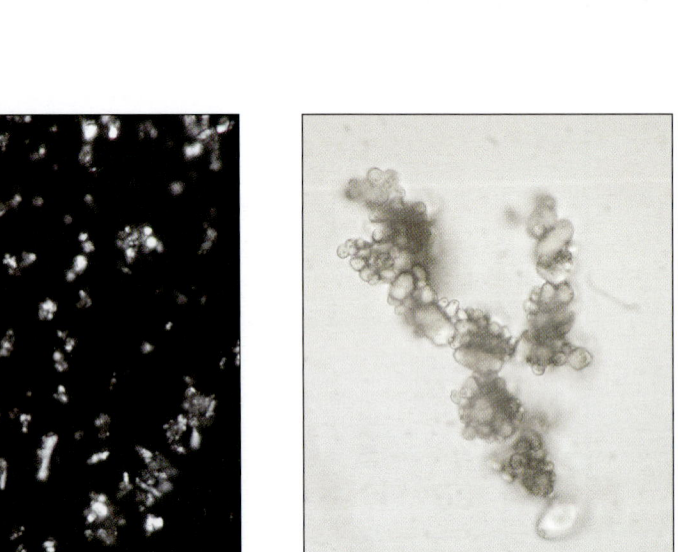

1. P1429: PPL/40×. Euhedral crystals of lead(II) sulfate.

2. P1429: XPL/40×. The field of view as in Fig. 1 under crossed polars.

3. P1429: RPL/40×. The field of view as in Fig. 1 in reflected light.

4. P1429: PPL/100×. Subhedral and anhedral crystals of lead(II) sulfate.

5. P0899: PPL/100×. Rounded subhedral crystals of lead(II) sulfate.

6. P0899: XPL/100×. The field of view as in Fig. 5 under crossed polars.

Crystallinity	Optic sign	2V	n_α	n_β	n_γ	δ
Orthorhombic	+ve	68°–75.4°	1.878	1.883	1.895	0.0170
					(data for mineral from Mason, 1968; cf. Webmineral, 2003)	

Lead(II) sulfate is the synthetic analogue of the mineral anglesite. In plane-polarised light, lead(II) sulfate forms transparent, colourless particles, with high relief and RI much greater than that of the medium. Particles shapes vary from subhedral subrounded to rounded polyhedral, equant particles to sub-rounded prismatic or cigar-shaped particles. Framboidal aggregates of particles are not uncommon. Particle surfaces are smooth, domed and very clear. A few particles contain fluid inclusions. Particle size is medium to fine and particle size distribution is relatively narrow. Dispersion is relatively strong.

Under crossed polars, lead(II) sulfate has moderate birefringence and high first order to low second order interference colours are typically observed. Particles appear very bright. Extinction is complete and elongate particles have straight extinction. In samples examined, although elongated particles were not well formed, they appear to be length slow.

Appearing a white colour in reflected light, this compound also had a weak yellow to yellow-white fluorescence at the edges of particles with UV excitation.

The Pigment Compendium

LEAD SULFATE (OTHER TYPES)

[Various]

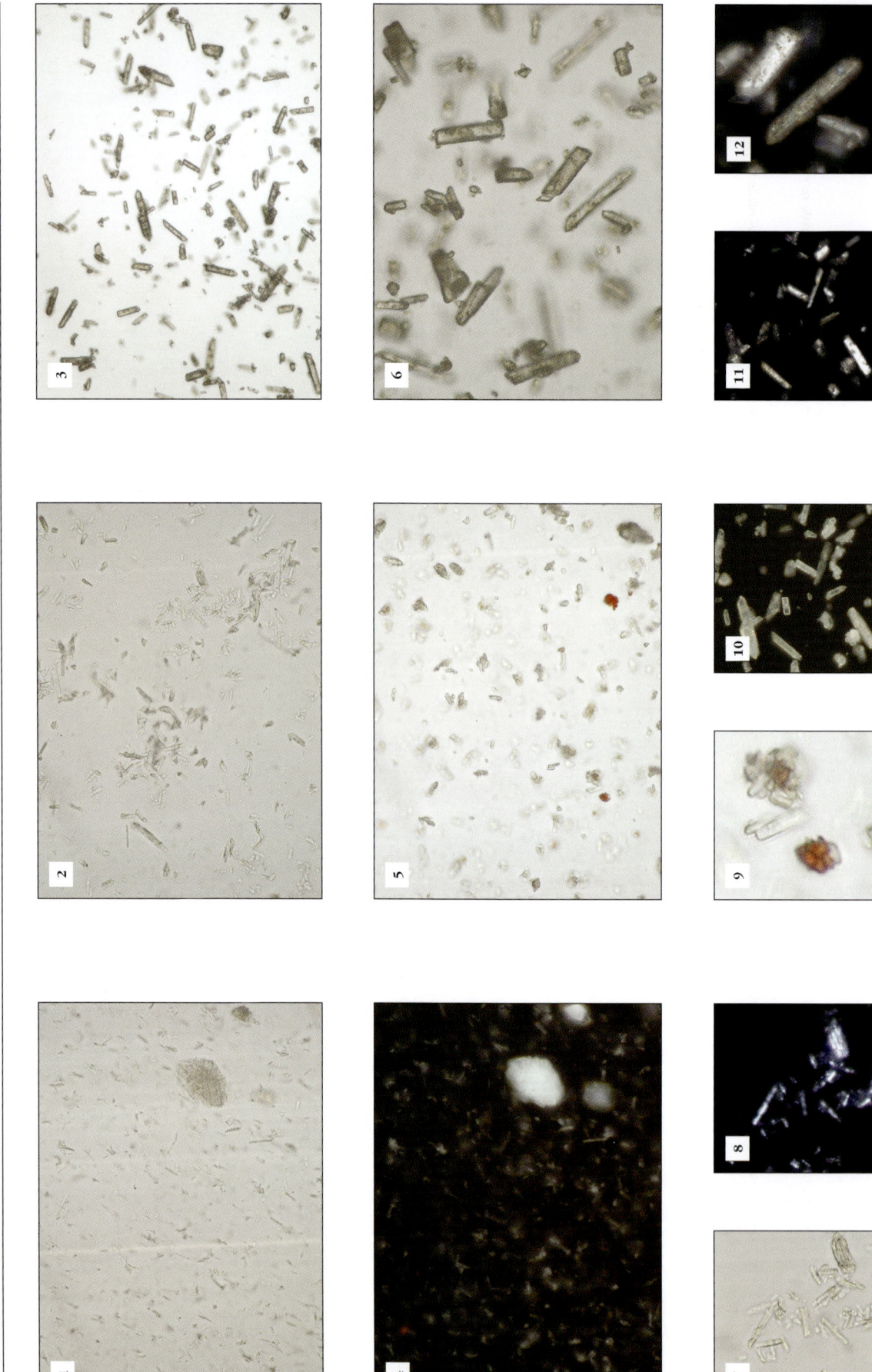

Key:

1. P1433: PPL/100×. Acicular particles and aggregate of monobasic lead sulfate.

2. P0116: PPL/100×. Bladed and acicular crystals of tribasic lead sulfate.

3. P1427: PPL/40×. Bladed, pale brown crystals of tetrabasic lead sulfate.

4. P1433: RPL/100×. The field of view as in Fig. 1 in reflected light.

5. P1437: PPL/100×. Aggregates and lath-shaped crystals of tribasic lead sulfate, includes red impurities.

6. P1427: PPL/100×. Euhedral bladed crystals of tetrabasic lead sulfate.

7. P0116: PPL. Detail of crystal morphology from Fig. 2.

8. P0116: XPL/●. Field of view in Fig. 7 under crossed polars.

9. P1437: PPL. Detail of crystal morphologies from Fig. 5.

10. P1427: RPL. Detail of crystals from Fig. 3 in reflected light.

11. P1427: XPL/●. The field of view in Fig. 10 under crossed polars.

12. P1427: XPL/●. Detail of crystals from Fig. 6 under crossed polars.

Crystallinity

	n	δ
PbSO$_4$.PbO: Monoclinic	$n_\alpha = 1.928$; $n_\beta = 2.007$; $n_\gamma = 2.036$	0.1080
PbSO$_4$.3PbO.H$_2$O: [Anisotropic]	>1.662	High
PbSO$_4$.4PbO: [Anisotropic]	>1.662	High
Pb$_4$O$_3$SO$_4$.H$_2$O	[Unknown]	[Unknown]
Pb$_4$(CO$_3$)$_2$(SO$_4$)(OH)$_2$: Trigonal	~1.96	[Unknown]

(data from Webmineral, 2003 and authors)

Under plane-polarised light, 'monobasic' lead sulfate, the synthetic analogue of the mineral lanarkite (PbSO$_4$.PbO; mineral also given as Pb$_2$(SO$_4$)O), forms transparent, colourless particles with low relief and RI just greater than that of the medium. In samples examined particles are euhedral, with finely acicular crystals; particle size is uniformly fine. Under crossed-polars, monobasic lead sulfate has low birefringence and low first order grey interference colours are observed. Particles have complete and straight extinction.

Under plane-polarised light, 'tribasic' lead sulfate (PbSO$_4$.3PbO.H$_2$O) forms transparent, colourless crystals with moderate relief and RI greater than that of the medium. Two particle morphologies are present in the sample shown here: acicular and lath-shaped crystals, with pyramidal terminations and finely polycrystalline aggregates. The latter are almost opaque. Particle size ranges from fine to medium. Under crossed polars, this form of lead sulfate has high birefringence and third and fourth order birefringence colours are typical. Acicular particles show complete and straight extinction. White in reflected light, this compound also had a dull yellow fluorescence with UV excitation.

Under plane-polarised light, 'tetrabasic' lead sulfate (PbSO$_4$.4PbO) forms colourless to pale yellow translucent crystals, with high relief and RI much greater than that of the medium. Particles are euhedral tabular and lath-shaped crystals, some with pyramidal terminations. Particle surfaces appear rough and etched and crystals are rich in fluid inclusions. Particle size ranges from fine to medium. Under crossed polars, this form of lead sulfate has high birefringence and fourth order and above interference colours are observed. Particles undergo complete and straight extinction. A yellow-white colour in reflected light, this compound also had a weak yellow to yellow-white fluorescence at the edges of particles with UV excitation.

A modern commercial lead sulfate pigment examined by the authors was found to be an intimate mixture of three phases: monobasic lead sulfate (as described above), lead oxide sulfate hydrate (Pb$_4$O$_3$SO$_4$.H$_2$O) and a lead carbonate sulfate hydroxide analogue of the mineral susannite Pb$_4$(CO$_3$)$_2$ (SO$_4$)(OH)$_2$. Differentiation of the phases optically is difficult.

The Pigment Compendium

MAGNESITE

MgCO$_3$

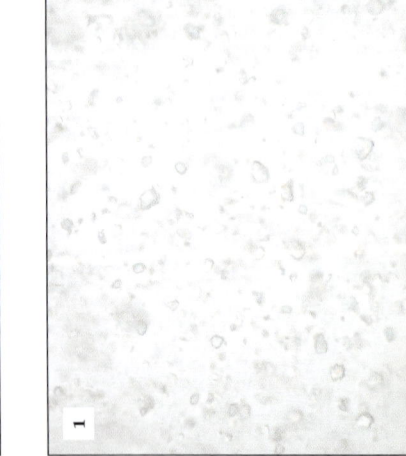

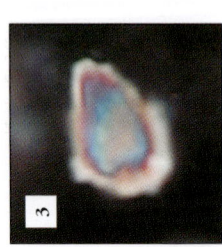

Key:
1. P1224: PPL/100×. Colourless particles of magnesite.
2. P1224: XPL/100×/⊙. The field of view as in Fig. 1 under crossed polars.
3. P1224: XPL ⊙. Detail of crystal from Fig. 2 showing birefringent crystal fringes.

Crystallinity	Optic sign	n_ω	n_ε	δ
Trigonal	–ve	1.700–1.782	1.509–1.563	0.190–0.218

(data from Deer *et al.*, 1992)

In plane-polarised light, magnesite forms translucent, colourless crystals. Relief is highly variable from moderate to very low with RI ranging from below that of the medium to above that of the medium. It should also be noted that refractive index increases dramatically with the degree of substitution of magnesium by other ions (Deer *et al.*, 1992). Particle morphology ranges from subhedral, hexagonal to anhedral plates. The majority of particles are angular and irregular. Particle surfaces are rough. Despite the derivation of magnesite from natural sources, particle size may be fine; particle size distribution may range from fine to very broad.

Under crossed polars, magnesite has high birefringence and fourth order and greater interference colours are typical. Many particles show concentric fringes of interference colours. Extinction is complete and straight.

Magnesite is difficult to distinguish optically from other carbonate minerals including calcite, dolomite and rhodochrosite. It is often found in metamorphosed ultrabasic rocks, in association with talc, chlorite and the serpentine group clay minerals.

Lit.: Deer *et al.* (1992)

The Pigment Compendium

TIN(IV) OXIDE AND CASSITERITE

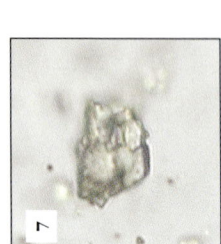

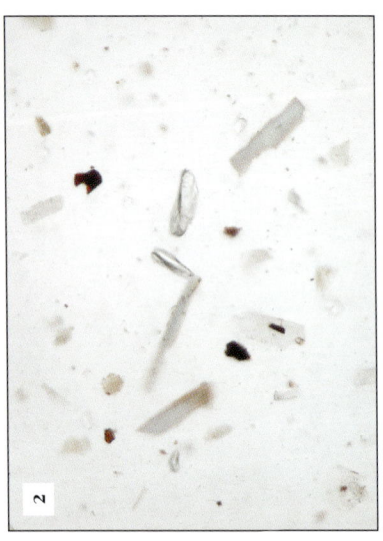

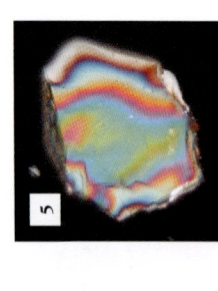

Key:

1. P1164: PPL/40×. Angular particles of cassiterite.

2. P1164: PPL/40×. High relief grains of cassiterite (centre) with low relief, blue and brown-coloured tourmaline.

3. P1164: XPL/40×. ⦿ The field of view as in Fig. 1 under crossed polars.

4. P1164: PPL. Detail of cassiterite particle.

5. P1164: XPL. ⦿ The crystal in Fig. 4 under crossed polars.

6. P1164: PPL. Cassiterite particles from Fig. 2 showing conchoidal fractures.

7. P1042: PPL. Particle of tin(IV) oxide from a sample of lead tin yellow.

Crystallinity	Optic sign	n_ω	n_ε	δ
Tetragonal	+ve	2.000–2.006	2.097–2.100	0.0940–0.0970

(data for mineral from Mason, 1968; cf. Webmineral, 2003)

In plane-polarised light tin(IV) oxide forms translucent colourless to dirty brown, almost opaque, particles. Colour is patchy and varies across particles. The relief is moderate to high, with an RI greater than that of the medium. Particle surfaces are distinctively rough and pitted. Particle size is moderate to coarse. Under crossed polars, particles are isotropic but have strong white or blue-white internal reflections.

Tin(IV) oxide may be encountered in certain synthetic pigments based on tin, notably lead tin oxide ('lead tin yellow') and related compounds as well as cobalt tin oxide ('cerulean blue').

Under plane-polarised light, cassiterite forms particles with very high relief and an RI that is much greater that that of the medium. Slight differences in RI have been reported in the literature; for example, McCrone *et al.* (1973–80) give $n_\varepsilon = 2.093$ and $n_\omega = 1.997$. Colour is variable and ranges from colourless to pale yellow and brown, some examples may show colour zoning. In strongly coloured particles, pleochroism is visible; McCrone *et al.* note that this is colourless for the ω direction, a salmon-pink to deep red for the ε direction. Particle shapes are typically angular shards, influenced in shape by the conchoidal fracture and one good cleavage, although McCrone *et al.* suggest that the cleavage is not visible in samples they examined. Euhedral particles have prismatic and acicular habits.

Under crossed polars, cassiterite has high birefringence. High third order and above interference colours are observed, often showing fringes of Newton's scale of colours on irregular-shaped particles. Cassiterite has straight extinction and acicular crystals are length fast.

Cassiterite is typically found with minerals including quartz, wolframite, tourmaline, topaz, fluorite and the micas (Deer *et al.*, 1992).

Lit.: Deer *et al.* (1992); McCrone *et al.* (1973–80)

The Pigment Compendium

TITANIUM(IV) OXIDE, ANATASE TYPE

TiO$_2$

Key:

1. P1484: PPL/100×. Angular pale brown translucent particles of titanium(IV) oxide, anatase type.

2. P1484: XPL/100×/ ○. The field of view as in Fig. 1 under crossed polars.

3. P1484: RPL/100×. The field of view as in Fig. 1 in reflected light.

4. P0046: PPL/100×. Titanium(IV) oxide, anatase type used as a substrate for an organic green dye, 'nettle green'.

5. P0046: ~XPL/100×. The field of view as in Fig. 4 under crossed polars.

6. P0045: PPL/100×. Titanium(IV) oxide, anatase type used as a substrate for an organic yellow dye.

7. P1484: PPL. Particle morphology of grains from Fig. 1.

8. P1484: PPL. Uneven and irregular particle morphologies from Fig. 1.

9. P1484: PPL. Particle morphologies from Fig. 1.

10. P1484: XPL/ ○. The grains from Fig. 9 under crossed polars, note that the grain margins are reflective but the grain interiors show very low first order interference colours.

11. P0045: PPL. An enlargement of grains from Fig. 6 showing particle morphology.

Crystallinity	Optic sign	n_ω	n_ε	δ
Tetragonal	−ve	2.54–2.55		0.0730
				(data from Laver, 1997)

In plane-polarised light, anatase-type titanium(IV) oxide ('titanium dioxide white') may be seen as translucent, yellow-brown particles that also appear white in reflected light. Particle size is very fine and close to that resolvable by the optical microscope, therefore particles invariably appear as rounded or bacterioid plates. Kampfer (1973) has published electron micrographs of the anatase form which shows a morphology that is generally cuboid or spheroid, with a particle size lying in the region of 0.2 µm. Relief in this pigment is high and the RI is much greater than the medium.

Under crossed polars the anatase variety of titanium dioxide white has high birefringence, but the extreme fineness of the particle size means that only low orders of interference colours are observed. The pigment also shows moderate grey internal reflections. The pigment is white in reflected light.

There is some fluorescence of this pigment from UV excitation; with excitation at 340 nm there is a strong emission peak centred around 540 nm, giving it a greenish colour (Laver, 1997). Samples examined by the authors gave weak white fluorescence under UV excitation.

Differentiation of the anatase from the rutile form of titanium(IV) oxide by optical means is difficult. However, careful observation of the birefringence (especially when directly judged against suitable reference samples) may permit the two forms to be separated since the value of δ for rutile is some four times that of anatase.

Pigments based on this compound may additionally contain barium sulfate or calcium sulfate as well as antimony(III) oxide ('antimony white'). The analyst should also be aware of the extent of production of coated titanium dioxide pigments, a technology that is both complex and sophisticated. Titanium dioxide whites have been reviewed by Laver (1997), including a discussion of microscopical appearance.

Lit.: Kampfer (1993); Laver (1997)

The Pigment Compendium

ANATASE

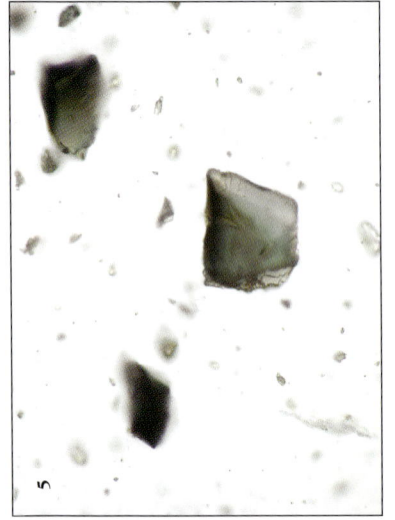

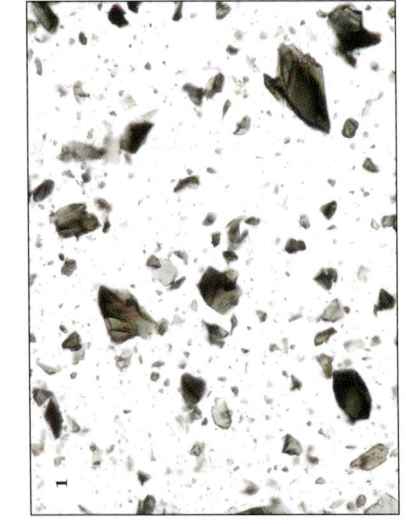

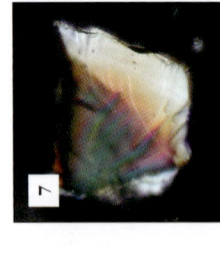

Key:

1. P1159: PPL/40×. Crushed angular anatase particles.

2. P1159: PPL/100×. Group of angular anatase particles.

3. P1159: PPL/100×. The group of particles in Fig. 2 rotated through 90°. Note change in body colour due to pleochroism.

4. P1159: XPL/40× ⭘ The same field of view as in Fig. 1 under crossed polars.

5. P1159: PPL/100×. An anhedral grain of anatase.

6. P1159: PPL. An enlargement of the grain in Fig. 5 showing conchoidal fractures.

7. P1159: XPL. The grain seen in Fig. 6 under crossed polars.

Crystallinity	**Optic sign**	n_ω	n_ε	δ
Tetragonal	−ve	2.561	2.488	0.0730

(data from Mason, 1968; cf. Deer et al., 1992)

In plane-polarised light, anatase forms translucent blue-grey to olive green crystals, which are pleochroic in these colours. Relief is high with RI much greater than that of the medium. Particle shapes in a pigment context are typically angular shards. Euhedral prismatic forms may be present. Particle surfaces have a glassy appearance and well-developed conchoidal fractures. Particle sizes may range from fine to coarse.

Under crossed polars anatase has high birefringence; however, the third and fourth order interference colours are strongly masked by the body colour, with anomalous bright blues sometimes observed. Extinction is inclined and may be complete or undulose. Coarse-grained particles of anatase may show twins.

In reflected light anatase is a weak yellow-brown colour. When viewed through the Chelsea filter, anatase appears green.

Anatase can be differentiated from rutile (q.v.) on the basis of their respective optic signs, anatase being negative and rutile positive; differentiation of the titanium(IV) oxide mineral brookite can be made on the basis of that mineral being biaxial.

Anatase, as a naturally occurring mineral analogue of one of the titanium dioxide whites, has been reviewed by Laver (1997).

Lit.: Deer et al. (1992); Laver (1997)

The Pigment Compendium

TITANIUM(IV) OXIDE, RUTILE TYPE

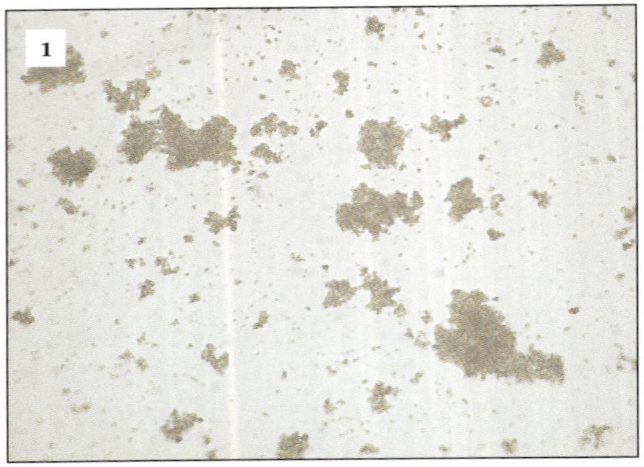

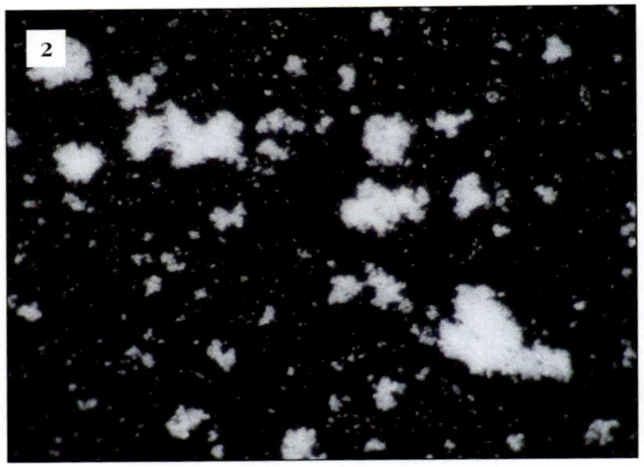

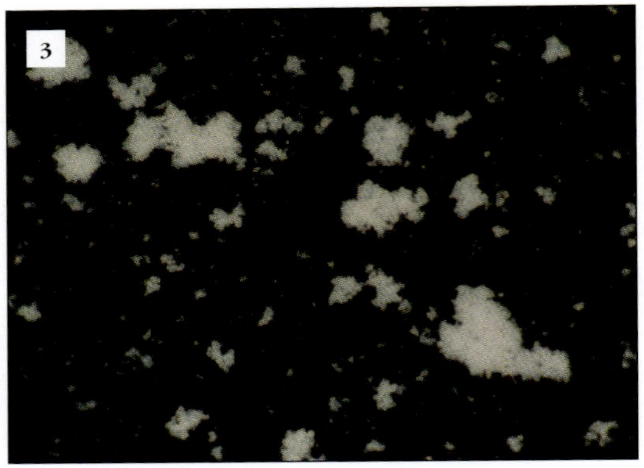

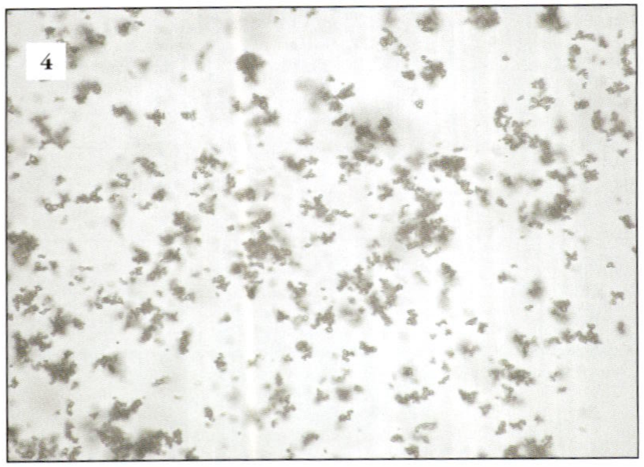

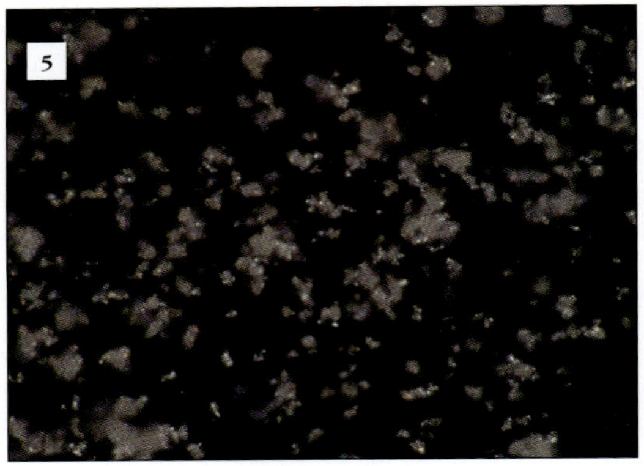

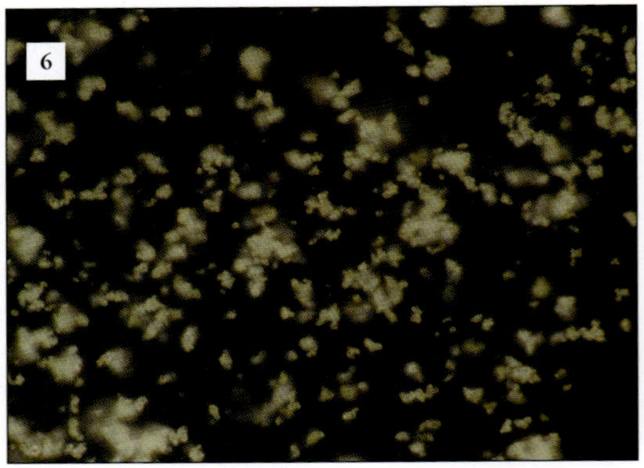

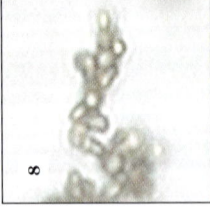

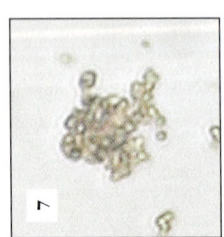

1. P0118: PPL/100×. Fine-grained aggregates and particles of synthetic rutile.

2. P0118: XPL/100×. ○. The field of view as in Fig. 1 under crossed polars.

3. P0118: RPL/100×. The field of view as in Fig. 1 in reflected light.

4. P0921: PPL/100×. Fine-grained particles of synthetic rutile

5. P0921: XPL/100×. The field of view as in Fig. 4 under crossed polars.

6. P0921: RPL/100×. The field of view as in Fig. 4 in reflected light.

7. P0118: PPL. Detail of particle morphology from Fig. 1.

8. P0921: PPL. Detail of particle morphology from Fig. 4.

Crystallinity	Optic sign	n_ω	n_ε	δ
Tetragonal	+ve	2.71–2.72		0.29

(data from Laver, 1997)

In plane-polarised light rutile-type titanium(IV) oxide ('titanium dioxide white') forms translucent, yellow-brown particles that also appear white in reflected light. Relief is high and RI is greater than the medium. Particle size is very fine and close to that resolvable by the optical microscope, therefore particles appear at best as rounded or bacterioïd plates. Kampfer (1973) has published electron micrographs of the rutile form that shows a morphology that is generally cuboid or spheroid with the particle size lying in the region of 0.2 μm, plus some rod-shaped particles that have an aspect ratio in the range 2:1 to 5:1.

Under crossed polars, the rutile variety of titanium dioxide white has extreme birefringence and high order interference colours are observed. Additionally this pigment can have strong, buff-coloured internal reflections, though this is not necessarily observed. Laver (1997) has commented that the fine particle size can mean that preparations have the appearance of a 'foggy cloud' that shows only low-level birefringence. The pigment is white in reflected light, although it has a more yellow cast than the anatase form of the pigment.

There is no functional fluorescence of this pigment from UV excitation; the principal UV excitation wavelength for rutile is 375 nm, but it emits only weakly from this at around 815 nm (Laver). Samples examined by the authors appeared grey or weakly white under UV excitation.

Differentiation of the anatase from the rutile form of titanium(IV) oxide by optical means is difficult. However, careful observation of the birefringence (especially when directly judged against suitable reference samples) may permit the two forms to be separated since the value of δ for rutile is some four times that of anatase.

Titanium dioxide white may additionally contain barium sulfate or calcium sulfate. The analyst should also be aware of the extent of production of coated titanium dioxide pigments, a technology that is both complex and sophisticated. Titanium dioxide whites have been reviewed by Laver (1997), including a discussion of microscopical appearance.

Lit.: Kampfer (1973); Laver (1997)

The Pigment Compendium

RUTILE

TiO$_2$

Key:

1. P1253: PPL/40×. Crushed angular shards of rutile, the range of colours present is due to the weak pleochroism.

2. P1253: XPL/40×. The same field of view as in Fig. 1 under crossed polars.

3. P1253: PPL/100×. Coarse grained anhedral particles.

4. P1253: RPL/40×. The same field of view as in Fig. 1 in reflected light.

5. P1253: RPL/40×. The same field of view as in Fig. 1 showing UV fluorescence.

6. P1253: PPL/○. Conchoidal fracture surfaces on rutile particle.

7. P1253: XPL/○. The crystals from Fig. 6 under crossed polars, slightly rotated.

8. P1253: PPL. Crystals of rutile, shows variation of colour within grains.

9. P1253: XPL. The grains in Fig. 8 under crossed polars.

Crystallinity	Optic sign	n_ω	n_ε	δ
Tetragonal	+ve	2.605–2.613	2.899–2.901	0.286–0.296

(data from Deer et al., 1992)

Under plane-polarised light, rutile forms transparent to translucent yellow-brown crystals that may be weakly pleochroic. Rutile has high relief and RI much greater than that of the medium. Rutile has strong dispersion, so refractive index varies with the light source used. Particle shapes range from euhedral prismatic forms, or more frequently angular shards produced by crushing. Particle surfaces vary from smooth and glassy appearing to those with conchoidal fractures. Some particles may be striated. Particle size may range from fine to very coarse.

Under crossed polars, rutile has very high birefringence with high fourth order and greater interference colours typical. These are somewhat masked by the high body colour. Particles have straight extinction. Simple twinning is common in rutile, but evidence of this may be lost on grinding.

This mineral appears orange in reflected light; some weak dull brown fluorescence was observed under UV excitation. When viewed through the Chelsea filter, rutile appears yellow.

Rutile can be differentiated from anatase (q.v.) on the basis of their respective optic signs, anatase being negative and rutile positive; differentiation of the titanium(IV) oxide mineral brookite can be made on the basis of that mineral being biaxial.

Rutile may be found in a great many rock types, typically in association with quartz and/or ilmenite. It is also found widely in earth pigments (Duval, 1992; Laver, 1997). Rutile, as a naturally occurring analogue of titanium dioxide whites, has been reviewed by Laver (1997).

Lit.: Deer et al. (1992); Duval (1992); Laver (1997)

ZINC(II) OXIDE

ZnO

1. P0190: PPL/100×. Fine-grained particles of zinc(II) oxide.

2. P0190: XPL/100×. The field of view as in Fig. 1 under crossed polars.

3. P0190: UV/100×. The field of view as in Fig. 1 in UV illumination.

4. P0190: RPL/100×. The field of view as in Fig. 1 in reflected light.

5. P0762: PPL/10×/H. Platy and crumb-like aggregates of zinc(II) oxide.

6. P0762: XPL/10× ● /H. The field of view as in Fig. 5 under crossed polars.

7. P0762: PPL/ ● /H. Detail of polycrystalline particle morphology from the sample in Fig. 5.

8. P0762: XPL/H. The field of view in Fig. 7 under crossed polars.

9. P0762: UV/H. The field of view in Fig. 7 showing UV fluorescence.

10. P0762: RPL/H. The field of view in Fig. 7 in reflected light.

Crystallinity Hexagonal	**Optic sign** +ve	n_ω 1.984–2.065	n_ε 2.000–2.081	δ 0.016
				(data from Winchell, 1931)

In plane-polarised light, zinc(II) oxide ('zinc white') forms transparent to translucent colourless or pale yellow-coloured particles. Particle size is typically fine or very fine, and particle size distribution is fairly narrow. As a consequence of particle size being close to that resolvable with the optical microscope, particles typically appear rounded or bacterioid. However, particle morphology is strongly affected by manufacturing process. Historical samples may be fine grained and in such examples, subhedral hexagonal plates and prisms may be discernible. Kühn (1986) has also noted, in addition to what he terms 'nodular' zinc oxide, the occurrence of acicular or 'boule'-shaped forms, often with three or four needles radiating from a single point. Occasionally fragments of this acicular form may be seen as curved particles, while particles may also clump together as irregular, polycrystalline aggregates. McCrone *et al.* (1973–80) and Mactaggart (2002) have sought to differentiate European and American manufacturing processes on the basis of occurrence of these morphologies in samples. However, as Kühn has pointed out, the particle size cf zinc(II) oxide is very fine and morphology is thus best studied by scanning electron microscopy; it would appear therefore that separation according to manufacturing process is probably an unresolved issue. Typical particle sizes for the 'nodular' form are in the range 0.25–1 μm, though there are apparently finer speciality grades; acicular particles may exhibit lengths of a few microns or more. Microphotographs of zinc(II) oxide have been published by Nelson (1940) and Vandemaele (1965).

Despite the RI implying that this pigment should have high relief, that observed in samples examined by the authors appeared low. RI is greater than that of the medium. Zinc(II) oxide has strong dispersion and so relief varies with differing light sources.

Under crossed polars, zinc(II) oxide has low birefringence and first order interference colours are typical. Very fine examples appear almost isotropic. Coarser particles are seen to be polycrystalline and show standing extinction crosses; elongation in these particles is length slow. Zinc(II) oxide has moderately strong grey-white internal reflections. It appears white or grey-white in reflected light.

The fluorescence of zinc(II) oxide under UV is well established, though the form and extent in individual cases is affected by the presence of impurities and admixtures. The fluorescence colour is a yellow-green, the peak emission lying around 520 nm. Zinc(II) oxide is also a strong UV absorber with an absorption edge lying in the 380 to 400 nm region, a property that can affect the colour of the pigment.

Zinc(II) oxide is commonly found in association with other white pigments such as lead carbonates and sulfates, barium sulfate and titanium dioxide whites. Commercial zinc oxides may also contain small amounts of zinc sulfate, sulfide or chloride (Kühn, 1986).

The chemical, physical and optical properties of this pigment have been reviewed by Kühn (1986).

Lit.: Kühn (1986); Mactaggart (2002); McCrone *et al.* (1973–80); Nelson (1940); Vandemaele (1965); Winchell (1931)

The Pigment Compendium

ZINC SULFIDE AND SPHALERITE

ZnS

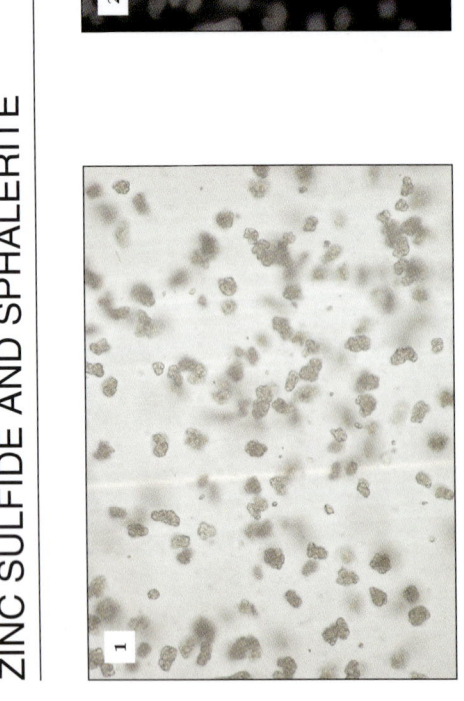

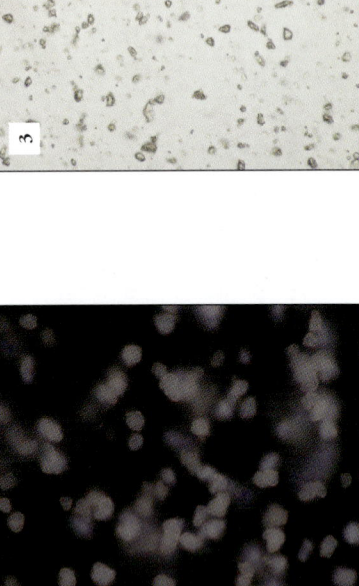

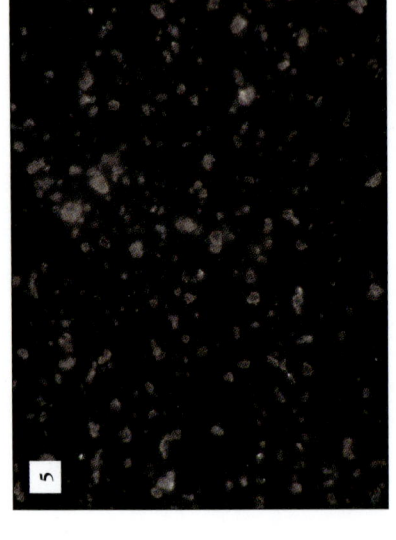

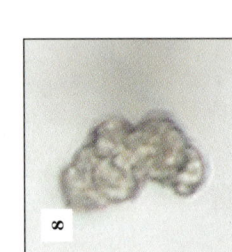

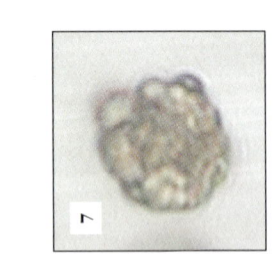

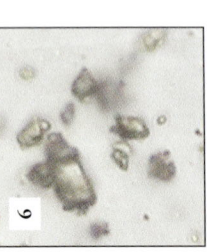

Crystallinity
Cubic

δ
Isotropic

n
2.37

(data for mineral from Deer et al., 1992)

In plane-polarised light, both sphalerite (Zn, Fe)S and the close synthetic analogue zinc sulfide form transparent, colourless or very pale yellow crystals. Relief is high and RI is much greater than that of the medium. Natural and synthetic compounds are readily distinguished by their particle size. Natural sphalerite typically has broad particle size distributions ranging from fine to coarse. Synthetic preparations on the other hand have a uniform, fine grain size. Particle morphologies for sphalerite are anhedral, angular shards, with glassy surfaces while morphologies for synthetic zinc sulfide are fine framboidal grain aggregates. Mineral sphalerite has equant dodecahedral cleavage that may lead to rhomb-shaped or tetrahedral fragments. Grassmann and Clausen (1973) have published an electron photomicrograph of a commercial synthetic zinc sulfide pigment that shows it to consist of spherical particles approximately 0.25–0.4 µm in diameter.

Sphalerite and its synthetic analogue are cubic and therefore isotropic under crossed polars; however, the crystals have yellow internal reflections. McCrone et al. (1973–80) note that some mineral crystals may show

extremely weak birefringence; these particles have sweeping extinction. This may be as a result of substitution or strain in the crystal.

Examples of the mineral and the synthetic analogue both appear a pale yellow-white colour in reflected light. Dull yellow to yellow-white fluorescence is observable under UV excitation.

The most common context in which synthetic zinc sulfide is likely to be encountered is *lithopone*, a co-precipitate of barium sulfate and zinc sulfide often used as a white extender pigment. An electron micrograph of lithopone published by Buxbaum (1998) illustrates the intimate mixture achieved in lithopones as well as a difference in particles size; the barium sulfate phase in that case had a mean particle diameter of 1.0 µm and the zinc sulfide 0.3 µm.

Lit.: Buxbaum (1998) 75; Deer et al. (1992); Grassmann & Clausen (1973); McCrone et al. (1973–80)

ZINC SULFATE HEPTAHYDRATE

ZnSO$_4$·7H$_2$O

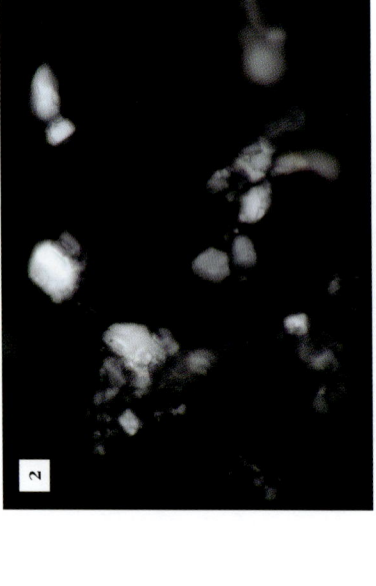

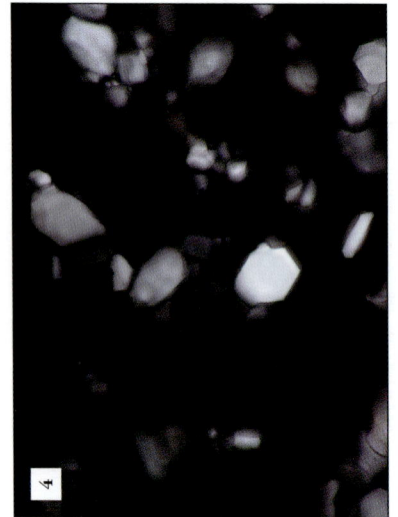

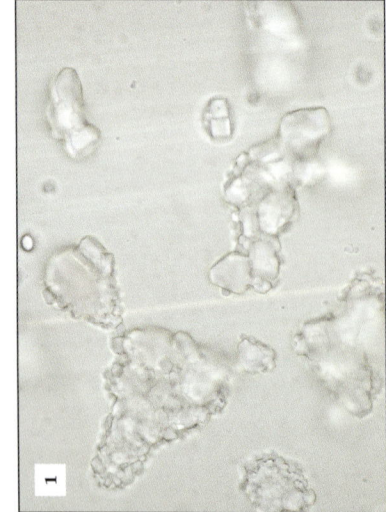

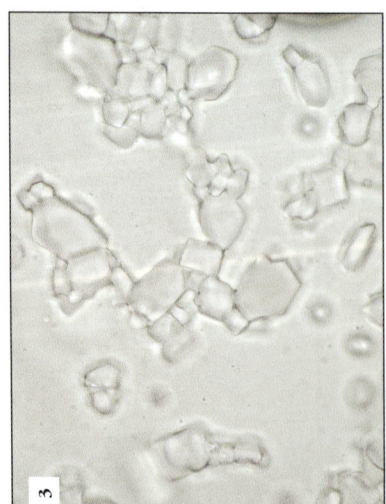

Crystallinity

Anisotropic

n	δ
<1.662	Low
	(data from authors)

In plane-polarised light zinc sulfate heptahydrate can be observed as transparent colourless crystals with moderate relief and an RI less than that of the medium. Crystals are typically euhedral, polyhedral plates, arranged in aggregates that have the appearance of bubbles in foam. Particle surfaces are smooth and some particles contain fluid inclusions. Particle size ranges from medium to coarse in material examined, though this property is probably more variable.

Under crossed polars, zinc sulfate heptahydrate has low birefringence with first order grey colours typical. The particles have straight and complete extinction.

Weak white reflections can be observed from some surfaces in reflected light; there was no observable UV fluorescence for examples studied by the authors.

The Pigment Compendium

QUARTZ

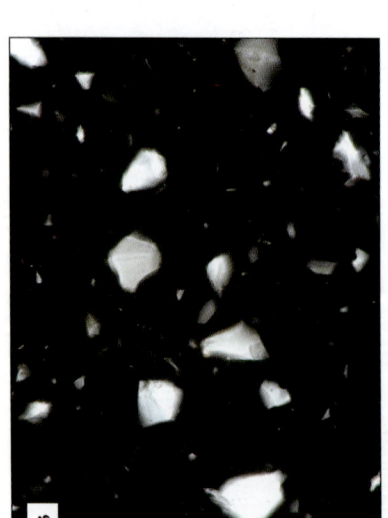

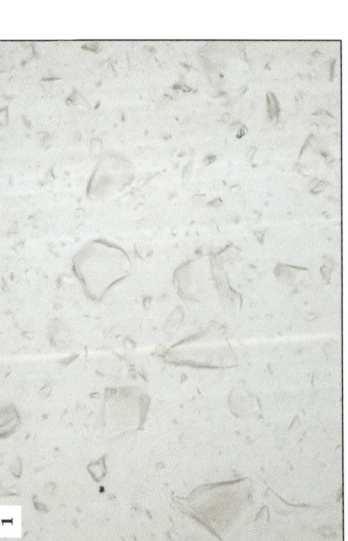

Key:
1. P1136: PPL/40×. Anhedral particles of quartz, note curved grain boundaries due to conchoidal fracture.

2. P0025: PPL/40×. Crushed quartz.

3. P1136: XPL/40×. ⭕ The field of view as in Fig. 1 under crossed polars.

4. P0025: XPL/40×. The field of view as in Fig. 2 under crossed polars.

Crystallinity	Optic sign	n_ω	n_ε	δ
Trigonal	+ve	1.544	1.553	0.0090
				(Rutley, 1988)

Quartz is a very common rock-forming mineral and may be observed as a naturally occurring minor component of mineral pigments or may have been added as a filler or even as a substrate for dyes. Quartz occurs as a wide variety of varicoloured minerals; however, when ground, it appears white and is always colourless when viewed using the polarising microscope.

In plane-polarised light, quartz has moderate to low relief (RI is less than the medium). In pigment contexts, crystals or massive samples of quartz will have been ground, and consequently, euhedral crystals will be unlikely. In rare cases subhedral fragments of hexagonal prismatic forms may be present. Quartz is normally colourless with few internal features, though mineral and fluid inclusions may be present. The latter are common and appear as small bubbles within the crystal. Quartz does not possess a cleavage but has a well-developed conchoidal fracture that is evident on most particles. Particle size and size range of pigmentary quartz can be highly variable (Katic, 1973).

Under crossed polars, quartz has low birefringence and first order greys through yellows are typical (the latter only in relatively coarse samples, >30 μm). Finely ground samples are almost isotropic and basal sections are isotropic. Quartz has straight extinction, but suitable crystallographic sections may be difficult to identify. In samples from some geological sources extinction may be sweeping.

So-called 'amorphous' silicas (in fact cryptocrystalline rather than amorphous) also exist; these may have gross particle sizes in the range 1–100 μm, but the small size of individual internal crystallites mean that they appear isotropic (Patton, 1973a).

Lit.: Katic (1973); Patton (1973a); Rutley (1988)

The Pigment Compendium

DIATOMITE

SiO$_2$(am)

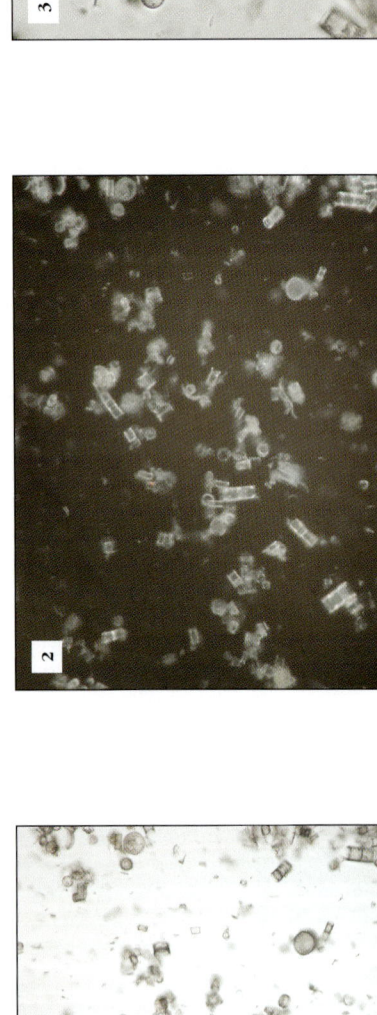

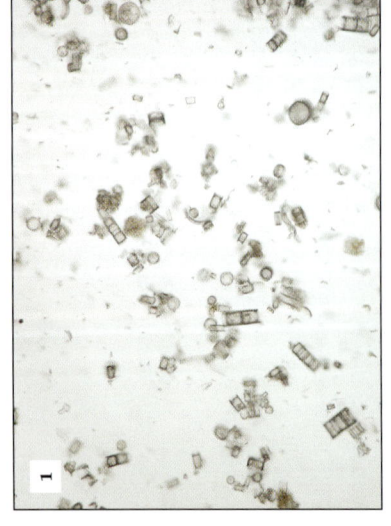

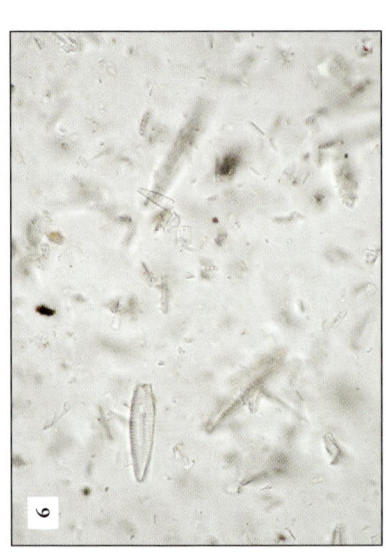

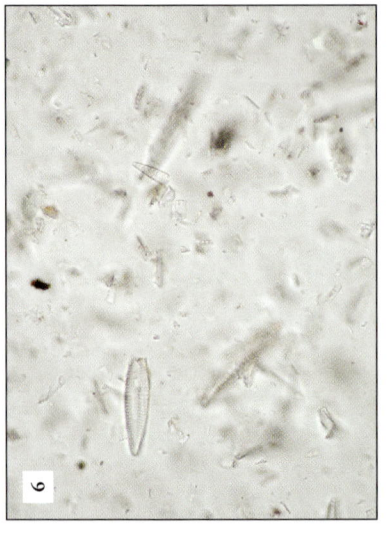

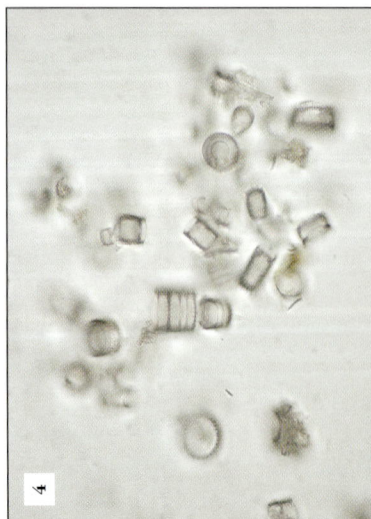

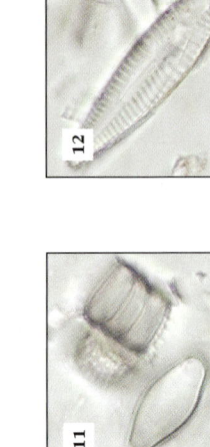

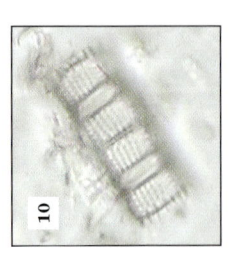

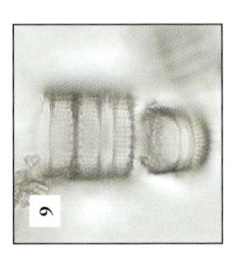

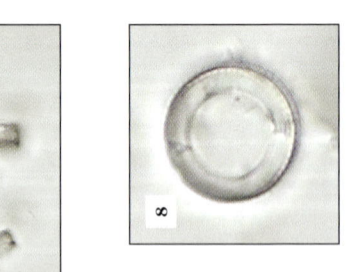

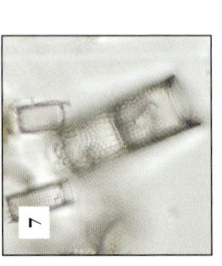

1. P0169: PPL/40×. Centric and colonial centric diatoms.

2. P0169: RPL/40×. The field of view as in Fig. 1 in reflected light.

3. P0169: PPL/100×. An enlarged view of the area in the centre of Fig. 1.

4. P0169: PPL/100×. Central and colonial centric diatoms.

5. P1377: PPL/100×/H. Pennate diatoms and fragments of diatoms and sponge spicules.

6. P1377: PPL/100×/H. Pennate diatoms and fragments of diatoms and sponge spicules.

7. P0169: PPL. Enlarged view of centric diatoms showing ornamented frustules.

8. P0169: PPL. Circular section of centric diatom.

9. P0169: PPL. Ornamented frustules of colonial centric diatoms.

10. P1184: PPL. Ornamented frustules of colonial centric diatoms.

11. P1184: PPL. Pennate and colonial centric diatoms.

12. P1377: PPL/H. Pennate diatom, the central groove is the raphe.

Crystallinity	n	δ
Amorphous	<1.662	0.0

Diatomites are rocks composed almost entirely of the tests ('frustules') of marine or lacustrine microorganisms of the Bacillariophyta, known as diatoms. Unusually, these unicellular phytoplankton secrete siliceous tests that are opaline (however, it should be noted that in fossilised varieties, the structure has reverted to that of quartz).

In plane-polarised light, diatoms have a clearly biological form. The most common varieties are 'pennate' or 'centric'. Pennate diatoms have bilateral symmetry, and are formed of two valves, while centric forms are radially symmetrical and form various circular and cylindrical forms. The frustules are variously ornamented with grooves, perforations (*puncta*), nodules and ribs. Identification of individual diatom species is based on the shape and ornamentation of the frustules and is an area of research in itself. Pigmentary grades of diatomite will have widely varying particle size and size range; they are also likely to contain a high proportion of fragments due to processing (Kranich, 1973).

In plane-polarised light, the diatom frustules are clearly observed and distinctive. They are transparent and colourless with moderate relief and RI less than that of the medium. Diatoms are of variable size, but most particles are medium to coarse. Siliceous sponge spicules may also present in marine-derived diatomites.

Being formed of opal (which is amorphous) diatoms are isotropic under cross-polarised light.

Lit.: Kranich (1973)

PALYGORSKITE GROUP CLAY MINERALS: PALYGORSKITE

$(Mg,Al)_2Si_4O_{10}(OH) \cdot 4H_2O$

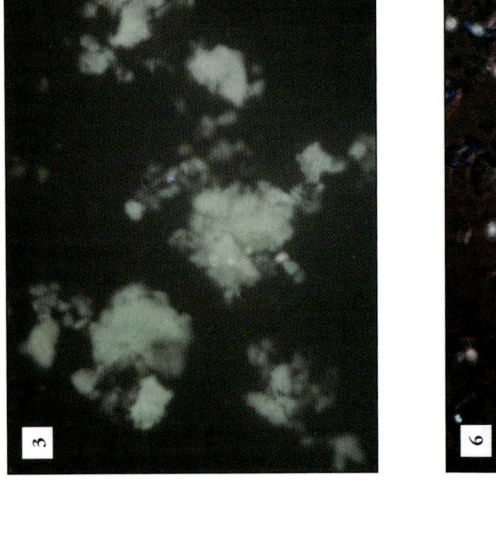

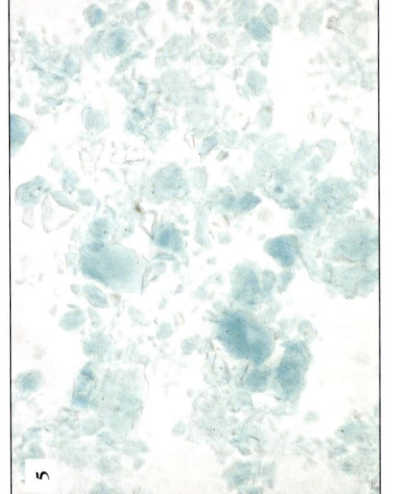

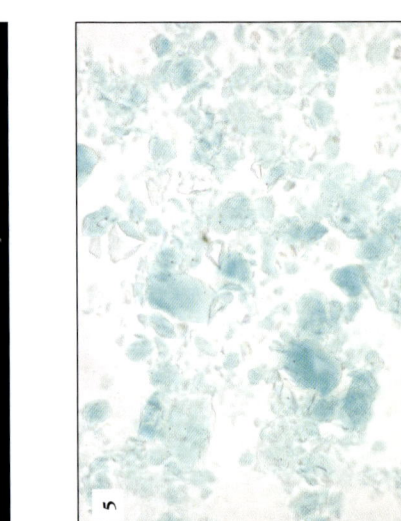

Key:

1. P0170: PPL/40×. Aggregates of palygorskite with higher relief grains of calcite.

2. P0170: XPL/40×. The same field of view as in Fig. 1 under crossed polars.

3. P0170: UV/40×. The same field of view as in Fig. 1 in UV illumination. Note the calcite does not fluoresce.

4. P0860: PPL/40×. Aggregates of palygorskite with higher relief grains of calcite. Note the brownish coloration of some particles.

5. P0181: PPL/100×. Maya blue with naturally occurring calcite particles.

6. P0181: ~XPL/100× ○. The same field of view as in Fig. 5 under crossed polars, note the anomalous blue interference colours. The white birefringent grains are calcite.

7. P0181: PPL. Showing an enlarged particle of Maya blue.

Crystallinity	Optic sign	2V	n_α	n_β	n_γ	δ
Monoclinic	−ve	30°–61°	1.522–1.528	1.530–1.546	1.533–1.548	0.020

(data from Bradley, 1940)

In plane-polarised light, natural palygorskite (formerly known as attapulgite) occurs as colourless, translucent particle aggregates. Pleochroism has been reported in coloured varieties as pale yellow to pale yellow-green. When combined with indigo (*q.v.*) in the preparation of the structural composite pigment known as *Maya blue* it can acquire colour ranging from a slightly greenish shade to a bright, almost turquoise blue colour. Aggregate particles often have fibrous or platy habits. Relief is moderate and RI is less than that of the medium. Particle surfaces appear rough and fibrous. Particle size varies from very fine to coarse and broad particle size distributions are frequently present. Haden (1973) describes pigmentary grades as consisting of needles about 1 μm in length and 0.01 μm in thickness.

Under crossed polars, palygorskite has low birefringence and very low first order interference colours are present. The finely polycrystalline nature of the particles is clear and complete extinction of particles cannot be attained. When formed as a composite with indigo, the colour of the dye masks the birefringence and blue interference colours are seen.

The mineral appears a weak white colour on some surfaces in reflected light. Variable levels of fluorescence were observed with UV excitation; where present, the colour was white.

Palygorskite is often found in association with calcite. Detrital minerals, such as quartz, may also be present. Analysts should be reminded that individual clay minerals, due to their fine particle size, are very difficult to differentiate using optical microscopy.

Lit.: Bradley (1940); Haden (1973)

The Pigment Compendium

KAOLINITE GROUP CLAY MINERALS: KAOLINITE

$Al_2Si_2O_5(OH)_4$

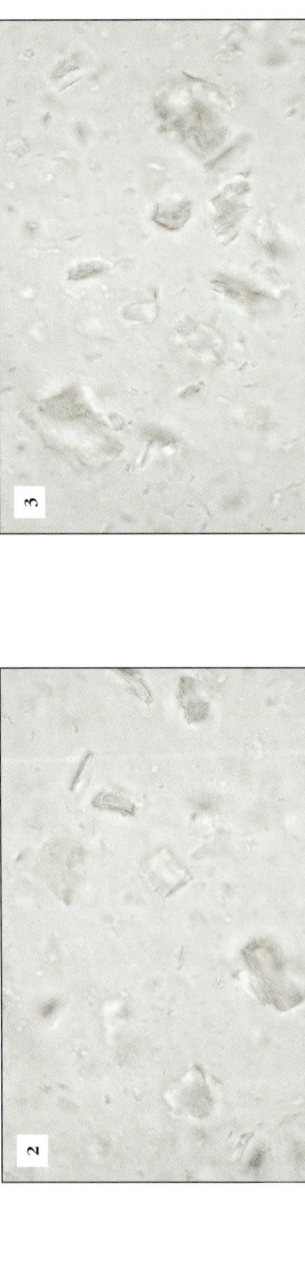

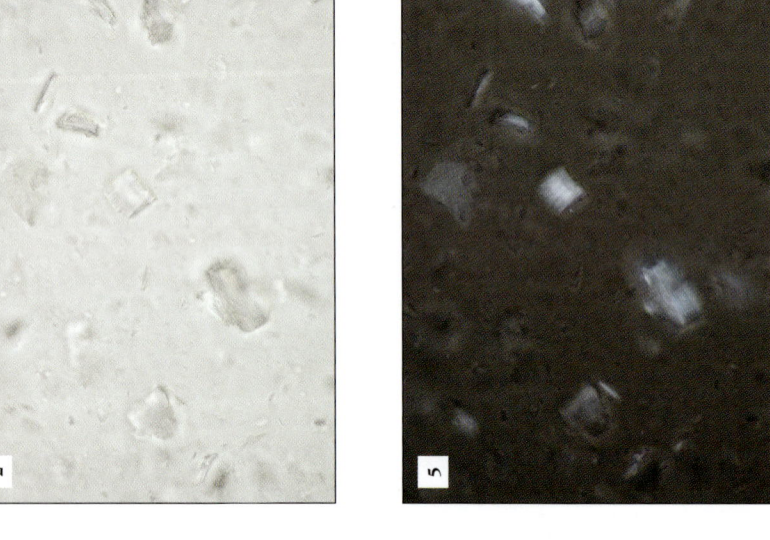

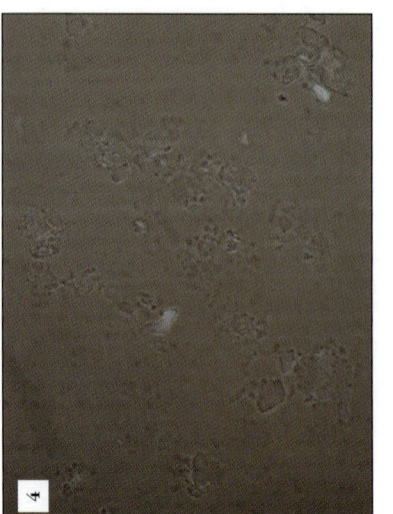

Key:

1. P0150: PPL/100×. Anhedral plates of kaolinite.

2. P1216: PPL/100×. Anhedral and lamellar particles of kaolinite.

3. P1216: PPL/100×. Lamellar particles of kaolinite.

4. P0150: ~XPL/100×. The same view as in Fig. 1 under partially crossed polars.

5. P1216: ~XPL/100×. The same view as in Fig. 2 under crossed polars.

6. P1216: ~XPL/100×/●. The same view as in Fig. 3 under crossed polars.

7. P1216: PPL. Kaolinite 'book' from Fig. 2.

8. P1216: ~XPL. The particle in Fig. 7 under crossed polars.

Crystallinity	Optic sign	2V	n_α	n_β	n_γ	δ
Triclinic/Monoclinic	−ve	24°–50°	1.553–1.565	1.559–1.569	1.560–1.570	0.006

(data from Deer *et al.*, 1992)

In plane-polarised light, kaolinite forms translucent, colourless particles. Relief is moderate to low and RI is less than that of the medium. Particles are polycrystalline aggregates that often occur as plates or lamellar 'books' (the latter case is particularly true for the kaolinite mineral dickite). Particle shapes are typically anhedral, with surfaces that are slightly rough in appearance. Particle size distribution is broad, with particles ranging from very fine to coarse. However, in commercial pigments processing may lead to a range of properties; washed pigments have particle sizes within controlled ranges between 0.1 and 25 μm, while those formed by delamination give highly laminar plates perhaps 0.25 μm thick and up to 10 μm in diameter (Brooks and Morris, 1973).

Under crossed polars kaolinite has low birefringence and very low first order interference colours are present. The finely polycrystalline nature of the particles is clear and complete extinction of particles cannot be attained. Many of the particles exhibit mottled or sweeping extinction.

Kaolinite appears a weak white colour in reflected light. There does not appear to be any fluorescence with UV excitation.

Kaolinite may be found in association with other clay minerals (particularly nacrite and dickite) as well as many other minerals including muscovite, quartz, tourmaline and iron oxides. Analysts should be reminded that individual clay minerals, due to their fine particle size, are very difficult to differentiate using optical microscopy.

During pigment processing kaolin may be calcined, a process that leads to loss of hydroxyl groups in the clay and a consequent change of structure. Particle sizes are larger, probably due to fusion of the original clay plates (Brooks and Morris).

Lit.: Brooks & Morris (1973); Deer *et al.* (1992)

KAOLINITE GROUP CLAY MINERALS: NACRITE

$Al_2Si_2O_5(OH)_4$

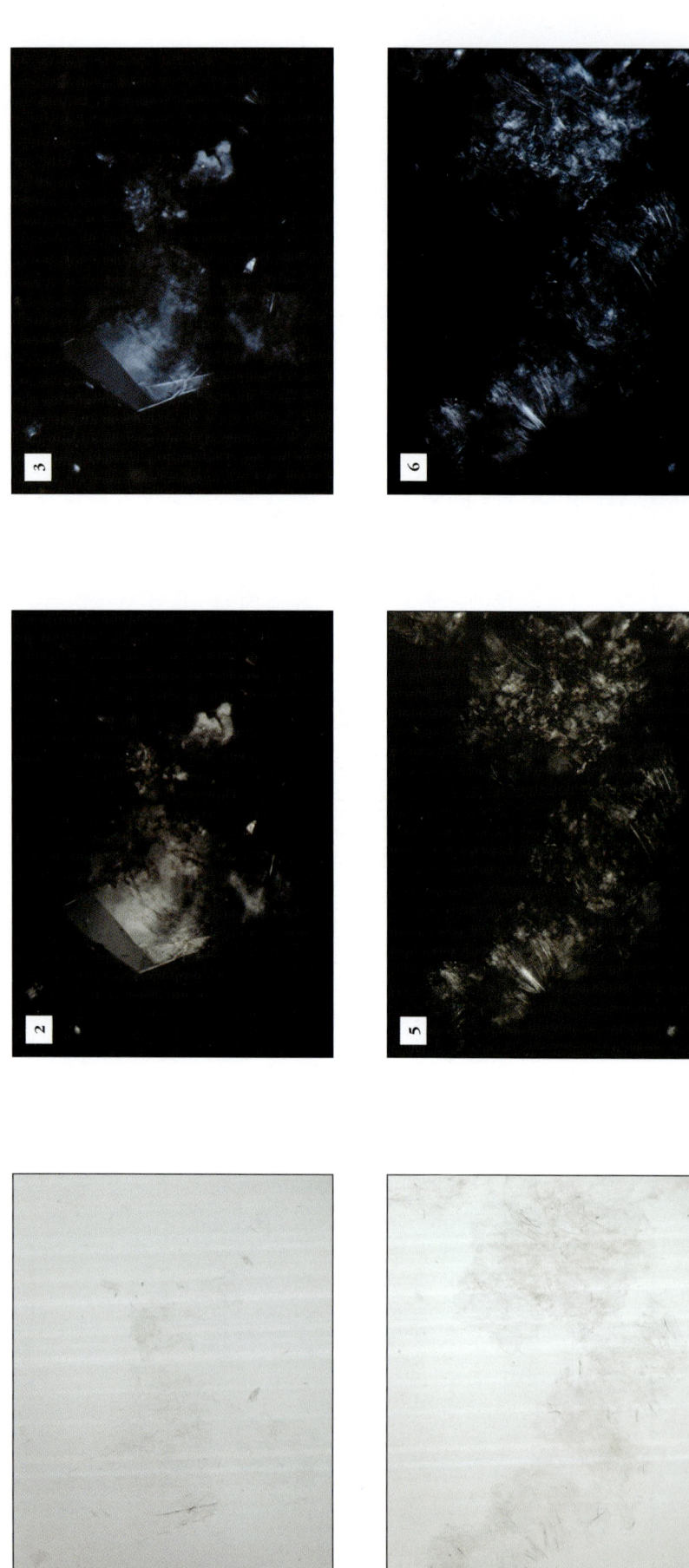

Key:

1. P1233: PPL/10×. Low relief colourless platy particles and aggregates of nacrite.

2. P1233: RPL/10×. White reflective surfaces of nacrite particles in reflected light. Same field of view as in Fig. 1.

3. P1233: XPL/10×. Low first order interference colours exhibited by nacrite plates and aggregates under crossed polars. Same field of view as in Fig. 1.

4. P1233: PPL/10×. Colourless fibrous and polycrystalline aggregates of nacrite.

5. P1233: RPL/10×. Weak white reflectance of fibrous nacrite aggregates in reflected light. Same field of view as in Fig. 4.

6. P1233: XPL/10×○. Fibrous and polycrystalline aggregates of nacrite exhibiting low first order interference colours under crossed polars. Same field of view as in Fig. 4.

Crystallinity	Optic sign	2V	n_α	n_β	n_γ	δ
Triclinic/Monoclinic	–ve	24°–50°	1.553–1.565	1.559–1.569	1.560–1.570	0.006

(data from Deer et al., 1992)

Nacrite is chemically identical to kaolinite, but has a slightly different structure.

In plane-polarised light, nacrite forms translucent, colourless particles with low relief, and RI just less than that of the medium. Particles have platy habits and appear to be formed of fibrous and lamellar crystals. All clay minerals can (and often do) form as pseudomorphs of pre-existing minerals, which may account for apparently euhedral particle shapes. Particle surfaces are rough and fibrous. Particle size distribution is typically broad, ranging from very fine to very coarse.

Under crossed polars, nacrite has low birefringence and very low first order interference colours are present. The polycrystalline nature of the particles is

clear and complete extinction of particles cannot be attained. Many particles of nacrite exhibit mottled or sweeping extinction.

Samples of nacrite examined were white in reflected light and also fluoresced green-white with UV excitation.

Nacrite is often found in association with kaolinite and also occurs in hydrothermally mineralised environments with calcite, dolomite, baryte or quartz. Analysts should be reminded that individual clay minerals, due to their fine particle size, are very difficult to differentiate using optical microscopy.

Lit.: Deer et al. (1992)

SMECTITE GROUP CLAY MINERALS: MONTMORILLONITE–BEIDELLITE
$(Na,Ca)_{0.3}(Al,Mg)_2Si_4O_{10}(OH)_2 \cdot nH_2O$ – $Na_{0.5}Al_2(Si_{3.5}Al_{0.5})O_{10}(OH)_2 \cdot nH_2O$

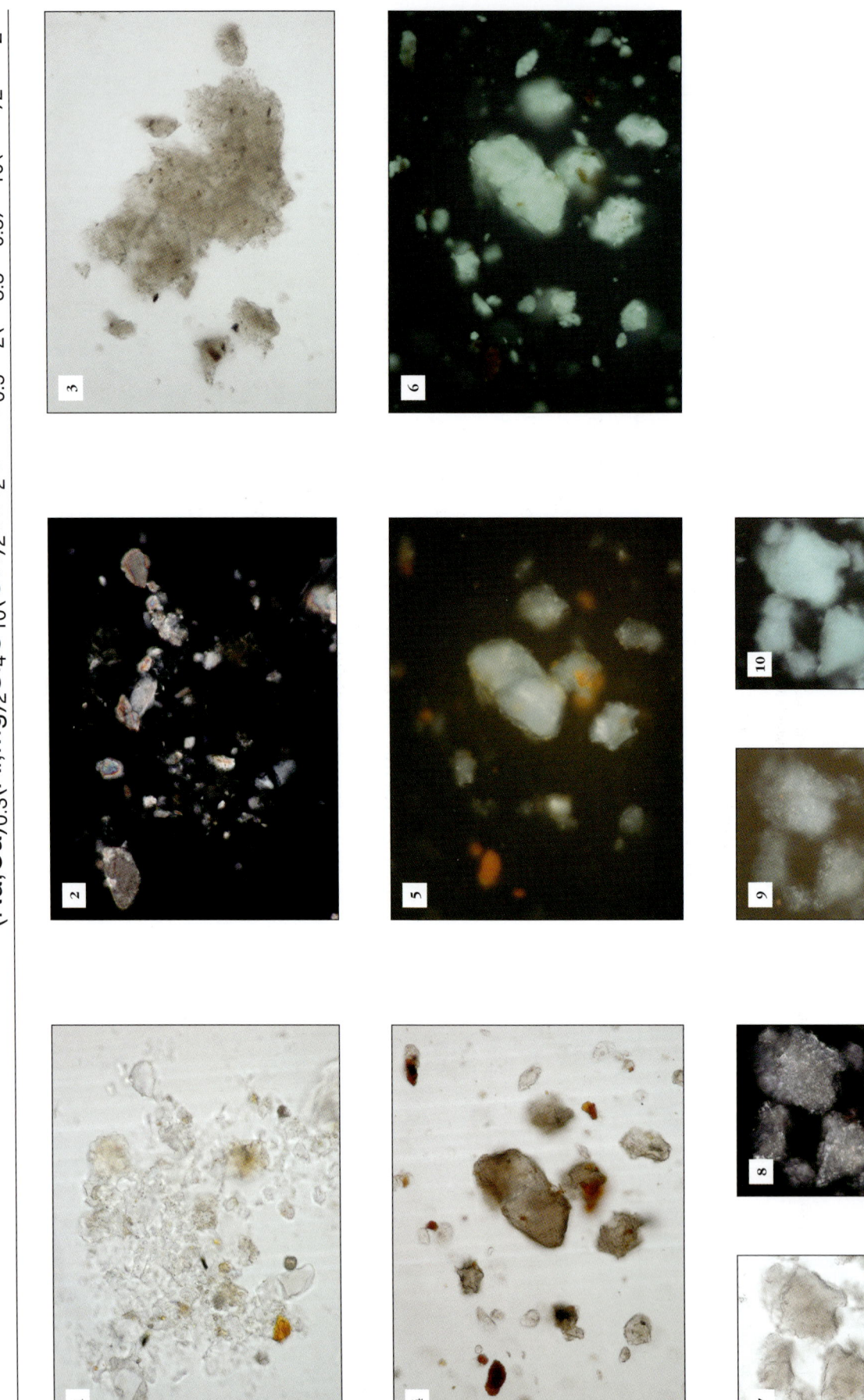

Key:

1. P0159: PPL/100×. Yellow montmorillonite and beidellite particles plus calcite.

2. P0159: XPL/100×/○. The field of view as in Fig. 1 under crossed polars, the highly birefringent material is calcite.

3. P1153: PPL/40×. Polycrystalline aggregate of beidellite.

4. P1439: PPL/40×. Rounded particles of beidellite and montmorillonite.

5. P1439: RPL/40×. The field of view as in Fig. 4 under crossed polars.

6. P1439: UV/40×. The field of view as in Fig. 4 in UV light.

7. P1231: PPL. Polycrystalline particles of montmorillonite.

8. P1231: XPL, ○. The particles in Fig. 7 under crossed polars.

9. P1231: RPL. The particles in Fig. 7 in reflected light.

10. P1231: UV. The particles in Fig. 7 in UV light.

Crystallinity	Optic sign	2V	n_α	n_β	n_γ	δ
Montmorillonite: Monoclinic	−ve	0°–30°	1.48–1.57	1.50–1.60	1.50–1.60	0.02–0.03
Beidellite: Monoclinic	−ve	9°–20°	1.494–1.503	1.525–1.532	1.526–1.533	0.0300–0.032

(data from Deer et al., 1992 and Webmineral, 2003)

In plane-polarised light, montmorillonite and beidellite form translucent brown-yellow particles with moderate relief and RI less than that of the medium. The colour is variable across particles, but least intense at the particle boundaries. Particles form aggregates of anhedral, angular grains. Particles surfaces are very rough, and size distributions may be broad, with particles ranging from very fine to very coarse.

Under crossed polars the polycrystalline nature of the particles is clear. Montmorillonite and beidellite have moderate birefringence with high first order to second order interference colours typical. However, individual particles are fine and appear speckled and polycrystalline.

These minerals appear white to yellow-white in reflected light. Dull yellow to white fluorescence was observed under UV excitation.

The montmorillonite–beidellite series is found in altered sedimentary rocks and may occur with other smectite and illite group clays, plus quartz and carbonate minerals. This series is the prime constituent of the bentonite deposits commonly also known as 'Fuller's earths'. These are weathered volcanogenic sediments and may additionally contain quartz, micas, feldspars, zircons, apatites and so forth. Analysts should be reminded that individual clay minerals, due to their fine particle size, are very difficult to differentiate using optical microscopy. It has also been suggested that staining techniques can be used to identify and differentiate these minerals optically (McCrone et al., 1973–80).

Lit.: Deer et al. (1992); McCrone et al. (1973–80)

The Pigment Compendium

SMECTITE GROUP CLAY MINERALS: NONTRONITE

$Na_{0.3}Fe^{3+}_2(Si,Al)_4O_{10}(OH)_2 \cdot nH_2O$

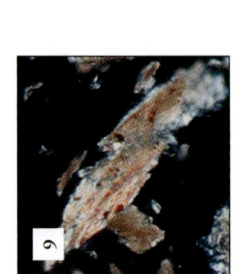

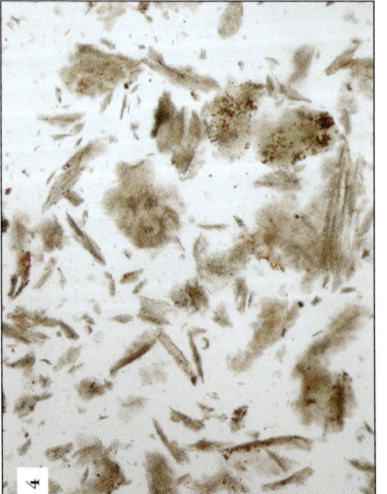

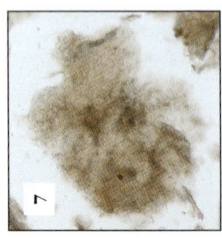

Key:

1. P1235: PPL/10×. Yellow-brown polycrystalline aggregates of nontronite.

2. P1235: XPL/10×/●. Same field of view as in Fig. 1 under crossed polars.

3. P1235: UV/10×. Same field of view as in Fig. 1, shows moderate green fluorescence of nontronite aggregates in UV light.

4. P1235: PPL/10×. Polycrystalline and fibrous aggregates of nontronite.

5. P1235: XPL/10×/●. Same field of view as in Fig. 4 under crossed polars.

6. P1235: RPL/10×. Same field of view as in Fig. 4 in reflected light.

7. P1235: PPL. Pale brown polycrystalline aggregate of nontronite.

8. P1235: XPL/●. Aggregate in Fig. 7 under crossed polars.

9. P1235: XPL/●. Moderate birefringence of a fibrous nontronite aggregate.

Crystallinity	Optic sign	2V	n_α	n_β	n_γ	δ
Monoclinic	–ve	25°–70°	1.56–1.61	1.57–1.64	1.57–1.64	0.030–0.045
					(data from Deer *et al.*, 1992 and Webmineral, 2003)	

In plane-polarised light, nontronite forms translucent to weakly translucent yellow-brown to green particles. The crystals are pleochroic, but this may be difficult to see in the finely polycrystalline particles. Particles are polycrystalline aggregates of fine fibrous crystals. Relief is very low and just less than that of the medium. Particle surfaces appear rough and fibrous. Shape is of anhedral, ragged appearing particles. Particle size distribution may be very broad, ranging from very fine to very coarse particles.

Under crossed polars, the polycrystalline nature of the particles is clear. Nontronite has moderate birefringence with high first order to second order interference colours typical. These may be masked by the strong body colour. The fibrous nature of the particles is clear and extinction is sweeping and twinkling.

Samples examined were a yellow-brown in reflected light, fluorescing a dull yellow-green with UV excitation.

Nontronite is found in marine sediments and as an alteration product of basaltic igneous rocks. It is often found with iron oxides and hydroxides, plus other smectite group clays and quartz. Analysts should be reminded that individual clay minerals, due to their fine particle size, are very difficult to differentiate using optical microscopy.

Lit.: Deer *et al.* (1992)

The Pigment Compendium

VERMICULITE GROUP CLAY MINERALS

$(Mg,Fe^{2+},Al)_3(Al,Si)_4O_{10}(OH)_2 \cdot 4H_2O$

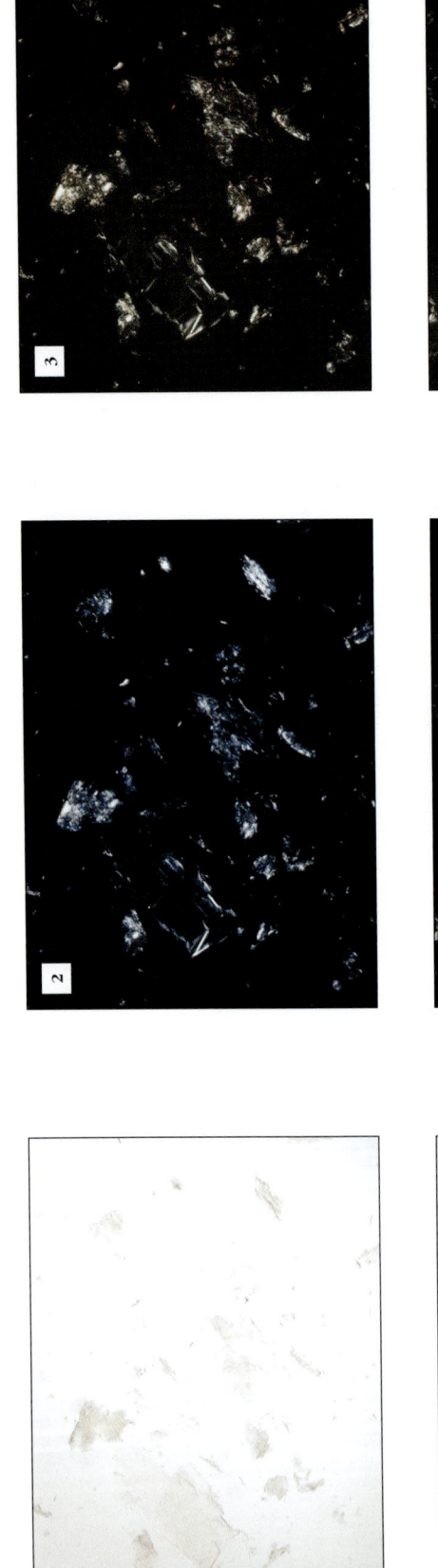

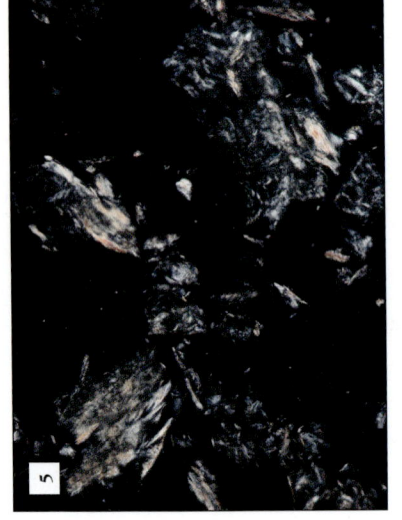

Crystallinity	Optic sign	2V	n_α	n_β	n_γ	δ
Monoclinic	–ve	0°–18°	1.520–1.564	1.530–1.583	1.530–1.583	0.02–0.03
					(data from Deer *et al.*, 1992 and Webmineral, 2003)	

In plane-polarised light, vermiculite may be seen as translucent colourless to weakly coloured yellow-brown particles. Relief is very low and RI is just less than that of the medium. Particles may be found as very fine grains or as large plates that pseudomorph other minerals. In the latter case (which is common), surfaces have the appearance of fibrous mats. In reflected light, colour can be seen on some surfaces to be white to yellowish white.

Under crossed polars, vermiculite has low birefringence, with first order to very low second order interference colours typical. Particles are clearly poly-crystalline and composed of randomly orientated clusters of fibres. Extinction is sweeping and mottled. Fibres are length slow.

Samples of vermiculite examined by the authors were white to yellowish white in reflected light and had a dull green-yellow fluorescence with UV excitation.

Vermiculite often forms as an alteration product of mica, chlorite or pyroxene, which accounts for the platy nature of the particles that is frequently observed. It is often found in earth deposits in association with minerals including the micas, pyroxenes, serpentinite group clays, chlorite and talc. Analysts should be reminded that individual clay minerals, due to their fine particle size, are very difficult to differentiate using optical microscopy.

Lit.: Deer *et al.* (1992)

The Pigment Compendium

SERPENTINE GROUP MINERALS

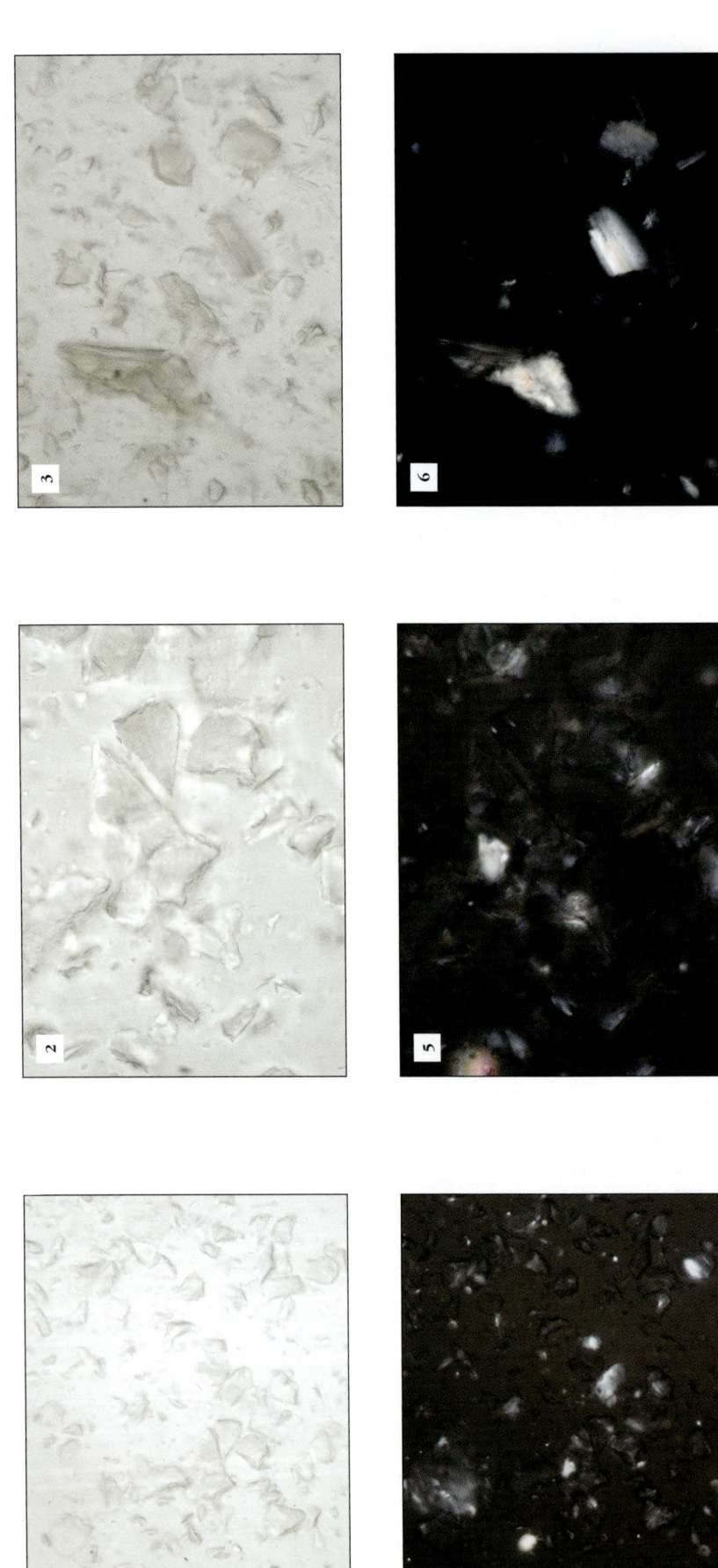

Key:

1. P1144: PPL/40×. Platy crystals of antigorite.
2. P1144: PPL/100×. Enlarged view of crystals in Fig. 1.
3. P1255: PPL/40×. Polycrystalline flakes of the serpentine group minerals.
4. P1144: ~XPL/40× **O** The field of view as in Fig. 1 under crossed polars.
5. P1144: ~XPL/100× **O** The field of view in Fig. 2 under crossed polars.
6. P1255: XPL/40×. The field of view in Fig. 3 under crossed polars.

Crystallinity	n_ω	n_ε	n_α	n_β	n_γ	δ
Antigorite: Monoclinic-Orthorhombic			1.558–1.567	1.566	1.562–1.574	0.004–0.007
Chrysotile: Trigonal	1.545–1.556	1.532–1.549				0.013–0.017
Lizardite: Trigonal	1.546–1.560	1.538–1.554				0.006–0.008

(data from Deer et al., 1992)

The serpentine group of minerals are lizardite, antigorite and chrysotile. The name serpentinite is generally applied to rocks composed of a mixture of these minerals. The three are thus often found intimately mixed and are often impossible to differentiate using the optical microscope. Chrysotile, because of its strongly fibrous habit and high relief, is easily distinguishable from antigorite and lizardite and therefore is discussed elsewhere under Asbestiform minerals.

In plane-polarised light, antigorite and lizardite are colourless or weakly coloured yellow-green. Red colours may be developed with iron oxide-rich environments. Particles are transparent with moderate to low relief and RIs less than that of the medium; these properties are insufficient to differentiate the two minerals. Lizardite and antigorite form platy crystals.

Under crossed polars all serpentine group minerals can be seen to have low birefringence. Antigorite and lizardite have very low first order interference colours, whereas chrysotile is more distinctive with high first order interference colours. The polycrystalline nature of all particles is clear and all show sweeping or mottled extinction. For antigorite, McCrone et al. (1973–80) describe 'incomplete and fan-like' extinction in equant particles, while blades and needles show parallel extinction; however, this phenomenon is more common in intergrowths of antigorite or lizardite with chrysotile. Antigorite is length slow.

In reflected light these minerals can appear transparent (black) to white or yellowish white. They are also fluorescent, exhibiting a weak yellow-white colour with UV excitation.

The serpentine group form as the alteration of olivine and pyroxene-rich rocks and relicts of these minerals may occur in association with the clays. Also present and common are iron oxides, especially hematite and magnetite, as well as talc and chlorite.

Lit.: Deer et al. (1992); McCrone et al. (1973–80)

STARCH [1]

[Complex; based on C₆H₁₀O₅]

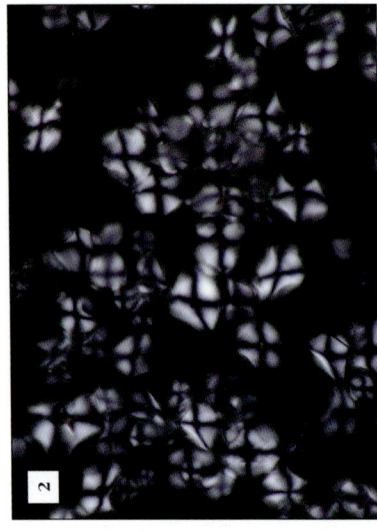

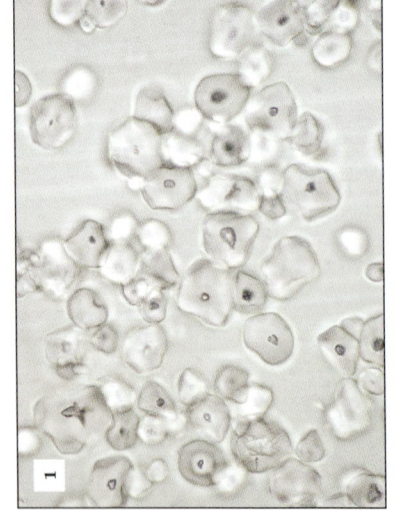

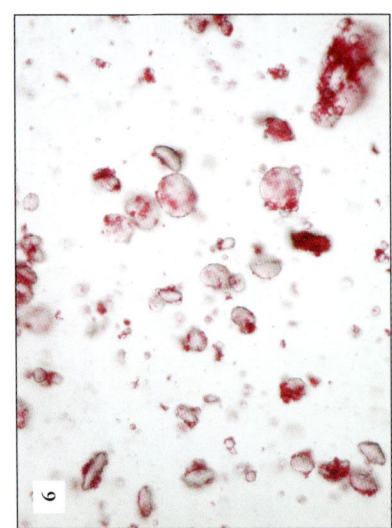

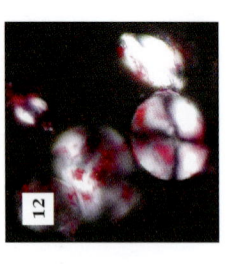

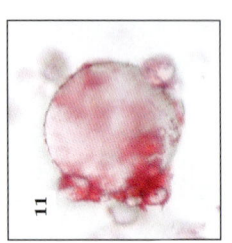

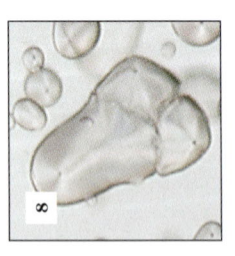

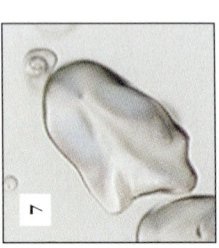

1. P1491: PPL/100×. Maize starch.

2. P1491: XPL/100×. Maize starch under crossed polars, note centred extinction cross.

3. P1493: PPL/10×. Potato starch.

4. P1493: XPL/40×. Potato starch under crossed polars, note off-centre extinction cross.

5. P1493: /40×. Same field of view as in Fig. 4 with the sensitive tint plate inserted.

6. P0579: PPL/40×/H. Wheat starch used as a substrate for cochineal.

7. P1493: PPL. Potato starch: an 'oyster shell'-shaped grain.

8. P1493: PPL. Potato starch: a 'triplet' grain. Note the three *bila* towards the extremities of the grain.

9. P1493: XPL. Potato starch under crossed polars, note the prominent *bilum* at the centre of the extinction cross.

10. P1493: XPL/STP. The grain in Fig. 9 with the sensitive tint plate inserted.

11. P0579: PPL/H. Enlarged view of wheat starch with particles of cochineal clinging to it.

12. P0579: XPL/●/H. Wheat starch under crossed polars. The red particles are cochineal.

Crystallinity
Composite structure

n	≪1.662
δ	Anisotropic (data from authors)

Starches from different sources are differentiable using optical microscopy and are described in terms of their shape (globular, lenticular, ellipsoidal, ovoid, truncated or polygonal), degree of aggregation, size and the *bilum*, or organic centre of the grains. The *bilum* is conspicuous in some starches, but difficult to observe (certainly in plane-polarised light) in others. *Hila* may be spots, small rings, elongate, or even dark clefts in various starches. *Hila* may also be situated centrally in the grain or off-centre. Under crossed polars, starch grains exhibit a distinctive extinction cross or 'Maltese cross', the centre of which is the location of the *bilum* (NB, legumes which have large, cleft-shaped *bila* have an extinction 'cross' with a central bar). The appearance of extinction crosses may again be diagnostic and varies with the degree of crystallisation of the starch grains. Starch grains also frequently show concentric rings around the *bilum*. Again, these may be unobservable, faint or conspicuous; the rings are more distinct using oblique illumination or other microscopic techniques suited to the observation of biological materials. Observation of many starch grains with a sensitive tint or gypsum plate reveals a 'beautiful play of colours' (Winton, 1906) that may also aid in identification.

Maize starch (Zea mays L., Gramineae)

Maize starch is distinctive in that it is the only commonly occurring starch with polygonal grains greater than 15 μm diameter. Grain size is therefore coarse. The grains are clearly polygonal and sometimes form aggregates of several particles. Maize starch has a large and conspicuous *bilum*, which is situated approximately

centrally in the grain, but varies in shape from circular to oval or even branching. Rings are not visible. In crossed polars, extinction crosses are distinctive and symmetrical. Use of the sensitive tint plate shows the particles to be optically positive, divided clearly into yellow and blue segments with a pink extinction cross.

Potato starch (Solanum tuberosum L., Solanaceae)

Potato starches are those most commonly used in pigment and paper manufacture. Potato starches are coarse to very coarse grained (particles may be visible with the naked eye) and has ovoid, pear-shaped, or more irregular 'oyster shell-shaped' grains. The *bilum* is small and located at the narrow end of the grain. Some grains are 'twins' or 'triplets' with two or three *bila* respectively. Rings are distinctive in oblique illumination. Under crossed polars, the extinction crosses are distinct. With a sensitive tint plate inserted, the grains are observed to be optically positive, and demonstrate a play of colours as the stage is rotated.

Wheat starch (Triticum spp., Gramineae)

Wheat starch grains are predominantly circular (occasionally ellipsoid and elongate particles occur) and range in size from fine to very coarse. Grains have a small, inconspicuous, central *bilum* and indistinct rings. Under crossed polars, the extinction cross is not distinct and may often appear deformed. With a sensitive tint plate inserted, the grains are observed to be optically positive, and demonstrate a weak play of colours as the stage is rotated.

The Pigment Compendium

STARCH [2]

[Complex; based on C$_6$H$_{10}$O$_5$]

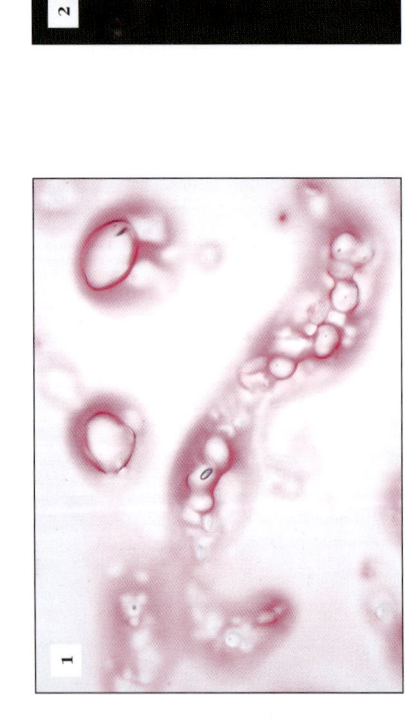

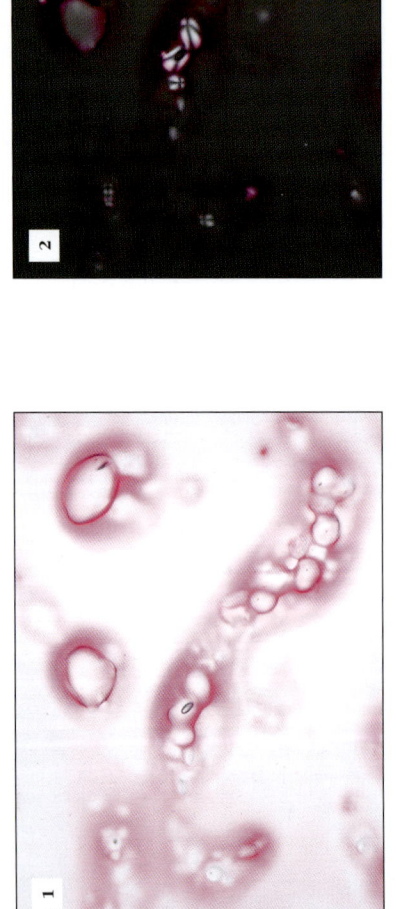

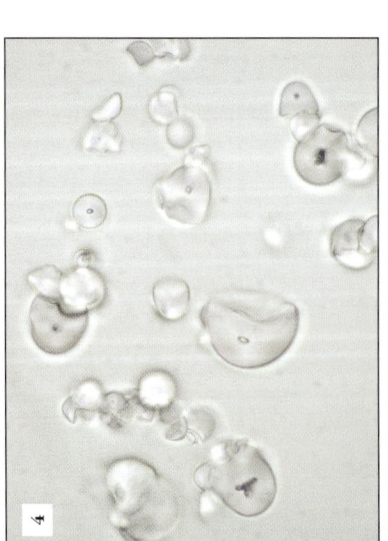

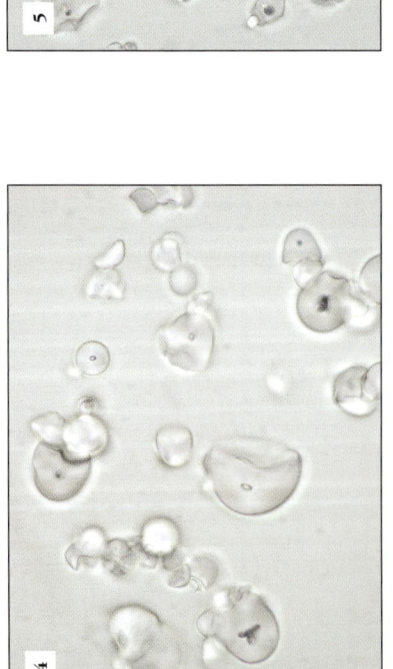

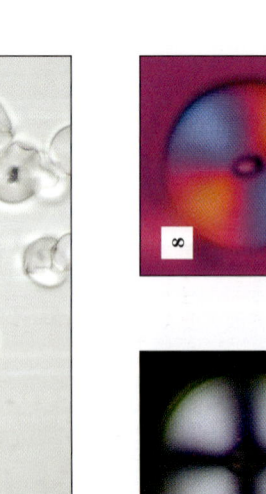

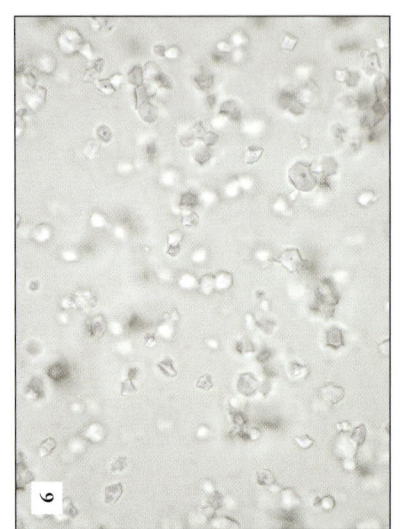

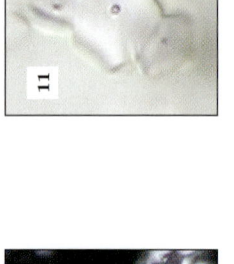

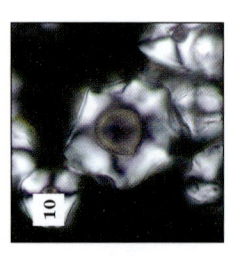

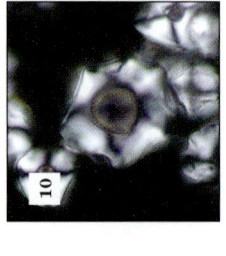

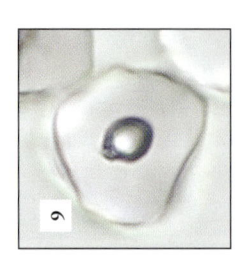

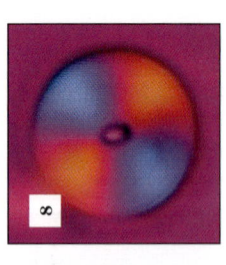

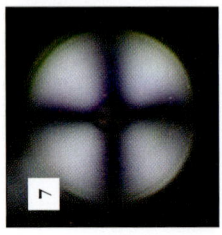

Key:

1. P0167: PPL/40×. Rhodamine B on arrowroot starch.

2. P0167: ~XPL/40×. The same field of view as in Fig. 1 under crossed polars. Note the distorted extinction crosses.

3. P0167: UV/40×. The same field of view as in Fig. 1 under UV illumination. The fluorescence is due to the Rhodamine B.

4. P1492: PPL/100×. Tapioca starch.

5. P1490: PPL/100×. Amioca (maize hybrid) starch.

6. P1489: PPL/100×. Rice starch.

7. P1492: XPL. Tapioca starch grain under crossed polars. Note centred extinction cross.

8. P1492: XPL/STP. The particle in Fig. 7 observed with the sensitive tint plate inserted.

9. P1490: PPL. Amioca starch grain; note prominent hilum.

10. P1490: XPL/●. Amioca starch under crossed polars.

11. P1489: PPL. Rice starch, note the strongly polygonal grain shape.

Crystallinity
Composite structure

n	δ
≪1.662	Anisotropic (data from authors)

Arrowroot starch (Maranta arundinacea L., Marantaceae)
Arrowroot starch is similar in appearance to potato starch, and is also commonly used for the same purposes. Arrowroot starch forms ovoid to pear-shaped, rounded particles, with a prominent *hilum* at the broad end of the grain. *Hila* may also be associated with a wing-like fissure crossing it. Rings are distinctive in oblique illumination, but hard to see otherwise. Under crossed polars, the extinction crosses are distinct. With a sensitive tint plate inserted, the grains are observed to be optically positive, and demonstrate a play of strong blue and yellow colours as the stage is rotated.

Tapioca starch (Manihot esculenta Crantz, Euphorbiaceae)
Tapioca starch forms spherical particles and aggregates of polygonal grains. Grain size is coarse. The *hilum* is conspicuous and ranges from a small speck to large, irregular marks. Under crossed polars, the extinction crosses are distinct and symmetrical; however, aggregates of grains have distorted extinction crosses. Use of the sensitive tint plate shows the particles to be optically positive, divided clearly into yellow and blue segments with a pink extinction cross.

Rice starch (Oryza sativa L., Gramineae)
Rice starch forms angular, polyhedral grains which exist as individuals and also in dense aggregates. Grain size is medium to fine. Aggregates are coarse grained. The hilum is conspicuous and situated centrally in the grain. Under crossed polars, grains have low first order grey interference figures, but the extinction cross is distinct. Use of the sensitive tint plate shows the particles to be optically positive, divided into yellow and blue segments with a pink extinction cross.

Apart from use as fillers, various starches have been used as substrates for dye-based pigments. In their natural state starch grains are colourless in plane-polarised light. The refractive index of starch grains is moderate and the RI is less than that of the medium.

The crystal structure of starch may be distorted or destroyed by heating.

Lit.: Winton (1906)

RED LAKE PIGMENTS

[Complex]

Key:

1. P0248: PPL/100×. Madder on fine-grained substrate.

2. P0964: PPL/100×/H. Madder on Fe-Ca-Si substrate.

3. P0982: PPL/100×/H. Madder on aluminium hydroxide substrate.

4. P1382: PPL/40×/H. Madder on amorphous substrate.

5. P0048: PPL/40×. Carmine on gibbsite.

6. P1329: PPL/100×/H. Carmine on maize starch.

7. P0982: UV/H. Particles from Fig. 3 in UV light.

8. P1382: UV/H. Particles from Fig. 4 in UV light.

9. P1329: XPL/●/H. Particles from Fig. 6 under crossed polars.

10. P0994: PPL/H. Particles of carmine adhering to crystal of aluminium hydroxide.

11. P0578: PPL/H. Intensely coloured carmine on aluminium hydroxide substrate.

12. P0571: PPL/H. Carmine adulterated with crushed vermilion which appears as dark spots and individual crystals.

Madder and Alizarin

Identification of madder (dyestuffs containing hydroxyanthraquinones such as alizarin and purpurin derived from *Rubia* species and other plants of the Rubiaceae) using optical microscopy is not conclusive, and it shows many similar characteristics with other red organic dyes. The dye has a crimson red colour and is usually seen to occur as very fine particles clinging to a substrate or as a stain on a substrate. Substrates are typically aluminium hydroxides or other Al-based compounds, including clays. More rarely organic substrates (starches) are encountered. For this reason particle size, size range and shape are all highly variable. It has also been suggested that the RI for madder-aluminium hydroxide lakes is ~1.66–1.70 and for alizarin lakes ~1.66–1.70 (McCrone *et al.*, 1973–80), the higher value for alizarin lakes perhaps being due to the RI of pure alizarin being about 1.70 (Merwin, 1917); on this basis it is implied that a differentiation may be achieved. On the other hand differentiation can be achieved using fluorescence. Madder fluoresces with a dull orange when viewed using UV excitation while alizarin lakes do not (Schweppe and Winter, 1997), making this a more reliable diagnostic procedure. Madder may be extended or adulterated with carmine and with ochres (of various types,

including burnt ochres and burnt siennas), powdered tile and copper hexa-cyanoferrates. Modern preparations have substituted alizarin compounds for madder (Schweppe and Winter). The identification of madder and alizarin-based pigments has been discussed in detail by Schweppe and Winter (1997: 1312, 1321).

Carmine

Insect-derived dyestuffs, commonly known as carmine and including cochineal, kermes and lac derived dyes, are not readily determined using optical microscopy. They are pink through orange-red in colour and appear either as a stain coating a substrate or more frequently as fine particles clinging to substrates. Substrates are typically aluminium hydroxides, clays or starches. For this reason particle size, size range and shape can all be highly variable. Cochineal and kermes have been noted as fluorescing bright pink using UV light, while lac lakes reputedly do not (Schweppe and Winter). Carmines may be extended or adulterated with madder and with other red pigments, including vermilion (mercury sulfide).

Lit.: McCrone *et al.* (1973–80); Merwin (1917); Schweppe & Winter (1997)

The Pigment Compendium

YELLOW LAKE PIGMENTS

[Complex]

Key:

1. P0044: PPL/100×. Weld on amorphous substrate.

2. P0222: PPL/100×. *Rhamnus* on amorphous substrate.

3. P0049: PPL/100×. Yellow lake on amorphous substrate.

4. P0547: PPL/40×/H. Orange-brown lake on clay-rich substrate.

5. P0840: PPL/100×. Yellow lake on amorphous substrate.

6. P0545: PPL/100×/H. Pale yellow lake on chalk.

7. P0049: RPL. Particles from Fig. 3 in reflected light.

8. P0547: XPL/H. Particles from Fig. 4 under crossed polars.

9. P0840: RPL. Particles from Fig. 5 in reflected light.

10. P0840: XPL. Particles from Fig. 5 in crossed polars.

11. P0545: XPL ● /H. Planktonic foraminifera from Fig. 6 under crossed polars.

12. P0545: RPL/H. Foraminifera from Fig. 11 in reflected light.

Yellow lakes are composed of a yellow dyestuff deposited onto a white substrate to produce a pigment. The dyes (in some cases historically referred to as 'pink') typically include dyes derived from the berries of the *Rhamnus* species, 'Persian berries'; those from weld (*Reseda luteola* L.) and the quercitron dyes derived from oak bark (*Quercus* spp.). All belong to the flavonoid group of compounds. In addition, yellow dyes may also conceivably be based on lichen, turmeric, gamboge and woad.

The substrates are typically chalk (and other ground limestones), aluminium hydroxides or starch. Less commonly substrates such as ground cuttlefish bone (sepia) may also be used.

Identification of components is generally straightforward for the substrate composition, and the main compounds used have been described elsewhere. The colourants may be identified by fluorescence and other analytical techniques.

Yellow lake pigments may be adulterated with or used as substitutes for pigments such as orpiment.

RED OCHRE

[Complex]

Key:

1. P0059: PPL/40×. Red ochre containing fine-grained hematite (red), quartz (colourless) and minor clays.

2. P0357: PPL/40×/H. Fine-grained intimate mixtures of hematite and smectite clays.

3. P0358: PPL/100×/H. Hematite-coated quartz grains.

4. P0059: XPL/40×/ **O**. The field of view as in Fig. 1 under crossed polars.

5. P0357: XPL/40×/H. The field of view as in Fig. 2 under crossed polars.

6. P0788: PPL/100×. Fine-grained plates of hematite.

7. P0059: PPL. Hematite-coated quartz grain from Fig. 1.

8. P0357: PPL/H. Finely crystalline mat of hematite and acicular clays from Fig. 2.

9. P0357: XPL/ **O** /H. A particle from Fig. 2 under crossed polars, the clays appear white.

10. P0358: XPL/ **O** / H. Hematite coated grains from Fig. 3 under crossed polars.

11. P0788: PPL. Enlargement of particles from Fig. 6 showing rounded plate-like morphologies.

12. P0788: RPL. The field of view as in Fig. 11 under reflected light.

Red ochres are composed primarily of the iron(III) oxide hematite, although other iron oxide minerals such as goethite, hematite, lepidocrocite and magnetite may also be present. As 'earths', these pigments are also almost invariably impure, containing variable amounts of accessory phases that typically include quartz, feldspars, calcite, dolomite, various clay minerals, and even gypsum, baryte or rutile (Duval, 1992). However, some may be especially pure, in which case differentiation from synthetic analogues must be based on the typically broader particle size distribution. Variation in the mineralogy and particle size present is strongly dependent on the formation environment of the ochre and processes used subsequently in its preparation as a pigment.

Lit.: Duval (1992)

YELLOW OCHRE

[Complex]

Key:

1. P0632: PPL/40×/H. Goethite-coated quartz grains.

2. P0053: PPL/40×. Plates of goethite with colourless quartz.

3. P0056: PPL/40×. Sienna; plates of goethite and goethite-coated quartz grains with minor opaque phases.

4. P1355: PPL/40×. Plates of finely disseminated goethite and clays with opaque minerals.

5. P0225: PPL/100×. Yellow ochre containing goethite, quartz, clays and feldspars.

6. P0271: PPL/100×. Sienna; plates of yellow goethite.

7. P0053: RPL. Enlargement of grains from Fig. 2 in reflected light.

8. P0056: RPL. Enlargement of grains from Fig. 3 in reflected light.

9. P0225: PPL. Enlargement of goethite mat from Fig. 5 showing fibrous crystal morphology.

10. P0271: PPL. An enlarged view of Fig. 6 showing particle morphologies.

11. P1283: PPL. Goethite-coated quartz grain.

12. P1085: PPL. Plates of goethite from a yellow ochre.

Yellow ochres are composed primarily of the iron oxide hydroxides goethite and, less frequently, lepidocrocite. For these minerals particle morphology may vary considerably from well-developed, usually fibrous and acicular crystals, to aggregates of finely divided particles. Fine particles of iron oxide hydroxides may also be observed as coating grains of other minerals such as quartz. Yellow ochres *sensu latu* may otherwise be rich in the minerals jarosite or natrojarosite.

Derived as they are from iron-rich soils and the decomposition of ore deposits, yellow ochres will be impure, the other phases present varying considerably from source to source. However, it is reasonable to expect to find quantities of quartz, feldspar, clays or carbonate minerals in association with the iron oxide hydroxides. The red iron oxide hematite may also be present in minor amounts, as well as opaque phases. Though not readily identifiable using optical methods, these are likely to be pyrite, magnetite or the manganese oxides. The occurrence of the latter in quantities up to 10% implies that the material is a sienna earth; siennas are otherwise optically indistinguishable from 'typical' yellow ochres. Greater amounts of manganese oxides present represent compositions more characteristic of umbers (*q.v.*). Some yellow ochre-type assemblages may contain humic material, which is again also typical of some siennas.

BROWN OCHRE AND BURNT SIENNA

[Complex]

Key:

1. P0348: PPL/40×/H. Brown ochre containing goethite, gypsum and calcite.

2. P0734: PPL/100×/◐/H. Brown ochre containing goethite, hematite, quartz and clays.

3. P0726: PPL/100×/H. A burnt sienna adulterated with fibrous colourless clays.

4. P0055: PPL/40×. A brown ochre containing goethite, opaque phases, feldspar, quartz and clays.

5. P0734: RPL/100×/H. Field of view as in Fig. 2 in reflected light.

6. P0272: PPL/100×. Burnt sienna containing hematite and calcite.

7. P0348: PPL/H. Enlarged view of particles from Fig. 1 showing orange goethite and colourless fragments of calcite.

8. P0734: PPL/◐/H. Enlargement of particles of goethite from Fig. 2.

9. P0726: ~XPL/◐/H. Particles from Fig. 3 under crossed polars, the clays show anomalous interference colours.

10. P0055: PPL. Perfect hexagonal plate of hematite enlarged from Fig. 4.

11. P0272: PPL. Enlargement of particles from Fig. 6 showing grain morphologies.

12. P0272: XPL/◐. Hematite-coated calcite particles from Fig. 6 under crossed polars.

Samples labelled 'brown ochre' have been found by the authors to typically contain goethite, hematite and also the black iron oxide, magnetite. Ochres derived from soils may also have humic earths within them. Theoretically the siennas fall into this category and are also said to contain 'significant' (~10%) amounts of manganese oxides.

Yellow ochres are also used as starting material for the production of pigments in a range of colours from orange through to black. These are manufactured by roasting to various temperatures, converting phases such as goethite to hematite and, eventually, magnetite. Such pigments may be difficult to differentiate optically from naturally occurring red ochres because accessory phases such as quartz and calcite often require temperatures higher than those used for processing to dissociate or fuse (calcite requires 898°C to break down to calcium oxide and quartz fuses at temperatures in excess of 1600°C).

Heating goethite at temperatures below 850°C leads to the formation of so-called 'disordered' hematite (Helwig, 1997; Béarat and Pradell, 1997). This is characterised by being poorly crystalline and it has been suggested that this process can take place naturally ('taphonically') as well as artificially ('anthropically'), but most authors consider that occurrences as pigments are of the latter (Masson, 1986; Béarat, 1996). In the examples illustrated above, those originally labelled 'burnt sienna' were found to contain hematite by X-ray diffraction, while in those called 'brown ochre' no hematite could be detected, even though such a phase was evidently present by optical microscopy; this may be considered to be a manifestation of the thermal alteration conditions of the various samples.

Lit.: Béarat (1996); Béarat & Pradell (1997); Helwig (1997); Masson (1986)

UMBERS AND WADS

[Complex]

Key:

1. P1486: PPL/40×. A sample of wad, note the high proportion of opaque manganese oxide phases present.

2. P0063: PPL/100×. Particles of goethite, quartz, manganese oxides and gypsum.

3. P0064: PPL/100×. Burnt umber; note the presence of bright red-orange particles of hematite.

4. P0354: PPL/100×/H. Rounded plates of iron hydroxides and clays.

5. P0268: PPL/100×. Coarse-grained rounded plates of iron hydroxides, manganese oxides, quartz and clays.

6. P0064: RPL/100×. The same view as in Fig. 3 in crossed polars.

7. P0354: PPL/H. An enlarged view of particle morphologies from Fig. 4.

8. P0268: RPL. Particles from Fig. 5 in reflected light.

9. P1109: PPL. A hexagonal opaque crystal of manganese oxide plus iron oxides and hydroxides.

10. P0265: PPL. A fine-grained umber.

11. P1069: PPL. Coarse-grained plates of goethite from an umber.

Geologically, umbers are very fine-grained sedimentary rocks composed of manganese hydroxides and oxides (primarily the minerals manganite and pyrolusite) plus iron hydroxide (primarily goethite but also hematite) particles. Manganese compounds should be present in amounts between 5 and 20% and iron oxides typically represent 45–70% of the rock (Ford, 2001; Buxbaum, 1998). As rocks, these are not pure substances and other minerals may be present in varying amounts depending on the mode and environment of formation.

Opaque phases observable optically are predominantly manganese oxides and hydroxides, but may also contain magnetite. Other associated minerals include silica (which may be amorphous or in the form of quartz), clay minerals, calcite, baryte and gypsum may also be present.

Raw umbers may additionally be adulterated with various extenders, including humic earths.

Burnt umbers will typically contain hematite rather than goethite, but traces of the latter phase are not unusual. Magnetite and a variety of manganese oxide minerals and silica are also likely to be present. Clay minerals, calcite, baryte and gypsum may also be components. As with the thermal alteration of ochres (*q.v.*), the crystal structure of hematite is likely to be disordered.

Wads contain significantly higher levels of manganese compounds than umbers.

Lit.: Buxbaum (1998); Ford (2001)

GREEN EARTHS

[Complex]

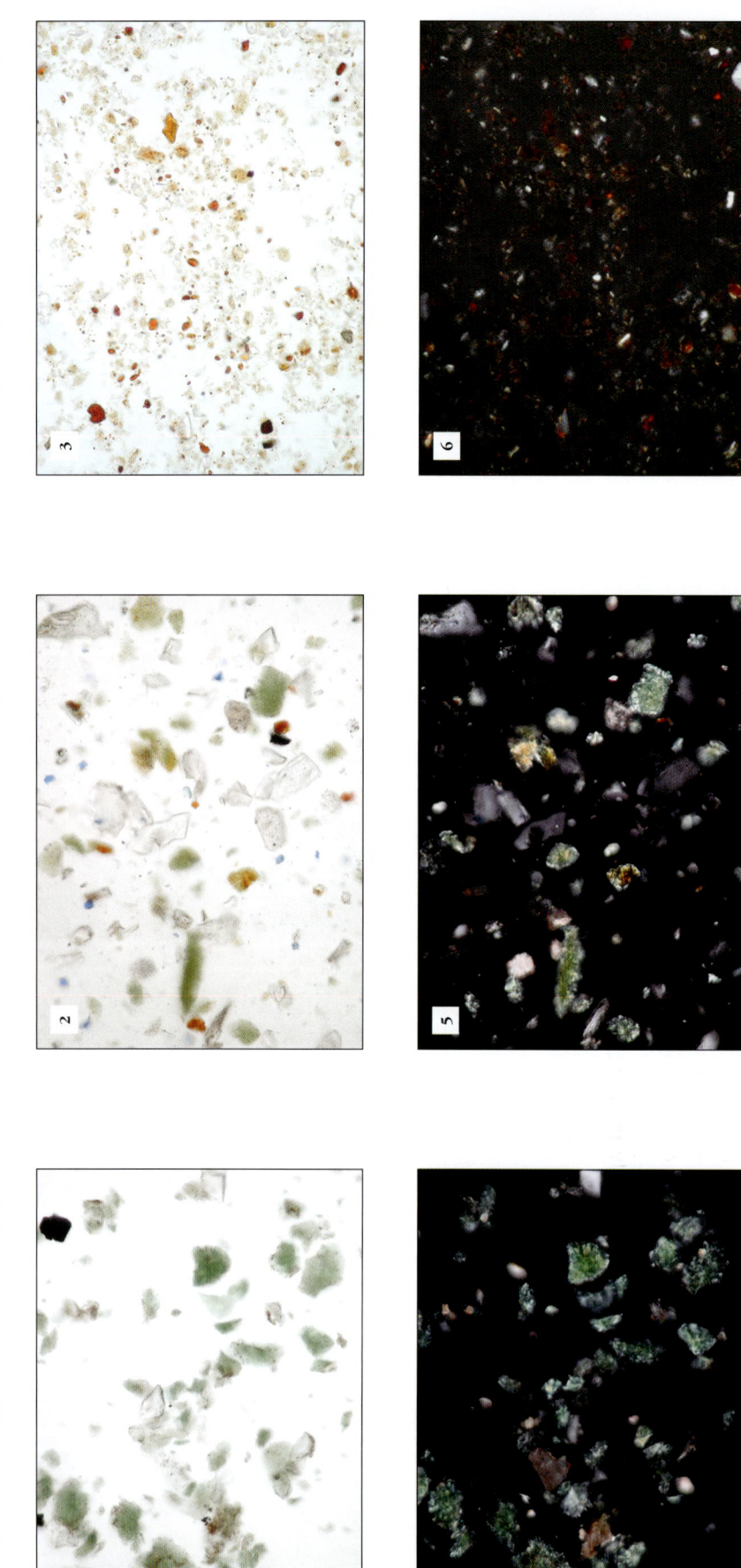

Key:

1. P1457: PPL/40X. Green glauconite with particles of calcite and opaque grains.

4. P1457: XPL/40X. The same field of view as in Fig. 1 under crossed polars.

2. P1469: PPL/40X. Green glauconite with quartz, calcite, micas and iron oxides. This sample has been adulterated with cobalt blue.

5. P1469: XPL/40X. The same field of view as in Fig. 2 under crossed polars.

3. P0273: PPL/100X. Burnt green earth; composed predominantly of brown iron oxides plus quartz.

6. P0273: ~XPL/100X/○. The same field of view as in Fig. 3 under partially crossed polars. The white birefringent particles are quartz.

The optical properties for minerals typically found in green earths are discussed elsewhere, for which see entries for glauconite and celadonite, chlorite, quartz, feldspars, calcite and the iron oxides and oxide hydroxides. Pigments are known to have been adulterated with blue and green dyes and with other pigments including cobalt blue. Green earths are also known in 'burnt' form; these are likely to contain brown iron oxides and other alteration products.

Green earth pigments have been reviewed as a group by Grissom (1986).

Lit.: Grissom (1986)

FELDSPAR MINERALS

$CaAl_2Si_2O_8–NaAlSi_3O_8–KAlSi_3O_8$

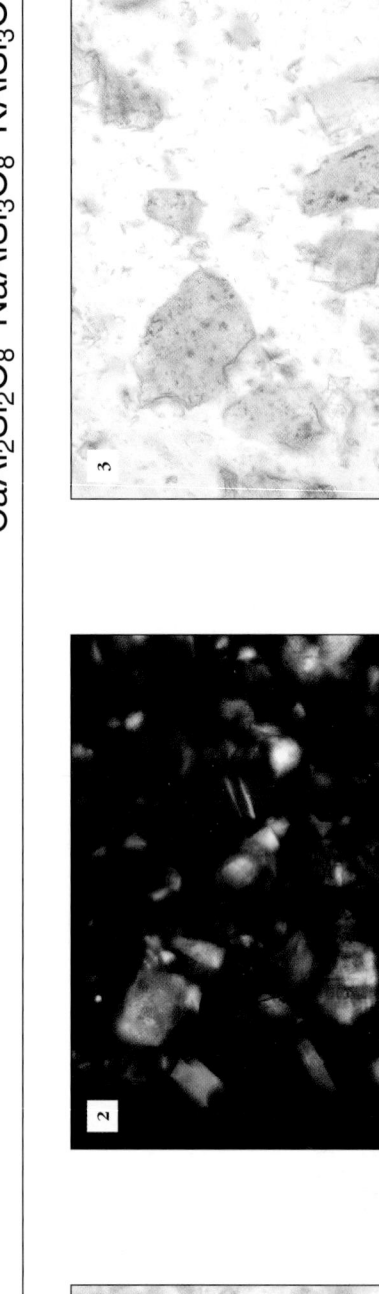

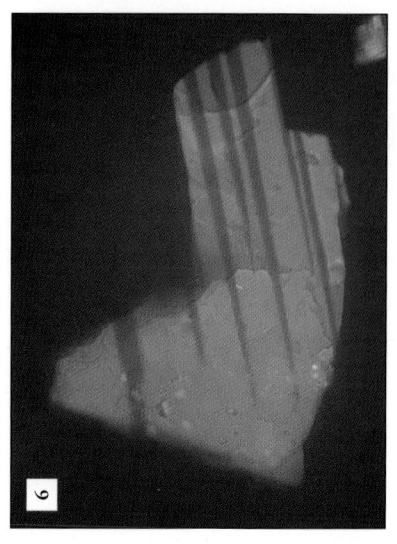

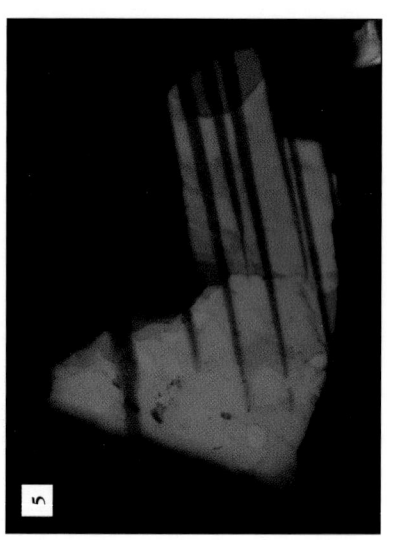

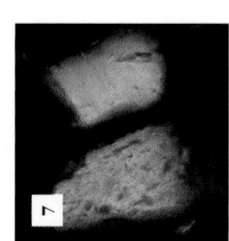

1. P1229: PPL/40×. Crushed angular particles of microcline.

2. P1229: XPL/40×/ **O**. The field of view as in Fig. 1 under crossed polars; the large particle towards the lower left shows cross-hatched twinning.

3. P1240: PPL/40×. Cloudy, inclusion-rich particles of orthoclase.

4. P1134: PPL/40×. Angular fragment of albite.

5. P1134: XPL/40×. The particle in Fig. 4 under crossed polars, note the multiple twins.

6. P1134: RPL/40×. The particle in Fig. 4 in reflected light.

7. P1240: XPL **O**. Particles of orthoclase from Fig. 3 under crossed polars.

Crystallinity	n_α	n_β	n_γ	δ
Albite: Triclinic	1.528–1.533	1.532–1.537	1.538–1.542	0.0090–0.0100
Microcline: Triclinic	1.518	1.522	1.525	0.0070
Orthoclase: Monoclinic	1.518	1.522	1.523–1.524	0.0050–0.0060

(data from Webmineral, 2003)

The feldspar group is a large and important category of rock forming minerals. They occur in most rock types and are therefore common components of earth pigments and may occur as traces in other mineral pigments. Two main varieties of feldspars exist, the potassic feldspars and the plagioclases. The varieties may be optically distinguished by the pattern of twinning which they exhibit.

In plane-polarised light all feldspars appear very similar and are hard to differentiate. They form transparent, colourless crystals, with moderate relief and RI less than that of the medium. Feldspars are extremely susceptible to weathering and break down to form clay minerals. In such cases, the crystals will appear cloudy and rich in inclusions. These feldspars are referred to as 'sieve textured'. When crushed, grain morphologies are angular shards. However, feldspars have three strong cleavages and this influences crushed grain shape by producing some trapezoidal grains. Unaltered particles have smooth, glassy surfaces. Particle size may range from fine through to coarse.

Under crossed polars, Feldspars have low birefringence and low first order greys are observed. The most diagnostic feature under crossed polars is the presence of twins, which will have different crystallographic orientations to their neighbours, and therefore one twin will appear in extinction when the other is fully illuminated. However, in pigments, feldspars may be crushed so finely that the particles are smaller than the twin repeat units and these may not be observed in all particles. Orthoclase potassic feldspars have simple twins or remain untwinned. The microcline potassic feldspars have fine cross-hatched multiple twins at right angles to each other; a phenomena referred to as 'tartan-twinning'. Plagioclases have one set of multiple twins. Unaltered feldspars have complete and inclined extinction.

The optical properties of feldspars are discussed in detail by Deer et al. (1992).

Lit.: Deer et al. (1992)

| The Pigment Compendium

FELDSPATHOID MINERALS [Various]

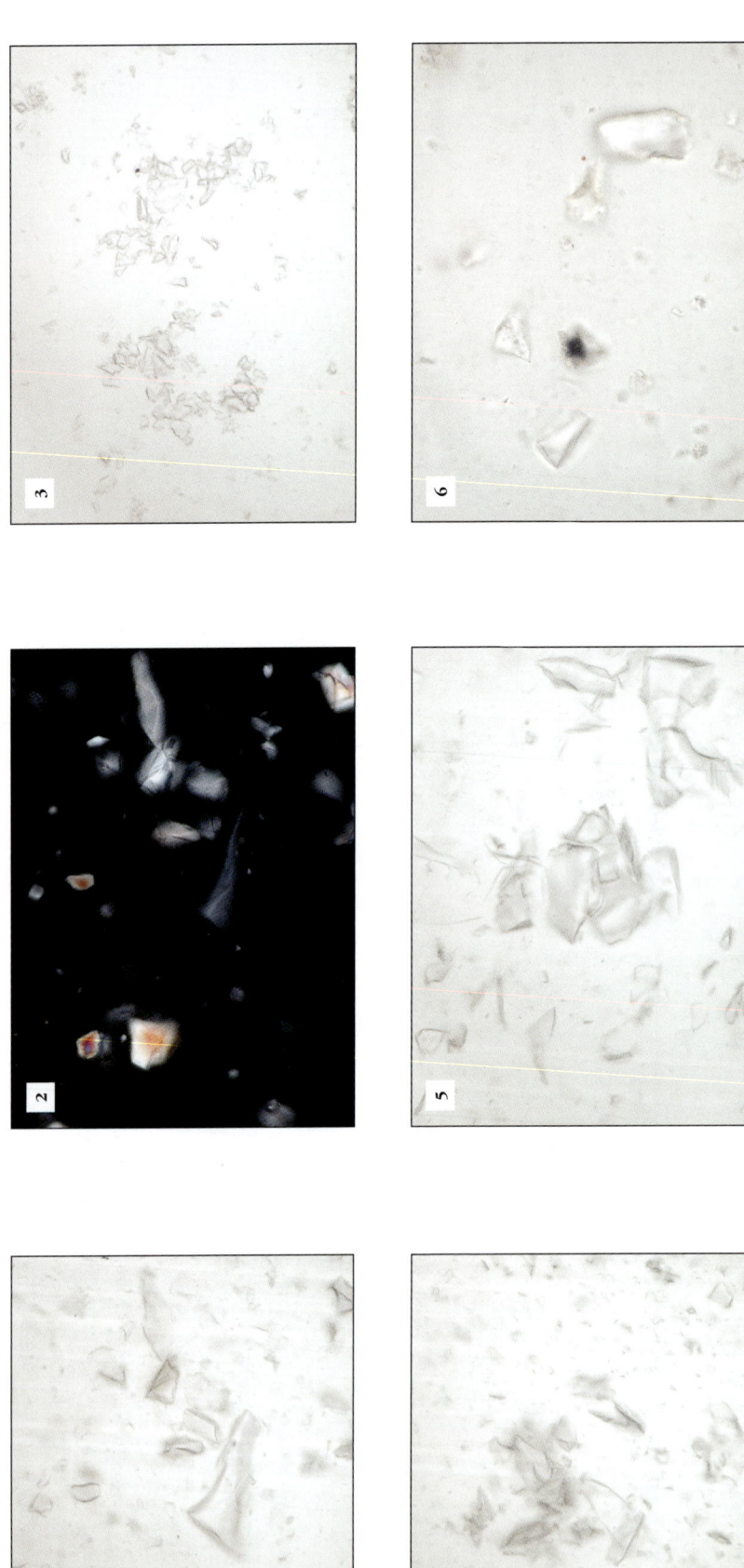

Key:

1. P1204: PPL/40×. Anhedral colourless fragments of 'hackmanite' showing conchoidal fracture on some grains.

2. P1204: ~XPL/40× / **o**. Weakly anisotropic hackmannite crystals in partially crossed polars.

3. P0008: PPL/100×. Colourless anhedral crystals of sodalite with feldspars as impurities (particularly the larger rounded colourless grain to the left of the image).

4. P1257: PPL/40×. Anhedral colourless crystals of sodalite.

5. P1137: PPL/40×. Colourless anhedral analcite grains with conchoidal fracture observable at some crystal edges.

6. P1385: PPL/100×/H. Colourless anhedral to subhedral crystals of wairakite.

Crystallinity	n	n_α	n_γ	δ
Sodalite: Cubic	1.483–1.484			Isotropic
Haüyne: Cubic	1.4961			Isotropic
Analcime: Triclinic		1.479–1.493	1.480–1.494	0.0010
Wairakite: Monoclinic		1.498	1.502	0.0040

(data from Webmineral, 2003)

The feldspathoids are a group of silicate minerals. Though not used directly as pigments, they appear as traces in mineral derived pigments, and primarily natural ultramarine. The prime colourant in this pigment, lazurite (*q.v.*), is a member of the sodalite family.

The sodalite group includes the minerals sodalite and haüyne. Under plane-polarised light both minerals appear transparent; they may be colourless, but also take on other colours including blue (thus causing confusion with lazurite). Particles have moderate relief, with the RI much less than that of the medium. Under crossed polars these phases are isotropic. In rare cases, substitutions distort the crystals and they become weakly birefringent, with very low first order interference colours. Associated phases showing first order greys are probably feldspars (*q.v.*). Sodalite, hackmannite and haüyne are not easily distinguished optically.

The analcite group includes the minerals analcime (analcite) and wairakite. Both are colourless under plane-polarised light and have moderate relief,

with RI less than that of the medium. Under crossed polars, though nominally of the cubic system, analcite and wairakite are frequently weakly anisotropic, showing first order greys. Both develop lamellar twins, which may be visible. The two are not easily distinguished optically.

The appearance of the colourless feldspathoid minerals is unaffected when viewed through the Chelsea filter. The blue feldspathoids transmit a deep red.

In addition to the pigment-bearing phases, feldspathoids may be found in association with (orthoclase) feldspar, calcite and opaque phases including iron pyrites. Feldspathoids do not occur in quartz-bearing rocks and the occurrence of the feldspathoids in earth pigments is unlikely.

Lit.: Deer *et al.* (1992)

MICA GROUP MINERALS

[Various]

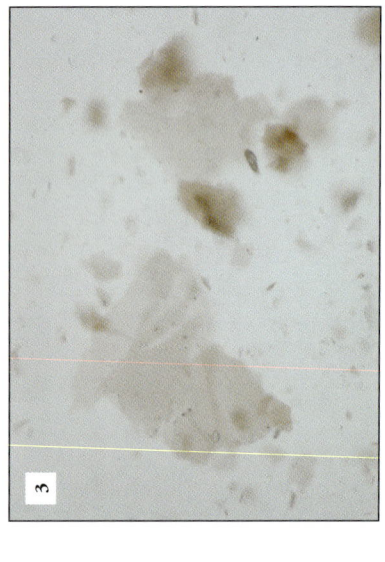

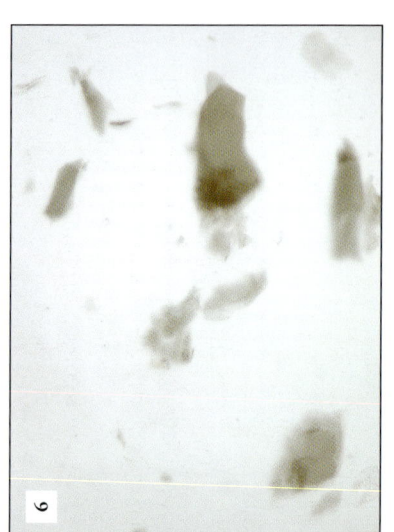

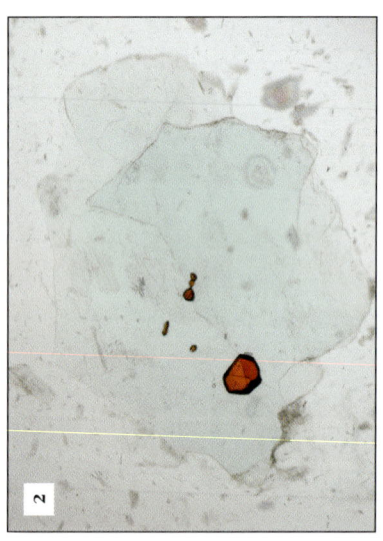

Key:

1. P1232: PPL/40×/ **O**. Colourless low relief plate of muscovite.

2. P0866: PPL/10×. Pale green plate of fuchsite containing orange rutile inclusions.

3. P1242: PPL/40×. Crystals of phlogopite.

4. P1232: XPL/40×/ **O**. The field of view as in Fig. 1 under crossed polars.

5. P0866: XPL/10×. The field of view as in Fig. 2 under crossed polars, note the partial illumination of the rutile crystal shows that it is twinned.

6. P1155: PPL/40×. Flakes of biotite.

Crystallinity	n_α	n_β	n_γ	δ
Fuchsite (Cr-Muscovite): Monoclinic	1.552–1.574	1.582–1.610	1.586–1.616	0.0340–0.0420
Phlogopite: Monoclinic	1.530–1.573	1.557–1.617	1.558–1.618	0.0280–0.0450
Biotite: Monoclinic	1.565–1.625	1.605–1.675	1.605–1.675	0.0400–0.0500
				(data from Webmineral, 2003)

In plane-polarised light, the mica minerals appear as transparent colourless (e.g. muscovite) or coloured brown or green crystals (e.g. biotite, phlogopite and fuchsite). The extraordinary cleavage of mica allows particles to be divided into extremely thin plates parallel to the basal section of the prism, and particles thus often fall on their basal sections. Occasionally, edge on, lamellar 'books' of mica are observed (and for coloured micas, these sections exhibit strong pleochroism; pleochroism is absent on basal sections). Relief for all micas is low to moderate and RI is less than that of the medium. Grain surfaces are smooth, while for lamellar particles the knife-sharp cleavage is visible. Micas are very prone to containing inclusions of other minerals and euhedral crystals of phases including apatite, zircon and rutile are common. Zircon inclusions may be associated with dark haloes of radiation damage.

Under crossed polars, micas have variable birefringence. On basal sections it is usually low and very thin plates show low first order interference colours or appear almost isotropic. On lamellar (prismatic) sections, third order interference colours are typical. For coloured micas, interference colours are masked by the body colour. Micas show mottled or sweeping extinction.

The Pigment Compendium

ASBESTIFORM MINERALS [Various]

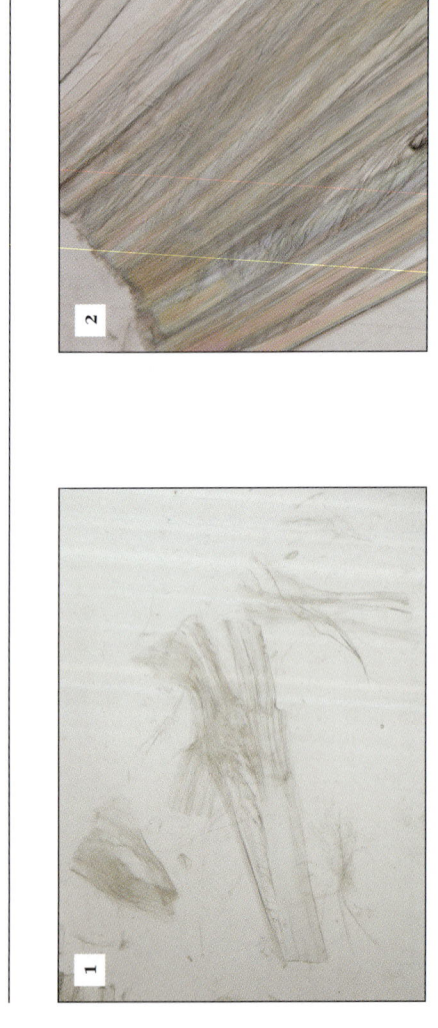

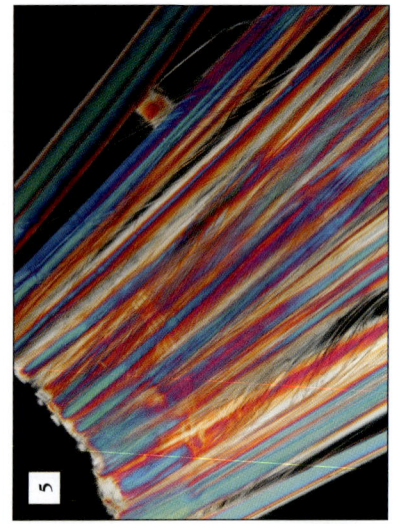

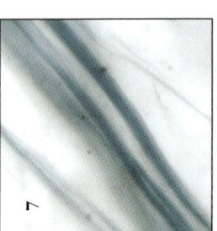

Key:

1. P1172: PPL/10×. Fibrous sheaf of chrysotile.

2. P1172: PPL/10×. The left-hand end of the chrysotile sheaf in Fig. 1 showing detail of fibrous structure.

3. P1504: PPL/10×. Fibrous blue crocidolite.

4. P1172: XPL/10× **○**. The field of view as in Fig. 1 under crossed polars.

5. P1172: XPL/10×. The field of view as in Fig. 2 under crossed polars.

6. P1504: XPL/10×. The field of view as in Fig. 3 under crossed polars.

7. P1504: PPL. Enlarged image of blue fibrous crocidolite.

8. P1504: PPL. The fibres in Fig. 7 with the polariser rotated through 90° to show pleochroism.

Crystallinity

	n_α	n_β	n_γ	δ
Tremolite–Actinolite series: Monoclinic	1.599–1.628	1.613–1.644	1.625–1.655	0.0250–0.0270
Chrysotile: Monoclinic	1.569		1.570	0.0010
Crocidolite (fibrous Riebeckite): Monoclinic	1.680–1.698	1.683–1.700	1.685–1.706	0.0050–0.0080

(data from Webmineral, 2003)

The term asbestiform refers to the fibrous nature of these minerals and is independent of the chemical or structural groupings. However, the majority of asbestiform minerals belong to the amphibole group and typically to the tremolite–actinolite solid solution series. Additionally the serpentine group clay mineral chrysotile also develops asbestiform habits. Crocidolite, commonly known as 'blue' asbestos, is related to the riebeckite–glaucophane solid-solution series (sodic amphiboles). Non-fibrous habits of these minerals are discussed under the entry for Glaucophane.

Under the microscope, these minerals are distinctive and appear as textile fibres or hairs. The most distinctive feature of these minerals is their habit, which is very fine and fibrous and crystals can be very long. Tremolite and chrysotile fibres commonly occur in bundles and are colourless in plane-polarised light. Crocidolite is blue (and remains blue when viewed through the Chelsea filter). Crocidolite is strongly pleochroic from blue to lavender or colourless. With substitutions in the solid-solution series of magnesium for iron RI decreases and birefringence correspondingly increases. For the riebeckite end-member of the solid solution series pleochroism is blue-yellow-colourless. All phases have moderate relief, with RI less than that of the medium.

Under crossed polars, all phases have moderate to high birefringence; however, thick mats and tufts of fibres may show anomalously high interference colours. For crocidolite extinction angle varies with composition and the fibres are length fast.

Appendix I: Samples

THE REFERENCE COLLECTION

A key element in the preparation of this book and its companion volume has been a review showing the range of pigments encountered in artefacts through research into documentary sources and the scientific literature. From that work an extensive list of the compounds that form the building blocks of pigments has been put together. To illustrate the microscopy, a substantial collection (now in excess of 1500 specimens) was assembled with the intention of including as many of these compounds as possible. Each of the samples had to be of good provenance and confirmed by analysis using multiple techniques – polarised light microscopy (PLM), scanning electron microscopy-energy dispersive X-ray spectrometry (SEM-EDX), X-ray diffraction (XRD) and Raman spectroscopy were used – to ensure clear characterisation of the collection wherever possible.

Material is derived from both historical and modern sources, though none are from artefacts. Some of these collections and information about the pigments that they contain have been published previously. In particular the original set of Turner studio pigments has been discussed in Townsend (1993) and Eastaugh (1995), the Hafkenscheid Collection has been the subject of a series of papers by Pey (1987a, 1987b, 1989, 1998), examples of cobalt violet pigments are contained in the recent paper by Corbeil (2003), some of the chromate pigments are included in Burnstock *et al.* (2003), numerous copper-based pigments are described in Scott (2002), pigments from Papua New Guinea are covered in Hill (2001), yeast cokes are included in Winter (1983), green earths formed the basis of Grissom (1986) and 'Hansa' yellows are discussed by Lake in the fourth volume of *Artists' Pigments*.

Specific sources for pigment examples as noted in the summary table are as follows:

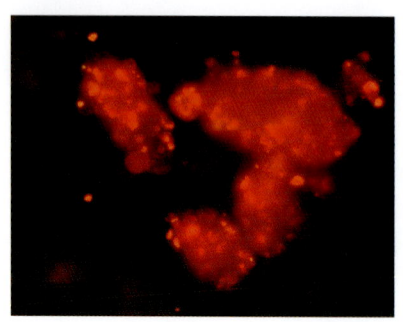

Ambers	Dr Janet Ambers, The British Museum, London, UK
Carlyle	Dr Leslie Carlyle,
Clearwell	Clearwell Caves, Forest of Dean, UK
Corbeil	Dr Marie-Claude Corbeil, Canadian Conservation Institute, Ottawa, Canada
Cornelissen	L. Cornelissen, London, UK
Courtauld	Courtauld Institute of Art, University of London, London, UK
De Kat	De Kat, Amsterdam, Netherlands
Eastaugh	Dr Nicholas Eastaugh, *The Pigmentum Project*, London, UK
Edwards	Keith Edwards, UK
Grissom	Carol Grissom, Washington DC, USA
Hafkenscheid	The Hafkenscheid Collection, Teylers Museum, Haarlem, Netherlands
Hatton	Gareth Hatton, RLAHA, University of Oxford, Oxford, UK
Hill	Rowena Hill, UK
Kremer	Kremer Pigmente, Germany
Lake	Susan Lake, Hirschorn Collection, Washington DC, USA
Middleton	Juliette Middleton, *The Pigmentum Project*, London, UK
Molyn	The Molyn Collection, Teylers Museum, Haarlem, Netherlands
Ovenstone	Dr James Ovenstone, University of Greenwich, London, UK
Pompeii	Museo Archaeologico, Pompeii, Italy
ROM	Royal Ontario Museum, Toronto, Canada
Scott	Dr David Scott, (formerly) Getty Conservation Institute, Los Angeles, USA
Seccaroni	Dr Claudio Seccaroni, Rome, Italy
Siddall	Dr Ruth Siddall, Department of Earth Sciences, University College, London, UK
Sigma-Aldrich	Sigma-Aldrich Chemical Co., Poole, UK
Steele	Dr Ian Steele, University of Chicago, Chicago, USA
Tayler	Richard Tayler Minerals, UK
Turner	J.M.W. Turner studio pigments, ex Courtauld Institute of Art, London, UK
von Imhoff	H.C. von Imhoff,
Walsh	Valentine Walsh, *The Pigmentum Project*, London, UK
Winter	Dr John Winter, Freer Gallery of Art, Washington DC, USA

P0001	*Smalte extra fein*	Kremer Pigmente, Germany, ref. K-1001
P0005	*Malachit* ('Malachite')	Kremer Pigmente, Germany, ref. K-1030063
P0007	*Eisenblau, vivianite* ('Vivianite')	Kremer Pigmente, Germany, ref. K-1040
P0008	*Sodalith* ('Sodalite')	Kremer Pigmente, Germany, ref. K-10420
P0009	*Lapis lazuli einfache qualitat* ('Lapis lazuli, poor quality')	Kremer Pigmente, Germany, ref. K-10500
P0010	*Zinnober aus mineral* ('Cinnabar as mineral')	Kremer Pigmente, Germany, ref. K-10620
P0011	*Auripigment echt* ('True orpiment')	Kremer Pigmente, Germany, ref. K-10700
P0012	*Bleiglanz schwarzgrau glanzend* ('Lead glance, black-grey, lustrous')	Kremer Pigmente, Germany, ref. K-10900
P0013	*Pyrite* ('Pyrite')	Kremer Pigmente, Germany, ref. K-1092
P0014	*Antimony stibnite* ('Stibnite')	Kremer Pigmente, Germany, ref. K-1094
P0015	*Purpurit eisenmang an phosphate* ('Purple iron manganese phosphate')	Kremer Pigmente, Germany, ref. K-10960
P0025	*Bergkristall* ('Mountain crystal')	Kremer Pigmente, Germany, ref. K-11400
P0026	*Marienglas* ('Marienglass (Selenite)')	Kremer Pigmente, Germany, ref. K-11800
P0027	*Elfenbein-schwarz echt* ('True ivory black')	Kremer Pigmente, Germany, ref. K-1200
P0028	*Bistre aus glanzruss* ('Bistre')	Kremer Pigmente, Germany, ref. K-1210
P0029	*Sepia* ('Sepia')	Kremer Pigmente, Germany, ref. K-12400
P0030	*Cadmiumgelb nr1 zitron* ('Cadmium yellow #1 lemon')	Kremer Pigmente, Germany, ref. K-21010
P0033	*Cadmiumrot nr1 hell* ('Cadmium red #1 light')	Kremer Pigmente, Germany, ref. K-21120
P0035	*Heliogengrun* ('Helio green')	Kremer Pigmente, Germany, ref. K-23000
P0036	*Heliogengrun gelbstihig* ('Helio green yellowish')	Kremer Pigmente, Germany, ref. K-23010
P0039	*Permanent gelb* ('Permanent yellow')	Kremer Pigmente, Germany, ref. K-2332
P0042	*Brilliant gelb PY74 Hansa* ('Brilliant yellow Pigment Yellow 74 Hansa')	Kremer Pigmente, Germany, ref. K-2365
P0044	*Weld* ('Weld')	Kremer Pigmente, Germany, ref. K-3626
P0045	*Brennessel pflanzenfarbe* ('Stinging nettle plant colour')	Kremer Pigmente, Germany, ref. K-36316
P0046	*Blattgrun pflanzenfarbe* ('Leaf green plant colour')	Kremer Pigmente, Germany, ref. K-36318

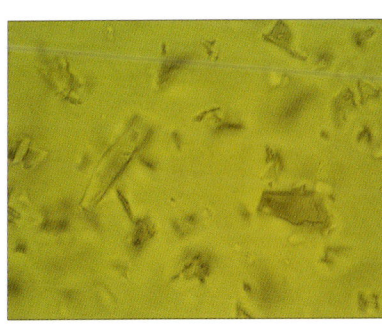

P0048	*Krapplack aus wurzeln* ('Madder root')	Kremer Pigmente, Germany, ref. K-3720C
P0049	*Schuttgelb stil de grain* ('Stil de grain, Rhamnus')	Kremer Pigmente, Germany, ref. K-37390A
P0053	*Goldocker H84 deutsch* ('German golden ochre (H84)')	Kremer Pigmente, Germany, ref. K-40210
P0055	*Satinober Deutsch* ('German ochre')	Kremer Pigmente, Germany, ref. K-40250
P0056	*Terra di Siena Italianish natur* ('Italian raw sienna')	Kremer Pigmente, Germany, ref. K-40400
P0059	*Bolus rot* ('Red bole')	Kremer Pigmente, Germany, ref. K-4050
P0062	*Manganbraun intensiv* ('Manganese brown intense')	Kremer Pigmente, Germany, ref. K-40623
P0063	*Umbragrunisch dunkel* ('Dark greenish umber')	Kremer Pigmente, Germany, ref. K-40630
P0064	*Umbra gebr. Italianisch* ('Italian burnt umber')	Kremer Pigmente, Germany, ref. K-4070
P0070	*Casselerbraun* ('Cassel brown')	Kremer Pigmente, Germany, ref. K-4100
P0072	*Ultramarin rot (altrosa)* ('Ultramarine red')	Kremer Pigmente, Germany, ref. K-42600
P0073	*Massicot bleiglatte* ('Massicot lead oxide')	Kremer Pigmente, Germany, ref. K-43010
P0074	*Neapelgelb hell* ('Light Naples yellow')	Kremer Pigmente, Germany, ref. K-4310
P0075	*Naples yellow*	Kremer Pigmente, Germany, ref. K-4311
P0081	*Kobaltgelb, aureolin* ('Aureolin, cobalt yellow')	Kremer Pigmente, Germany, ref. K-43500
P0083	*Chromgelb rotlich blh* ('Reddish chrome yellow')	Kremer Pigmente, Germany, ref. K-4380
P0085	*Zincgelb* ('Zinc yellow')	Kremer Pigmente, Germany, ref. K-4390
P0086	*Barytgelb barium chromat* ('Barium chromate')	Kremer Pigmente, Germany, ref. K-43940
P0091	*Chromoxidgrun stumpf* ('Opaque chrome oxide')	Kremer Pigmente, Germany, ref. K-44200
P0096	*Grunspan* ('Verdigris')	Kremer Pigmente, Germany, ref. K-44450
P0097	*Cadmiumgrun dunkel* ('Dark cadmium green')	Kremer Pigmente, Germany, ref. K-44510
P0100	*Ultramarinviolett rotlich* ('Reddish ultramarine violet')	Kremer Pigmente, Germany, ref. K-45120
P0101	*Miloriblau* ('Milori blue (Prussian blue)')	Kremer Pigmente, Germany, ref. K-45200
P0102	*Mangancolinblau aus altbestand* ('Manganese cerulean blue (old stock)')	Kremer Pigmente, Germany, ref. K-4530
P0103	*Manganviolett* ('Manganese violet')	Kremer Pigmente, Germany, ref. K-45350
P0105	*Kobaltblau mittel deckend* ('Cobalt blue, middle covering')	Kremer Pigmente, Germany, ref. K-45710

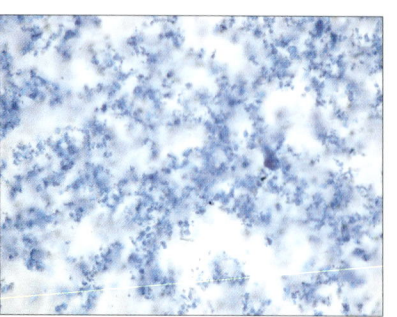

P0108	*Kobaltblau hell* ('Light cobalt blue')	Kremer Pigmente, Germany, ref. K-4572
P0112	*Kobaltviolett hell* ('Light cobalt violet')	Kremer Pigmente, Germany, ref. K-45810
P0113	*Kobaltviolett dunkel* ('Dark cobalt violet')	Kremer Pigmente, Germany, ref. K-45800
P0114	*Kobaltviolett hell* ('Light cobalt violet')	Kremer Pigmente, Germany, ref. K-45820
P0115	*Kremserweiss* ('Krems white (lead white)')	Kremer Pigmente, Germany, ref. K-4600
P0116	*Bleisulfat* ('Lead sulfate')	Kremer Pigmente, Germany, ref. K-4605
P0117	*Lithopone silber* ('Silver lithopone')	Kremer Pigmente, Germany, ref. K-4610
P0118	*Titanweiss* ('Titanium white')	Kremer Pigmente, Germany, ref. K-46200
P0121	*Gofun shirayuki* ('Gofun, shiryauki (oyster shells)')	Kremer Pigmente, Germany, ref. K-46400
P0123	*Beinschwarz aus knochen* ('Bone black from bones')	Kremer Pigmente, Germany, ref. K-4710
P0125	*Spinelschwarz tiefstschwarz* ('Spinel black, deepest black')	Kremer Pigmente, Germany, ref. K-47400
P0126	*Manganschwarz* ('Manganese black')	Kremer Pigmente, Germany, ref. K-47500
P0127	*Asphalt-graphit amerikanisch* ('American asphalt-graphite')	Kremer Pigmente, Germany, ref. K-47600
P0128	*Graphitsilberpuder* ('Graphite silver powder')	Kremer Pigmente, Germany, ref. K-47700
P0130	*Eisenoxid gelb 920* ('Iron oxide yellow 920')	Kremer Pigmente, Germany, ref. K-4800
P0131	*Eisenoxid orange* ('Iron oxide orange')	Kremer Pigmente, Germany, ref. K-4806
P0132	*Eisenoxid rot* ('Iron oxide red')	Kremer Pigmente, Germany, ref. K-4820
P0135	*Bismut metallpulv* ('Bismuth metal powder')	Kremer Pigmente, Germany, ref. K-5400
P0137	*Eisenoxid rot natur* ('Iron oxide red natural')	Kremer Pigmente, Germany, ref. K-48600
P0146	*Campagner kreide* ('Champagne chalk')	Kremer Pigmente, Germany, ref. K-5800
P0147	*Rugener kreide* ('Rugener chalk')	Kremer Pigmente, Germany, ref. K-58010
P0150	*China clay, Englisch* ('English china clay')	Kremer Pigmente, Germany, ref. K-5820
P0153	*Alabaster gips ungebrannt* ('Alabaster plaster, unburnt')	Kremer Pigmente, Germany, ref. K-5834
P0158	*Blancfix schwerspat natur* ('Blanc fixe')	Kremer Pigmente, Germany, ref. K-5870
P0159	*Bentonit* ('Bentonite')	Kremer Pigmente, Germany, ref. K-5890
P0160	*Knochenasche weiss* ('Bone ash white')	Kremer Pigmente, Germany, ref. K-5892

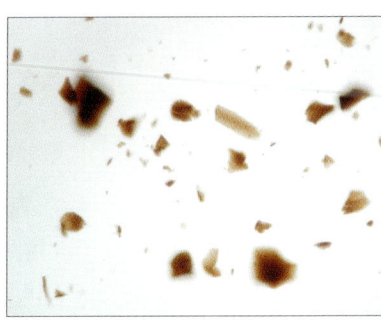

P0161	*Aluminium hydroxid* ('Aluminium hydroxide')	Kremer Pigmente, Germany, ref. K-58941

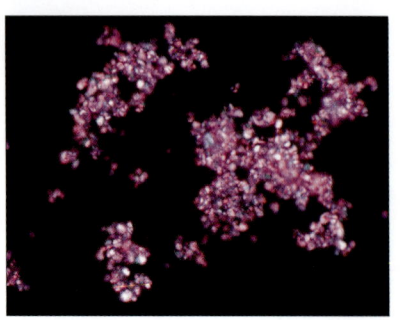

P0162	*Carrara weiss* ('Carrara white (marble)')	Kremer Pigmente, Germany, ref. K-5960
P0164	*Barium carbonat* ('Barium carbonate')	Kremer Pigmente, Germany, ref. K-6409
P0167	*Rhodamine B* ('Rhodamine B')	Kremer Pigmente, Germany, ref. K-9490
P0169	*Tripel rotlich, gutfullend* ('Reddish tripoli')	Kremer Pigmente, Germany, ref. K-99920
P0170	*Attapulgite*	Kremer Pigmente, Germany, ref. K-9994
P0176	*Green verditer*	Kremer Pigmente, Germany, ref. GRBICE
P0177	*Stinkspat* ('Antozonite (Fluorite)')	Kremer Pigmente, Germany, ref. -
P0178	*Tyrian purple*	Kremer Pigmente, Germany, ref. -
P0180	*Egyptian blue*	Kremer Pigmente, Germany, ref. -
P0181	*Maya blue*	Kremer Pigmente, Germany, ref. -
P0183	*Graphite*	H.-C. von Imhoff personal collection, ref. 006 101
P0185	*Ivory black*	H.-C. von Imhoff personal collection, ref. 006 103
P0186	*Asphalt*	H.-C. von Imhoff personal collection, ref. 006 702
P0190	*Zinc white*	H.-C. von Imhoff personal collection, ref. 130 101
P0191	*Cremnitz white*	H.-C. von Imhoff personal collection, ref. 182 101
P0192	*Manganese violet*	H.-C. von Imhoff personal collection, ref. 225 101
P0194	*Caput mortuum violet*	H.-C. von Imhoff personal collection, ref. 226 102
P0196	*Cobalt violet*	H.-C. von Imhoff personal collection, ref. 227 102
P0197	*Monastral fast blue*	H.-C. von Imhoff personal collection, ref. 306 101
P0198	*Indigo (synthetic)*	H.-C. von Imhoff personal collection, ref. 306 102
P0199	*Indigo (natural)*	H.-C. von Imhoff personal collection, ref. 306 103
P0201	*Ultramarine green*	H.-C. von Imhoff personal collection, ref. 316 102
P0202	*Manganese blue*	H.-C. von Imhoff personal collection, ref. 325 101
P0204	*Parisien blue*	H.-C. von Imhoff personal collection, ref. 326 102
P0205	*Cerulean blue*	H.-C. von Imhoff personal collection, ref. 327 101
P0206	*Cobalt blue*	H.-C. von Imhoff personal collection, ref. 327 102

P0207	*Copper blue*	H.-C. von Imhoff personal collection, ref. 329 101
P0209	*Oxide of chromium*	H.-C. von Imhoff personal collection, ref. 424 101
P0210	*Viridian*	H.-C. von Imhoff personal collection, ref. 424 102
P0213	*Cobalt green*	H.-C. von Imhoff personal collection, ref. 427 101
P0214	*Schweinfurt green*	H.-C. von Imhoff personal collection, ref. 429 101
P0215	*Copper green*	H.-C. von Imhoff personal collection, ref. 429 102
P0217	*Scheele's green*	H.-C. von Imhoff personal collection, ref. 429 104
P0219	*Hansa yellow*	H.-C. von Imhoff personal collection, ref. 506 101
P0220	*Hansa yellow*	H.-C. von Imhoff personal collection, ref. 506 102
P0222	*Persian berries yellow*	H.-C. von Imhoff personal collection, ref. 506 403
P0224	*Chrome yellow*	H.-C. von Imhoff personal collection, ref. 524 101
P0225	*Yellow ochre*	H.-C. von Imhoff personal collection, ref. 526 101
P0226	*Iron oxide yellow*	H.-C. von Imhoff personal collection, ref. 526 102
P0232	*Strontium yellow*	H.-C. von Imhoff personal collection, ref. 538 101
P0233	*Cadmium yellow*	H.-C. von Imhoff personal collection, ref. 548 101
P0236	*Cadmium yellow lithopone*	H.-C. von Imhoff personal collection, ref. 548 201
P0242	*Massicot*	H.-C. von Imhoff personal collection, ref. 582 101
P0243	*Molibdat orange ('Molybdate orange')*	H.-C. von Imhoff personal collection, ref. 642 101
P0244	*Cadmium orange*	H.-C. von Imhoff personal collection, ref. 648 101
P0245	*Cadmium orange*	H.-C. von Imhoff personal collection, ref. 648 102
P0248	*Dark madder*	H.-C. von Imhoff personal collection, ref. 706 401
P0253	*Iron oxide red*	H.-C. von Imhoff personal collection, ref. 726 101
P0257	*Indian red*	H.-C. von Imhoff personal collection, ref. 726 105
P0258	*Iron oxide red ('Mars red')*	H.-C. von Imhoff personal collection, ref. 726 106
P0260	*Cadmium red*	H.-C. von Imhoff personal collection, ref. 748 102
P0262	*Zinnober ('Cinnabar')*	H.-C. von Imhoff personal collection, ref. 780 101
P0263	*Minium*	H.-C. von Imhoff personal collection, ref. 782 101

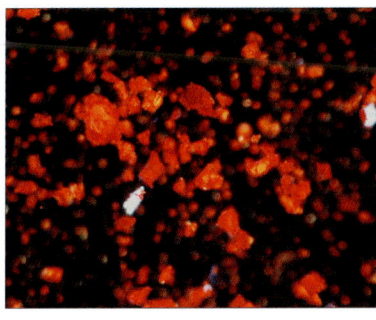

P0264	*Cassel earth*	H.-C. von Imhoff personal collection, ref. 806 701
P0265	*Raw umber green*	H.-C. von Imhoff personal collection, ref. 825 101
P0268	*Umber clear*	H.-C. von Imhoff personal collection, ref. 825 104
P0271	*Sienna*	H.-C. von Imhoff personal collection, ref. 826 103
P0272	*Burnt sienna*	H.-C. von Imhoff personal collection, ref. 826 104
P0273	*Bohemian green earth burnt*	H.-C. von Imhoff personal collection, ref. 826 105
P0278	*Copper compound, Scott 4*	Dr David Scott, Getty Conservation Institute, ref. 4
P0280	*Copper compound, Scott 6*	Dr David Scott, Getty Conservation Institute, ref. 6
P0281	*Copper compound, Scott 7*	Dr David Scott, Getty Conservation Institute, ref. 7
P0283	*Copper compound, Scott 9*	Dr David Scott, Getty Conservation Institute, ref. 9
P0285	*Copper compound, Scott 11*	Dr David Scott, Getty Conservation Institute, ref. 11
P0286	*Copper compound, Scott 12*	Dr David Scott, Getty Conservation Institute, ref. 12
P0287	*Copper compound, Scott 13*	Dr David Scott, Getty Conservation Institute, ref. 13
P0288	*Copper compound, Scott 14*	Dr David Scott, Getty Conservation Institute, ref. 14
P0289	*Copper compound, Scott 15*	Dr David Scott, Getty Conservation Institute, ref. 15
P0290	*Copper compound, Scott 16*	Dr David Scott, Getty Conservation Institute, ref. 16
P0292	*Copper compound, Scott 18*	Dr David Scott, Getty Conservation Institute, ref. 18
P0294	*Copper compound, Scott 20*	Dr David Scott, Getty Conservation Institute, ref. 20
P0298	*Copper compound, Scott 24*	Dr David Scott, Getty Conservation Institute, ref. 24
P0299	*Copper compound, Scott 25*	Dr David Scott, Getty Conservation Institute, ref. 25
P0300	*Copper compound, Scott 26*	Dr David Scott, Getty Conservation Institute, ref. 26
P0301	*Copper compound, Scott 27*	Dr David Scott, Getty Conservation Institute, ref. 27
P0303	*Copper compound, Scott 29*	Dr David Scott, Getty Conservation Institute, ref. 29
P0306	*Copper compound, Scott 32*	Dr David Scott, Getty Conservation Institute, ref. 32
P0307	*Copper compound, Scott 33*	Dr David Scott, Getty Conservation Institute, ref. 33
P0311	*Copper compound, Scott 37*	Dr David Scott, Getty Conservation Institute, ref. 37
P0312	*Copper compound, Scott 38*	Dr David Scott, Getty Conservation Institute, ref. 38

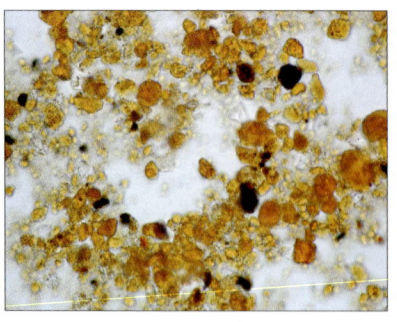

P0315	*Copper compound, Scott 41*	Dr David Scott, Getty Conservation Institute, ref. 41
P0317	*Char, yeast*	Dr John Winter, Freer Gallery of Art, Washington D.C., ref. 1
P0318	*Char, yeast*	Dr John Winter, Freer Gallery of Art, Washington D.C., ref. 2
P0319	*Char, yeast*	Dr John Winter, Freer Gallery of Art, Washington D.C., ref. 3
P0321	*Lapis lazuuli* ('Lapis lazuli')	Hafkenscheid Collection, Teylers Museum, Haarlem, ref. HI1
P0322	*Antimonium* ('Antimony')	Hafkenscheid Collection, Teylers Museum, Haarlem, ref. HI3 (H3)
P0323	*Cromiaat erts Americaansch* ('American chromium ore')	Hafkenscheid Collection, Teylers Museum, Haarlem, ref. HI4 (H4)
P0332	*Oprement Chineesch* ('Chinese orpiment')	Hafkenscheid Collection, Teylers Museum, Haarlem, ref. HI14 (H13)
P0333	*Oprement Levantsch of Aurumpigmentum* ('Levantine orpiment or auripiment')	Hafkenscheid Collection, Teylers Museum, Haarlem, ref. HI15 (H14)
P0348	*Fraaie dunker bruin oker* ('Beautiful dark brown ochre')	Hafkenscheid Collection, Teylers Museum, Haarlem, ref. HII5-B (H28)
P0354	*Omber* ('Umber')	Hafkenscheid Collection, Teylers Museum, Haarlem, ref. HII11 (H34)
P0356	*Terrasiena Geele* ('Yellow sienna earth')	Hafkenscheid Collection, Teylers Museum, Haarlem, ref. HII13 (H36)
P0357	*Bolus Armenische* ('Armenian bole')	Hafkenscheid Collection, Teylers Museum, Haarlem, ref. HII14 (H37)
P0358	*Bolus Roode* ('Red bole')	Hafkenscheid Collection, Teylers Museum, Haarlem, ref. HII15 (H38)
P0360	*Keulscheaarde* ('Cologne earth')	Hafkenscheid Collection, Teylers Museum, Haarlem, ref. HII17 (H40)
P0361	*Casselsche Aarde* ('Cassel earth')	Hafkenscheid Collection, Teylers Museum, Haarlem, ref. HII18 (H41)
P0369	*Puimsteen witte* ('Pumice stone white')	Hafkenscheid Collection, Teylers Museum, Haarlem, ref. HIII3 (H49)
P0370	*Puimsteen Grijze zeer ligt en fraai* ('Pumice stone, very light grey and beautiful')	Hafkenscheid Collection, Teylers Museum, Haarlem, ref. HIII4 (H50)

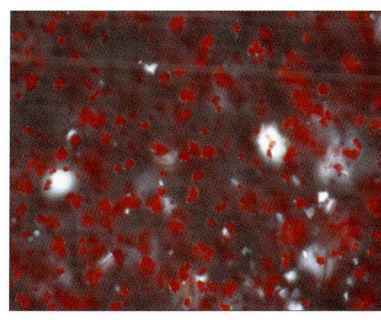

P0371	*Falcum* ('Talcum')	Hafkenscheid Collection, Teylers Museum, Haarlem, ref. HII5 (H51)
P0372	*Fransch Krÿt* ('French chalk')	Hafkenscheid Collection, Teylers Museum, Haarlem, ref. HIII6 (H52)
P0408	*Kurkuma Chineesche* ('Chinese curcuma')	Hafkenscheid Collection, Teylers Museum, Haarlem, ref. HV15 (H96)
P0410	*Kurkuma Bengalsche* ('Benegal curcuma')	Hafkenscheid Collection, Teylers Museum, Haarlem, ref. HV17 (H98)
P0429	*Indigo*	Hafkenscheid Collection, Teylers Museum, Haarlem, ref. HVI ? (H117)
P0482	*Drakenbloed* ('Dragon's blood')	Hafkenscheid Collection, Teylers Museum, Haarlem, ref. HIX1 (H172)
P0483	*Gum Guttae* ('Gamboge')	Hafkenscheid Collection, Teylers Museum, Haarlem, ref. HIX3 (H174)
P0493	*Arsenicum Witte* ('Arsenic white')	Hafkenscheid Collection, Teylers Museum, Haarlem, ref. HIX14 (H184)
P0503	[*Un-named*]	Hafkenscheid Collection, Teylers Museum, Haarlem, ref. HIX ? (H194)
P0507	*Loodwit Perebooms* ('Lead white from Perebooms')	Hafkenscheid Collection, Teylers Museum, Haarlem, ref. HX2 (H196C)
P0509	*Schulpwit* ('Shell white')	Hafkenscheid Collection, Teylers Museum, Haarlem, ref. HX2 (H197A)
P0510	*Schulpwit Engelsch* ('Shell white, English')	Hafkenscheid Collection, Teylers Museum, Haarlem, ref. HX3-B (H197B)
P0532	*Papagaai Groen* ('Parrot green')	Hafkenscheid Collection, Teylers Museum, Haarlem, ref. HX18-A, #1 (H211)
P0534	*Minraal Groen* ('Mineral green')	Hafkenscheid Collection, Teylers Museum, Haarlem, ref. HX18-B (H211)
P0535	*Minraal Groen Donker* ('Mineral green, dark')	Hafkenscheid Collection, Teylers Museum, Haarlem, ref. HX18-B, 2 (H211)
P0537	*Berggroen* ('Mountain green')	Hafkenscheid Collection, Teylers Museum, Haarlem, ref. HX19-B (H212B)
P0538	*Cromaat geel* ('Chrome yellow')	Hafkenscheid Collection, Teylers Museum, Haarlem, ref. HX ? (H213)
P0545	*Schÿtgeel Englisch* ('Pink, English')	Hafkenscheid Collection, Teylers Museum, Haarlem, ref. HXI3 (H221)

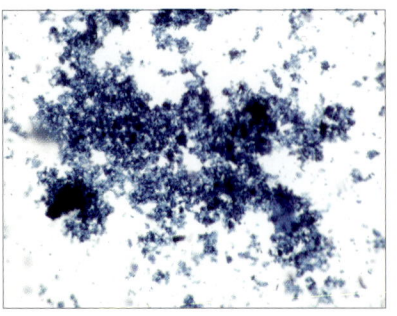

P0547	*Schÿtgeel Bruin Donker Extra fraai* ('Brown pink, dark, very beautiful')	Hafkenscheid Collection, Teylers Museum, Haarlem, ref. HXI5 no. 1 (H223)
P0549	*Vermilloen Heele* ('Vermilion, light')	Hafkenscheid Collection, Teylers Museum, Haarlem, ref. HXI8 (H225)
P0551	*Vermilloen Inlandsch 2× Gemalen* ('Vermilion, Dutch, ground 2×')	Hafkenscheid Collection, Teylers Museum, Haarlem, ref. HXI9-B (H226 2×)
P0553	*Vermilloen inlandsch 4× gemalen* ('Vermilion, Dutch, ground 4×')	Hafkenscheid Collection, Teylers Museum, Haarlem, ref. HXI9-D (H226 4×)
P0571	*Carmÿn* ('Carmine')	Hafkenscheid Collection, Teylers Museum, Haarlem, ref. HXI11 (H228 #20B)
P0578	*Florentÿnsche lak > cochenille* ('Florentine lake, cochineal')	Hafkenscheid Collection, Teylers Museum, Haarlem, ref. HXI13-A1 (H230)
P0579	*Florentÿnsche lak > cochenille* ('Florentine lake, cochineal')	Hafkenscheid Collection, Teylers Museum, Haarlem, ref. HXI13A-2 (H230-2)
P0584	*Ultramarÿn/Cobalt blaauw ?* ('Ultramarine or cobalt blue?')	Hafkenscheid Collection, Teylers Museum, Haarlem, ref. HXI17A (H234)
P0601	*Oost Indiesche Inkt* ('East Indian ink')	Hafkenscheid Collection, Teylers Museum, Haarlem, ref. HXII11 (H250)
P0613	*Blaauwsel Noordsche* ('Nordic Saxony blue')	Hafkenscheid Collection, Teylers Museum, Haarlem, ref. HXIII2 (H264)
P0618	*Strooi Blauw* ('Strewing blue')	Hafkenscheid Collection, Teylers Museum, Haarlem, ref. HXIII4 (H266)
P0620	*Blaauwsel vit de Kiezerl. Oostenar Fabriek* ('Smalt from the Imperial Austrian factory')	Hafkenscheid Collection, Teylers Museum, Haarlem, ref. HXIII6 (H268)
P0621	*Blaauwsel vit de Hungarisch Fabriek* ('Smalt from the Hungarian factory')	Hafkenscheid Collection, Teylers Museum, Haarlem, ref. HXIII7 (H269)
P0632	*Geel Oker Zuiver Gemalen Fransche Oude Mÿn* ('French yellow ochre, ground, old mine')	Hafkenscheid Collection, Teylers Museum, Haarlem, ref. HXIV 1-A (H282)
P0638	*Venetiaansch Rood* ('Venetian red')	Hafkenscheid Collection, Teylers Museum, Haarlem, ref. HXIV-3 (H284)
P0656	*Loodaarde gemalen* ('Lead earth, ground')	Hafkenscheid Collection, Teylers Museum, Haarlem, ref. HXIV 11-C (H292)
P0658	*Oprement Groene* ('Green orpiment')	Hafkenscheid Collection, Teylers Museum, Haarlem, ref. HXIV 13-B (H294)
P0721	*Lood menie (kristalmenie)* Minium (Crystallised minium)	Molyn Collection, Teylers Museum, Haarlem

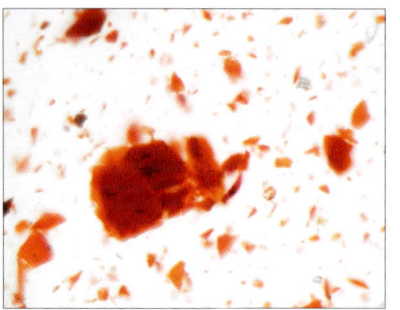

P0723	*Orangechromaat* ('Chrome orange')	Molyn Collection, Teylers Museum, Haarlem
P0726	*Terra di sienna* ('Sienna')	Molyn Collection, Teylers Museum, Haarlem
P0734	*Bruin oker* ('Brown ochre')	Molyn Collection, Teylers Museum, Haarlem
P0735	*Turkschrood* ('Turkish red')	Molyn Collection, Teylers Museum, Haarlem
P0740	*Barytgeel* ('Baryte yellow')	Molyn Collection, Teylers Museum, Haarlem
P0749	*Englisch blauw monastral* ('English monastral blue')	Molyn Collection, Teylers Museum, Haarlem
P0750	*Ultramarine blau (soda ultr)* ('Ultramarine blue')	Molyn Collection, Teylers Museum, Haarlem
P0751	*Smalt of strooi blauw* ('Smalt or strewer's blue')	Molyn Collection, Teylers Museum, Haarlem
P0754	*Vlamzwart* ('Flame black')	Molyn Collection, Teylers Museum, Haarlem
P0756	*Van dyksbruin* ('Van Dyck brown')	Molyn Collection, Teylers Museum, Haarlem
P0762	*Zinkwit (lood vky)* ('Zinc white')	Molyn Collection, Teylers Museum, Haarlem
P0764	*Gips* ('Plaster')	Molyn Collection, Teylers Museum, Haarlem
P0771	*Titanwit (standaard T)* ('Titanium white')	Molyn Collection, Teylers Museum, Haarlem
P0775	*Permanentwit (blank fixe)* ('Permanent white (*blanc fixe*)')	Molyn Collection, Teylers Museum, Haarlem
P0785	*Brown ochre*	De Kat, Amsterdam
P0786	*Curcumin*	De Kat, Amsterdam
P0788	*Brown ochre English*	De Kat, Amsterdam
P0796	*Bone black*	De Kat, Amsterdam
P0800	*Lamp black*	De Kat, Amsterdam
P0809	*Mummie Zwart Of Asfart Trinidad N.BL.6* ('Mummy black from Trinidad asphalt')	De Kat, Amsterdam
P0813	*Lead white Dutch process*	Dr Leslie Carlyle personal collection
P0814	*Lead white Kremer*	Dr Leslie Carlyle personal collection
P0815	*Lamp black Kremer*	Dr Leslie Carlyle personal collection
P0817	*Azurit dark* ('Natural azurite')	Kremer Pigmente, Germany, ref. K102002
P0818	*Azurite deep*	Kremer Pigmente, Germany, ref. K102002
P0826	*Helioblue* ('Helio blue PB15')	Kremer Pigmente, Germany, ref. K23050
P0833	*Drachenblut* ('Dragonsblood')	Kremer Pigmente, Germany, ref. K3700
P0836	*Gummi gutti* ('Gummi gutti')	Kremer Pigmente, Germany, ref. K3705
P0837	*Gummi gutti rohren stucke* ('Gummi gutti raw pieces')	Kremer Pigmente, Germany, ref. K3706

P0839	*Curcuma-pulver* ('Curcumin powder')	Kremer Pigmente, Germany, ref. K37220
P0840	*Schuttgelb* ('Stil de grain (Rhamnus)')	Kremer Pigmente, Germany, ref. K3739 A
P0846	*Aureolin* ('Aureolin')	Kremer Pigmente, Germany, ref. K4350
P0856	*Heliogen blau gold* ('Helio brown gold')	Kremer Pigmente, Germany, ref. K5090
P0860	*Attapulgite*	Kremer Pigmente, Germany, ref. K9994
P0862	*Kobalt senate natural* ('Cobalt arsenate, natural')	Kremer Pigmente, Germany
P0863	*Zinnober Monte Amiata* ('Cinnabar from Monte Amiata')	Kremer Pigmente, Germany
P0866	*Fuchsite*	Kremer Pigmente, Germany, ref. K11424
P0867	*Aerinite*	Kremer Pigmente, Germany
P0871	*Copper resinate*	Kremer Pigmente, Germany, ref. K1?70
P0874	*Antimony(III) oxide*	Sigma-Aldrich Co. Ltd, UK, ref. 23,089-8
P0876	*Barium manganate*	Sigma-Aldrich Co. Ltd, UK, ref. 21,019-6
P0880	*Calcium carbonate*	Sigma-Aldrich Co. Ltd, UK, ref. 23,921-6
P0882	*Calcium sulfate*	Sigma-Aldrich Co. Ltd, UK, ref. 23,713-2
P0883	*Calcium sulfate dihydrate*	Sigma-Aldrich Co. Ltd, UK, ref. C3771
P0884	*Calcium sulfate hemihydrate*	Sigma-Aldrich Co. Ltd, UK, ref. 30,766-1
P0888	*Copper(I) oxide*	Sigma-Aldrich Co. Ltd, UK, ref. 20,882-5
P0890	*Copper(II) hydroxide*	Sigma-Aldrich Co. Ltd, UK, ref. 28,978-7
P0891	*Copper(II) oxide*	Sigma-Aldrich Co. Ltd, UK, ref. 24,174-1
P0897	*Lead(II) carbonate basic*	Sigma-Aldrich Co. Ltd, UK, rcf. 24,358 2
P0898	*Lead(II) chromate*	Sigma-Aldrich Co. Ltd, UK, ref. 31,044-1
P0899	*Lead(II) sulfate*	Sigma-Aldrich Co. Ltd, UK, ref. 30,773-4
P0900	*Lead(II) sulfide*	Sigma-Aldrich Co. Ltd, UK, ref. 37,259-5
P0902	*Lead(IV) oxide*	Sigma-Aldrich Co. Ltd, UK, ref. 23,714-0
P0907	*Mercury(I) iodide*	Sigma-Aldrich Co. Ltd, UK, ref. 40,074-2
P0909	*Mercury(II) iodide*	Sigma-Aldrich Co. Ltd, UK, ref. 22,109-0
P0910	*Mercury(II) oxide*	Sigma-Aldrich Co. Ltd, UK, ref. 21,335-7

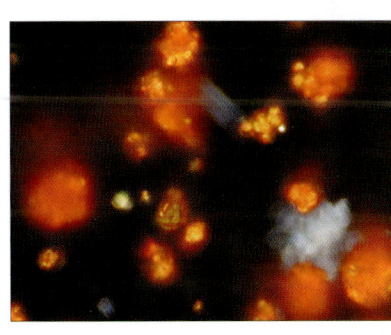

P0914	*Potassium hexachloroplatinate(IV)*	Sigma-Aldrich Co. Ltd, UK, ref. 20,606-7

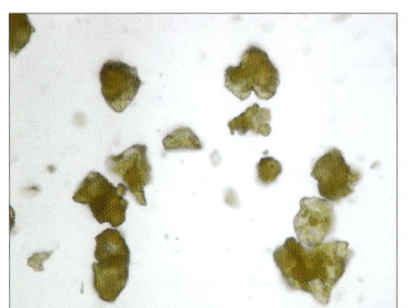

P0917	*Sulfur*	Sigma-Aldrich Co. Ltd, UK, ref. 21,519-8
P0921	*Titanium(IV) oxide*	Sigma-Aldrich Co. Ltd, UK, ref. 23,203-3
P0929	*Zinc sulfate heptahydrate*	Sigma-Aldrich Co. Ltd, UK, ref. 22,137-6
P0930	*Zinc sulfide*	Sigma-Aldrich Co. Ltd, UK, ref. 24,462-7
P0935	*Prussian blue ammonia III*	J. Middleton, The Pigmentum Project, ref. 2000310 (9)
P0936	*Prussian blue insoluble sodium*	J. Middleton, The Pigmentum Project, ref. 2000311 (10)
P0939	*Prussian blue antimony II*	J. Middleton, The Pigmentum Project, ref. 2000314 (13)
P0941	*Prussian blue copper b*	J. Middleton, The Pigmentum Project, ref. 2000316 (15)
P0942	*Prussian blue copper d*	J. Middleton, The Pigmentum Project, ref. 2000317 (16)
P0943	*Prussian blue from blood*	J. Middleton, The Pigmentum Project, ref. 17
P0946	*Prussian blue soluble I*	J. Middleton, The Pigmentum Project, ref. 20
P0962	*Gamboge*	J.M.W. Turner studio pigments, ref. 3
P0964	*Brown madder*	J.M.W. Turner studio pigments, ref. 5
P0965	*Prussian blue*	J.M.W. Turner studio pigments, ref. 6
P0976	*Cobalt blue*	J.M.W. Turner studio pigments, ref. 17
P0982	*Rose madder*	J.M.W. Turner studio pigments, ref. 23
P0985	*Light chrome yellow*	J.M.W. Turner studio pigments, ref. 26
P0994	*Cochineal*	J.M.W. Turner studio pigments, ref. 35
P1003	*Cobalt blue*	J.M.W. Turner studio pigments, ref. 44
P1004	*Calcium carbonate*	J.M.W. Turner studio pigments, ref. 45
P1005	*Lead tin yellow*	Dr Nicholas Eastaugh personal collection, ref. LTY01
P1016	*Lead tin yellow*	Dr Nicholas Eastaugh personal collection, ref. LTY12
P1018	*Lead tin yellow*	Dr Nicholas Eastaugh personal collection, ref. LTY14
P1041	*Lead tin yellow*	Dr Nicholas Eastaugh personal collection, ref. LTY37

P1042	*Lead tin yellow*	Dr Nicholas Eastaugh personal collection, ref. LTY38
P1044	*Lead tin yellow*	Dr Nicholas Eastaugh personal collection, ref. LTY40
P1045	*Lead tin yellow*	Dr Nicholas Eastaugh personal collection, ref. LTY41
P1046	*Lead tin yellow*	Dr Nicholas Eastaugh personal collection, ref. LTY42
P1052	*Original smalt*	L. Cornelissen, London
P1055	*Smalt. Coarse T2 200 mesh*	L. Cornelissen, London
P1065	*Indian yellow real*	Valentine Walsh personal collection
P1067	*Channel Black*	Valentine Walsh personal collection
P1068	*High jet black*	Valentine Walsh personal collection
P1069	*Raw umber*	Valentine Walsh personal collection
P1071	*Ivory black extra*	Valentine Walsh personal collection
P1072	*Mars black red*	Valentine Walsh personal collection
P1078	*NE2 char: peachstone 1*	Dr Nicholas Eastaugh personal collection, ref. 1
P1079	*NE2 char: peachstone 2*	Dr Nicholas Eastaugh personal collection, ref. 2
P1080	*NE2 char: peachstone 3*	Dr Nicholas Eastaugh personal collection, ref. 3
P1081	*ROM purple pigment stick A*	Royal Ontario Museum, Ottowa, ref. 960.243.75
P1082	*ROM purple pigment stick B*	Royal Ontario Museum, Ottowa, ref. 981 × 5.6
P1083	*Woad indigo*	Keith Edwards personal collection, ref. 01
P1085	*Yellow ochre*	Keith Edwards personal collection, ref. 03
P1087	*Antimony crimson*	Keith Edwards personal collection, ref. 05
P1088	*Bistre*	Keith Edwards personal collection, ref. 06
P1093	*Ultramarine green*	Keith Edwards personal collection, ref. 11
P1094	*Antimony vermilion*	Keith Edwards personal collection, ref. 12
P1095	*Molybdenum red-chrome*	Keith Edwards personal collection, ref. 13
P1097	*Spanish red*	Keith Edwards personal collection, ref. 15
P1103	*Bremen blue (blue ashes)*	Keith Edwards personal collection, ref. 21

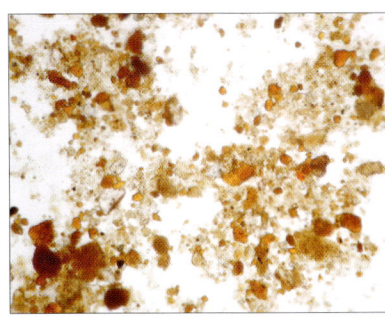

P1104	*Hematite red*	Keith Edwards personal collection, ref. 22
P1105	*Bideford black*	Keith Edwards personal collection, ref. 23
P1106	*Jarosite*	Keith Edwards personal collection, ref. 24
P1108	*Egyptian blue*	Keith Edwards personal collection, ref. 26
P1109	*Raw umber*	Keith Edwards personal collection, ref. 27
P1111	*Han blue and soda flux*	Keith Edwards personal collection, ref. 29
P1114	*Han blue*	Keith Edwards personal collection, ref. 32
P1120	*Cerussite*	Keith Edwards personal collection, ref. 38
P1132	*Aerinite*	Richard Tayler Minerals, UK
P1134	*Albite*	Richard Tayler Minerals, UK
P1136	*Amethyst*	Richard Tayler Minerals, UK
P1137	*Analcite*	Richard Tayler Minerals, UK
P1140	*Anhydrite*	Richard Tayler Minerals, UK
P1141	*Ankerite*	Richard Tayler Minerals, UK
P1144	*Antigorite*	Richard Tayler Minerals, UK
P1145	*Antlerite*	Richard Tayler Minerals, UK
P1146	*Aragonite*	Richard Tayler Minerals, UK
P1149	*Azurite*	Richard Tayler Minerals, UK
P1151	*Baryte*	Richard Tayler Minerals, UK
P1153	*Beidellite*	Richard Tayler Minerals, UK
P1155	*Biotite*	Richard Tayler Minerals, UK
P1157	*Bixbyite*	Richard Tayler Minerals, UK
P1159	*Brookite*	Richard Tayler Minerals, UK
P1162	*Calcite, var. coral*	Richard Tayler Minerals, UK
P1164	*Cassiterite*	Richard Tayler Minerals, UK
P1168	*Chalconatronite*	Richard Tayler Minerals, UK
P1171	*Chrysocolla*	Richard Tayler Minerals, UK
P1172	*Chrysotile*	Richard Tayler Minerals, UK
P1175	*Clinochlore*	Richard Tayler Minerals, UK
P1176	*Coal, cannel*	Richard Tayler Minerals, UK
P1177	*Coal, lignite*	Richard Tayler Minerals, UK
P1182	*Crocoite*	Richard Tayler Minerals, UK
P1183	*Cuprite*	Richard Tayler Minerals, UK
P1184	*Diatomite*	Richard Tayler Minerals, UK
P1187	*Dolomite*	Richard Tayler Minerals, UK
P1190	*Erythrite*	Richard Tayler Minerals, UK
P1192	*Fluorite*	Richard Tayler Minerals, UK
P1193	*Fluorite*	Richard Tayler Minerals, UK
P1200	*Goethite*	Richard Tayler Minerals, UK
P1201	*Graphite*	Richard Tayler Minerals, UK
P1202	*Gypsum*	Richard Tayler Minerals, UK
P1204	*Hackmanite*	Richard Tayler Minerals, UK
P1207	*Hausmannite*	Richard Tayler Minerals, UK
P1208	*Hematite, var. micaceous*	Richard Tayler Minerals, UK
P1212	*'Huntite' (Dolomite)*	Richard Tayler Minerals, UK
P1216	*Kaolin*	Richard Tayler Minerals, UK

P1219	*Lazurite*	Richard Tayler Minerals, UK
P1220	*Lepidocrocite*	Richard Tayler Minerals, UK
P1224	*Magnesite*	Richard Tayler Minerals, UK
P1225	*Magnetite*	Richard Tayler Minerals, UK
P1226	*Malachite*	Richard Tayler Minerals, UK
P1229	*Microcline, var. pink*	Richard Tayler Minerals, UK
P1231	*Montmorillonite*	Richard Tayler Minerals, UK
P1232	*Muscovite*	Richard Tayler Minerals, UK
P1233	*Nacrite*	Richard Tayler Minerals, UK
P1234	*Natrojarosite*	Richard Tayler Minerals, UK
P1235	*Nontronite*	Richard Tayler Minerals, UK
P1240	*Orthoclase*	Richard Tayler Minerals, UK
P1242	*Phlogopite*	Richard Tayler Minerals, UK
P1244	*Pseudomalachite*	Richard Tayler Minerals, UK
P1247	*Pyrite*	Richard Tayler Minerals, UK
P1248	*Pyrolusite*	Richard Tayler Minerals, UK
P1253	*Rutile*	Richard Tayler Minerals, UK
P1255	*Serpentine*	Richard Tayler Minerals, UK
P1257	*Sodalite*	Richard Tayler Minerals, UK
P1258	*Sphalerite*	Richard Tayler Minerals, UK
P1260	*Stibnite*	Richard Tayler Minerals, UK
P1262	*Talc*	Richard Tayler Minerals, UK
P1263	*Tenorite*	Richard Tayler Minerals, UK
P1267	*Vermiculite*	Richard Tayler Minerals, UK
P1268	*Vermiculite*	Richard Tayler Minerals, UK
P1270	*Witherite*	Richard Tayler Minerals, UK
P1283	*Luyckse geel oker*	Valentine Walsh personal collection, ref. -
P1285	*Chromeisenstein*	Valentine Walsh personal collection, ref. -
P1286	*Realgar echt* ('True realgar')	Valentine Walsh personal collection, ref. -
P1288	*French ultramarine from Barbes (?) …*	Courtauld Institute of Art, London, ref. C-001
P1289	*Best French Ultramarine Robersons 1853*	Courtauld Institute of Art, London, ref. C-002
P1291	*Ultramarine, W. Eatwell 49, Dorset Street […] Street*	Courtauld Institute of Art, London, ref. C-004
P1295	*Prepared by A.P. Laurie CERULEAN BLUE Presented by H.R.H. Woolford*	Courtauld Institute of Art, London, ref. C-008
P1299	*Cobalt Green from Roberson 3/oz*	Courtauld Institute of Art, London, ref. C-012
P1300	*Prepared by A.P. Laurie COBALT GREEN Presented by H.R.H. Woolford*	Courtauld Institute of Art, London, ref. C-013
P1302	*Emerald green*	Courtauld Institute of Art, London, ref. C-015
P1304	*Cadmium yellow from Field, 1852*	Courtauld Institute of Art, London, ref. C -017
P1305	*Cadmium yellow from Griffen Chemist Baker St 1851*	Courtauld Institute of Art, London, ref. C-018
P1307	*Cadmium yellow label circa 1850*	Courtauld Institute of Art, London, ref. C-020

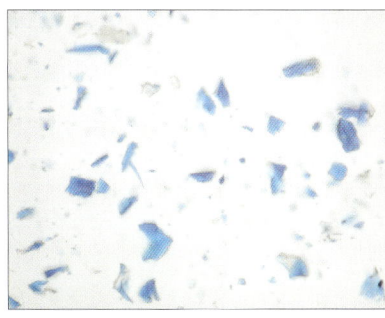

P1308	*Zinc tetroxy chromate, Shell Research, Egham Industrial Chemicals, Whitehall Lane, Surrey*	Courtauld Institute of Art, London, ref. C-021
P1310	*Chrome of lead from Field*	Courtauld Institute of Art, London, ref. C-023
P1312	*Orpiment*	Courtauld Institute of Art, London, ref. C-025
P1315	*Aureolin*	Courtauld Institute of Art, London, ref. C-028
P1316	*Aureolin*	Courtauld Institute of Art, London, ref. C-029
P1321	*Strontium yellow*	Courtauld Institute of Art, London, ref. C-034
P1322	*Gamboge*	Courtauld Institute of Art, London, ref. C-035
P1324	*Raw siena, goethite*	Courtauld Institute of Art, London, ref. C-037
P1326	*Mosaic? Sulfide, made Erner & Amend, New York*	Courtauld Institute of Art, London, ref. C-039
P1327	*Native cinnabar 2*	Courtauld Institute of Art, London, ref. C-040
P1329	*Carmine*	Courtauld Institute of Art, London, ref. C-042
P1341	*Mercure iodide*	Courtauld Institute of Art, London, ref. C-054
P1342	*Vermilion, Chang Dou*	Courtauld Institute of Art, London, ref. C-055
P1347	*Vermilion Dark BW7 Romney*	Courtauld Institute of Art, London, ref. C-060
P1351	*Manganese violet*	Courtauld Institute of Art, London, ref. C-064
P1353	*Vivianite Poroma, Southern Highland PNG, Oct. 1986*	Rowena Hill personal collection, ref. RH1
P1355	*Yellow ochre from Goroka, Eastern Highland Province, PNG*	Rowena Hill personal collection, ref. RH3
P1360	*PY73*	Susan Lake, Hirschorn Collection, Washington, ref. PY73
P1361	*PY97*	Susan Lake, Hirschorn Collection, Washington, ref. PY97
P1374	*Pompeii 9517*	Museo Archaeologico, Pompeii, ref. 9517
P1376	*Pompeii 9524*	Museo Archaeologico, Pompeii, ref. 9524
P1377	*Pompeii 9530*	Museo Archaeologico, Pompeii, ref. 9530
P1379	*Pompeii 9649*	Museo Archaeologico, Pompeii, ref. 9649
P1382	*Pompeii 18107*	Museo Archaeologico, Pompeii, ref. 18107

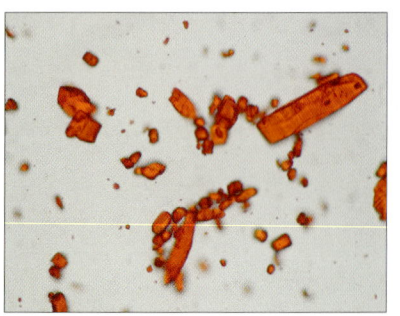

P1385	*Pompeii 18110B*	Museo Archaeologico, Pompeii, ref. 18110B
P1396	*Pigment yellow 4 (5G)*	Susan Lake, Hirschorn Collection, Washington, ref. SL 16
P1406	*Blue verditer*	Dr Leslie Carlyle personal collection
P1409	*Indian yellow*	Dr Leslie Carlyle personal collection
P1413	*Yellow A1*	Dr Claudio Seccaroni personal collection
P1414	*Yellow B*	Dr Claudio Seccaroni personal collection, ref. yellow B
P1415	*Yellow C1*	Dr Claudio Seccaroni personal collection, ref. yellow C 1
P1418	*Yellow G1*	Dr Claudio Seccaroni personal collection, ref. yellow G1
P1427	*Tetrabasic lead sulfate*	Dr Ian Steele, University of Chicago, ref. B
P1428	*Orthorhombic lead oxide*	Dr Ian Steele, University of Chicago, ref. C
P1429	*Lead sulfate*	Dr Ian Steele, University of Chicago, ref. D
P1430	*Beta-lead oxide*	Dr Ian Steele, University of Chicago, ref. E
P1432	*Tetragonal lead oxide*	Dr Ian Steele, University of Chicago, ref. G
P1433	*Monobasic lead sulfate*	Dr Ian Steele, University of Chicago, ref. H
P1434	*Hydrocerrusite*	Dr Ian Steele, University of Chicago, ref. I
P1435	*Plumbonacrite*	Dr Ian Steele, University of Chicago, ref. J
P1437	*Tribasic lead sulfate + mix*	Dr Ian Steele, University of Chicago, ref. L
P1439	*Bentonite*	Carol Grissom personal collection, ref. 55
P1440	*Green earth/Celadonite?*	Carol Grissom personal collection, ref. C3811
P1441	*Green earth/Celadonite*	Carol Grissom personal collection, ref. 4733
P1443	*Green earth/Glauconite*	Carol Grissom personal collection, ref. R4735
P1452	*Green earth/Galuconite*	Carol Grissom personal collection, ref. 91932B
P1453	*Green earth/Glauconite*	Carol Grissom personal collection, ref. 91933
P1457	*Green earth/Glauconite*	Carol Grissom personal collection, ref. 91937
P1458	*Green earth/Glauconite*	Carol Grissom personal collection, ref. 91938A

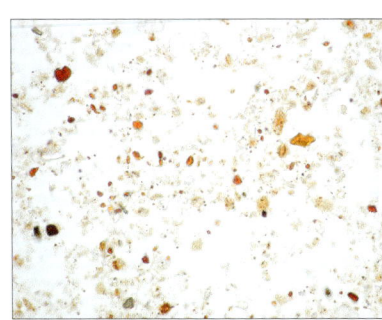

P1469	*Green earth*	Carol Grissom personal collection, ref. Forbes No. 28
P1470	*Cobalt violet light*	Dr Ruth Siddall personal collection
P1472	*Cinnabar*	Dr Ruth Siddall personal collection
P1474	*Realgar*	Dr Ruth Siddall personal collection
P1475	*Chrysocolla*	Dr Ruth Siddall personal collection
P1476	*Magnesium cobalt arsenate*	Dr Marie-Claude Corbeil, Canadian Conservation Institute, ref. CV A
P1478	*Cobalt phosphate octahydrate*	Dr Marie-Claude Corbeil, Canadian Conservation Institute, ref. CV C
P1480	*Lithium cobalt phosphate*	Dr Marie-Claude Corbeil, Canadian Conservation Institute, ref. CV E
P1484	*Titanium anatase/brookite*	Dr James Ovenstone, University of Greenwich
P1486	*Wad*	Keith Edwards personal collection
P1488	*Eggshell white*	Dr Ruth Siddall personal collection
P1489	*Rice starch*	Dr Ruth Siddall personal collection
P1490	*Amioca starch*	Dr Ruth Siddall personal collection
P1491	*Maize starch*	Dr Ruth Siddall personal collection
P1492	*Tapioca starch*	Dr Ruth Siddall personal collection
P1493	*Potato starch*	Dr Ruth Siddall personal collection
P1494	*Huntite*	Janet Ambers, British Museum
P1498	*Egyptian blue*	Gareth Hatton, R.L.A.H.A., University of Oxford, ref. GDH5c
P1500	*Turmeric*	Dr Nicholas Eastaugh personal collection
P1500	*Glaucophane*	Dr Ruth Siddall personal collection
P1503	*Sulfur*	Dr Ruth Siddall personal collection
P1504	*Blue asbestos/crocidolite*	Dr Ruth Siddall personal collection
P1507	*Dried blood*	Valentine Walsh personal collection
P1508	*Pink coral*	Valentine Walsh personal collection
P1511	*Bismuthinite*	Richard Tayler Minerals, UK

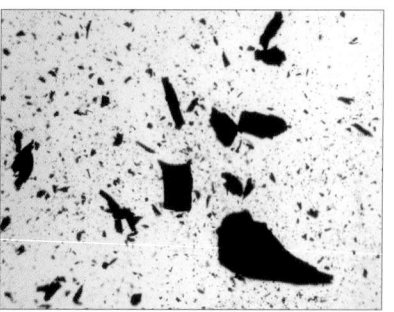

P1512	*Bismuthinite*	Richard Tayler Minerals, UK
P1513	*Brochantite*	Richard Tayler Minerals, UK, ref. S144
P1514	*Pararealgar*	National Museums and Galleries of Wales, ref. NMW 48.264.G.R498
P1516	*Botallackite*	National Museums and Galleries of Wales, ref. NMW 83.41G.M4102
P1517	*Metacinnabar*	National Museums and Galleries of Wales, ref. NMW 83.41G.M773
P1520	*Indigo carmine CI 73015 (Acid blue 74)*	Sigma-Aldrich Co. Ltd, UK
P1522	*Lead(II) iodide*	Sigma-Aldrich Co. Ltd, UK, ref. 20,360-2
P1523	*Aluminium oxide*	Sigma-Aldrich Co. Ltd, UK, ref. 20,260-6
P1524	*Chromium(III) phosphate hydrate*	Sigma-Aldrich Co. Ltd, UK, ref. 49519-0
P1525	*Egyptian green*	Kremer Pigmente, Germany

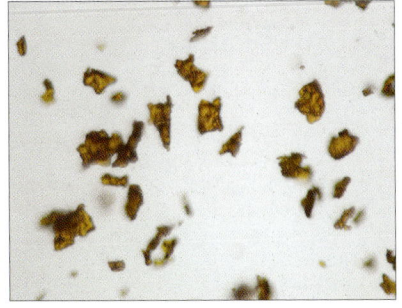

Appendix II: References

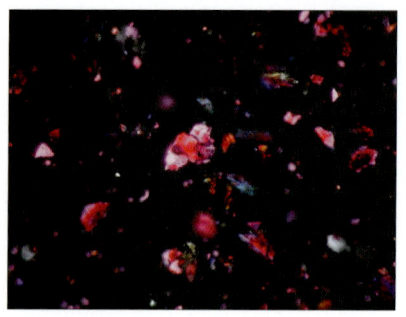

ADAMS *et al.* (1991)   Adams, A.E.; MacKenzie, W.S.; Guilford, C. *Atlas of Sedimentary Rocks under the Microscope* Longman Scientific & Technical (1991)

ARTISTS' PIGMENTS (1986)   Feller, R.L. (ed.) *Artists' Pigments. A Handbook of their History and Characteristics* **1** National Gallery of Art (Washington) & Cambridge University Press (1986)

ARTISTS' PIGMENTS (1993)   Roy, A. (ed.) *Artists' Pigments. A Handbook of their History and Characteristics* **2** National Gallery of Art, Washington & Oxford University Press (1993)

ARTISTS' PIGMENTS (1997)   FitzHugh, E.W. (ed.) *Artists' Pigments. A Handbook of their History and Characteristics* **3** National Gallery of Art, Washington & Oxford University Press (1997)

ARTISTS' PIGMENTS (FORTH.)   Berrie, B.H. (ed.) *Artists' Pigments. A Handbook of their History and Characteristics* **4** National Gallery of Art, Washington & Oxford University Press (forth.)

ASPEREN DE BOER (1974)   Asperen de Boer, J.R.J. van 'An Examination of Particle Size Distributions of Azurite and Natural Ultramarine in Some Early Netherlandish Paintings' *Studies in Conservation* **19** (1974) 233–243

BACCI & PICOLLO (1996)   Bacci, M.; Picollo, M. 'Non-destructive spectroscopic detection of cobalt (II) in paintings and glass' *Studies in Conservation* **41** #3 IIC, London (1996) 136–144

BOMFORD *et al.* (1980)   Bomford, D.; Dunkerton, J.; Gordon, D.; Roy, A. 'Three Panels from Perugino's Certosa di Pavia Altarpiece' *National Gallery Technical Bulletin* **4** (1980)

BAER *et al.* (1986)   Baer, N.S.; Joel, A.; Feller, R.L.; Indictor, N. 'Indian Yellow' *Artists' Pigments. A Handbook of their History and Characteristics* **1** Feller, R.L. (ed.) National Gallery of Art, Washington & Cambridge University Press (1986) 17–36

BÉARAT (1996a)   Béarat, H. 'Chemical and mineralogical analyses of Gallo-Roman wall painting from Dietikon, Switzerland' *Archaeometry* **38** #1 (1996) 81–95

Béarat & Pradell (1997) Béarat, H.; Pradell, T. 'Contribution of Mössbauer spectroscopy to the study of ancient pigment and paintings' *Roman Wall Painting. Materials, Techniques, Analysis and Conservation. Proceedings of the International Workshop Fribourg 7–9 March 1996* Béarat, H. *et al.* (eds.) Institute of Mineralogy and Petrography, Fribourg (1997) 239–256

Berrie (1997) Berrie, B.H. 'Prussian Blue' *Artists' Pigments. A Handbook of their History and Characteristics* **3** FitzHugh, E.W. (ed.) National Gallery of Art, Washington & Oxford University Press (1997) 191–217

Bomford *et al.* (1980) Bomford, D.; Brough, J.; Roy, A. 'Three Panels from Perugino's Certosa di Pavia Altarpiece' *National Gallery Technical Bulletin* **4** (1980) 3–31

Bown (1998) Bown, P.R. (ed.) *Calcareous Nannofossil Biostratigraphy*, British Micropalaeontological Society Publication Series, Chapman and Hall/Kluwer Academic Publishers (1998) 1–315

Bradley (1940) Bradley, W.F. 'The Structural Scheme of Attapulgite' *American Mineralogist* **25** (1940) 405–410

Brooks & Morris (1973) Brooks, L.E.; Morris, H.H. 'Aluminum Silicate (Kaolin)' *Pigment Handbook* **1** Patton, T.C. (ed.) John Wiley, New York (1973) 199–215

Burns (1993) Burns, R.G. *Mineralogical applications of crystal field theory* 2nd ed. *Cambridge Topics in Mineral Physics and Chemistry* **5** Cambridge University Press (1993)

Burnstock *et al.* (2003) Burnstock, A.R.; Jones, C.G.; Cressey, G. 'Characterisation of artists' chromium-based yellow pigments' *Zeitschrift für Kunsttechnologie und Konservierung* **17** (2003) 74–84

Buxbaum (1998) *Industrial Inorganic Pigments* Buxbaum, Gunter (ed.) Wiley-VCH (1998)

Buzzegoli *et al.* (2000) Buzzegoli, E.; Cardaropoli, R.; Kunzelman, D.; Moioli, P.; Montalbano, L.; Piccolo, M.; Seccaroni, C. 'Valerio Mariani da Pesaro, il trattato "Della Miniatura": primi raffronti con le analisi e le opere' *OPD Restauro* **12** (2000) 248–256

Casas & Llopis (1992) Palet Casas, A.; Andres Llopis, J. de 'The Identification of Aerinite as a Blue Pigment in the Romanesque Frescos of the Pyrenean Region' *Studies in Conservation* **37** (1992) 132–136

Cascales *et al.* (1986) Cascales, C.; Alonso, J.A.; Rasines, I. 'The New Pyrochlores $Pb_2(MSb)O_{6.5}$ (M = Ti, Zr, Sn, Hf)' *Journal of Materials Science Letters* **5** (1986) 675–677

Clay & Watson (1944) Clay, H.F.; Watson, V. 'Some Important Properties of Chromes' *Journal of the Oil and Colour Chemists' Association* **27** #283 (1944) 3–18

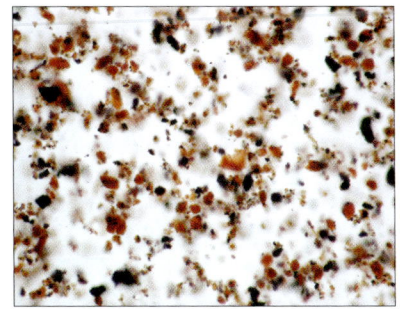

CLAY & WATSON (1948) Clay, H.F.; Watson, V. 'The Crystal Structure of Lead Chrome' *Journal of the Oil and Colour Chemists' Association* **31** #9 (1948) 418–422

COLOUR INDEX (1971) *Colour Index* 3rd ed. Bradford, UK: Society of Dyers & Colourists (1971)

CORBEIL & HELWIG (1995) Corbeil, M.C.; Helwig, K. 'An Occurrence of Pararealgar as an Original or Altered Artists' Pigment' *Studies in Conservation* **40** (1995) 133–138

CORBEIL *et al.* (2002) Corbeil, M.C.; Charland, J.P.; Moffatt, E.A. 'The characterization of cobalt violet pigments' *Studies in Conservation* **47** (2002) 237–249

CORBEIL *et al.* (2002) Corbeil, M.C.; Charland, J.P.; Moffatt, E.A. 'The Characterization of Cobalt Violet Pigments' *Studies in Conservation* **47** (2002) 237–249

CORNELL & SCHWERTMANN (1996) Cornell, R.M.; Schwertmann, U. *The Iron Oxides. Structure, Properties, Reactions, Occurrence and Uses* New York: VCH (1996)

CORNMAN (1986) Cornman, M. 'Cobalt Yellow (Aureolin)' *Artists' Pigments. A Handbook of their History and Characteristics* **1** Feller, R.L. (ed.) National Gallery of Art, Washington & Cambridge University Press (1986) 37–46

DANA (1944) Palache, C.; Berman, H.; Frondel, C. *Dana's System of Mineralogy* 7th ed., Vol. 1, John Wiley & Sons, New York (1944)

DANA (1997) Gaines, R.V.; Skinner, H.; Catherine, W.; Foord, E.E.; Mason, B.; Rosenzweig, A. (eds.) *Dana's New Mineralogy: The System of Mineralogy of James Dwight Dana and Edward Salsbury Dana* 8th ed., J. Wiley & Sons, New York (1997)

DEER *et al.* (1992) Deer, W.A.; Howie, R.A.; Zussman, J. *An Introduction to the Rock Forming Minerals* 2nd ed., Longman Scientific & Technical (1992)

DE LA RIE (1982) de la Rie, E. Rene 'Fluorescence of Paint and Varnish Layers (Parts 1–3)' *Studies in Conservation* **27** (1982) 1–3

DUNHAM (1937) Dunham, K.C. 'The paragenesis and color of fluorite in the English Pennines' *American Mineralogist* **22** (1937) 468–478

DUNKERTON & ROY (1996) Dunkerton, J.; Roy, A. 'The Materials of a Group of Late Fifteenth-century Florentine Panel Paintings' *National Gallery Technical Bulletin* **17** London: National Gallery (1996) 20–31

DUNN (1973) Dunn, E.J., Jr 'White Hiding Lead Pigments' *Pigment Handbook* **1** Patton, T.C. (ed.) John Wiley, New York (1973) 65–84

DUVAL (1992) Duval, A.R. 'Les préparations colorées des tableaux de L'Ecole Français des dix-septième et dix-huitième siècles' *Studies in Conservation* **37** #4 (1992) 239–258

EASTAUGH (1988) Eastaugh, N. *Lead Tin Yellow. Its History, Manufacture, Colour and Structure* Unpublished Ph.D. thesis, Courtauld Institute of Art, University of London (1998)

EASTAUGH (1995) Eastaugh, N. 'Some Dyes and Dye-based Pigments in Turner's Palette' *Turner's Painting Techniques in Context* UKIC, London (1995) 46–49

EASTAUGH (2002) Eastaugh, N. 'The Realities of Digital Image Acquisition, Integration and Access in a Large Database Project' *Digital Imaging for the Paintings' Conservator. Preprints for a conference held in the Starr Auditorium at Tate Modern on Friday 6th December 2002* UKIC, London (2002) 46–52

ERGUN (1968) Ergun, S. 'Optical Studies of Carbon' In: *Chemistry and Physics of Carbon* Vol. 3, P.L. Walker (ed.) Marcel Dekker Inc. (1968) 45–119

FELLER & BAYARD (1986) Feller, R.L.; Bayard, M. 'Terminology and Procedures used in the Systematic Examination of Pigment Particles with the Polarizing Microscope' *Artist Pigments. A Handbook of their History and Characteristics* **1** Feller, R.L. (ed.) National Gallery of Art, Washington & Cambridge University Press (1986) 285–298

FELLER (1986) Feller, R.L. 'Barium Sulfate – Natural and Synthetic' *Artists' Pigments. A Handbook of their History and Characteristics* **1** Feller, R.L. (ed.) National Gallery of Art, Washington & Cambridge University Press (1986) 47–64

FELLER & JOHNSTON-FELLER (1997) Feller, R.L.; Johnston-Feller, R.M. 'Vandyke Brown, Cassel Earth, Cologne Earth' *Artists' Pigments. A Handbook of their History and Characteristics* **3** FitzHugh, E.W. (ed.) National Gallery of Art, Washington & Oxford University Press, Oxford (1997) 157–190

FIEDLER & BAYARD (1986) Fiedler, I.; Bayard, M. 'Cadmium Yellows, Oranges and Reds' *Artists' Pigments. A Handbook of their History and Characteristics* **1** Feller, Robert L. (ed.) National Gallery of Art, Washington & Cambridge University Press (1986) 65–108

FIEDLER & BAYARD (1997) Fiedler, I.; Bayard, M.A. 'Emerald Green and Scheele's Green' *Artists' Pigments. A Handbook of their History and Characteristics* **3** FitzHugh, E.W. (ed.) National Gallery of Art, Washington & Oxford University Press, Oxford (1997) 219–271

FITZHUGH & ZYCHERMAN (1983) FitzHugh, E.W.; Zycherman, L.A. 'An Early Man-made Blue Pigment from China – Barium Copper Silicate' *Studies in Conservation* **28** #1 (1983) 15–23

FITZHUGH & ZYCHERMAN (1992) FitzHugh, E.W.; Zycherman, L.A. 'A Purple Barium Copper Silicate Pigment from Early China' *Studies in Conservation* **37** #3 (1992) 145–154

FITZHUGH (1986) FitzHugh, E.W. 'Red Lead and Minium' *Artists' Pigments. A Handbook of their History and Characteristics* **1**

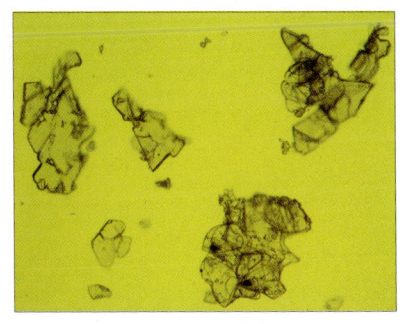

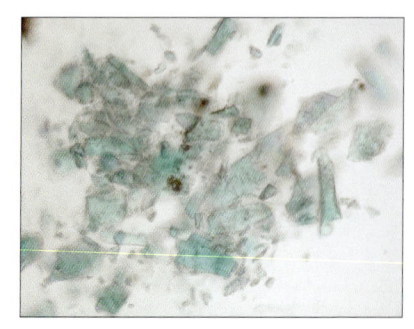

Feller, R.L. (ed.) National Gallery of Art, Washington & Cambridge University Press (1986) 109–140

FITZHUGH (1997)   FitzHugh, E.W. 'Orpiment and Realgar' *Artists' Pigments. A Handbook of their History and Characteristics* **3** FitzHugh, E.W. (ed.) National Gallery of Art, Washington & Oxford University Press, Oxford (1997) 47–79

FORD (2001)   Ford, T. 'Derbyshire wad and umber' *Mining History* **14** #5 (2001) 39–45

FREESTONE (1995)   Freestone, I.C. 'Ceramic Petrography' *American Journal of Archaeology* **99** (1995) 111–115

GATES (1995)   Gates, G. 'A Note on the Artists' Pigment Aureolin' *Studies in Conservation* **40** #3 (1995) 201–206

GETTENS & FITZHUGH (1993a)   Gettens, R.J.; FitzHugh, E.W. 'Azurite and Blue Verditer' *Artists' Pigments. A Handbook of their History and Characteristics* **2** Roy, A. (ed.) National Gallery of Art, Washington & Oxford University Press, Oxford (1993) 23–36

GETTENS & FITZHUGH (1993b)   Gettens, R.J.; FitzHugh, E.W. 'Malachite and Green Verditer' *Artists' Pigments. A Handbook of their History and Characteristics* **2** Roy, A. (ed.) National Gallery of Art, Washington & Oxford University Press, Oxford (1993) 183–202

GETTENS & STOUT (1966)   Gettens, R.J. & Stout, G.L. *Painting Materials. A Short Encyclopaedia* Dover: New York (1966)

GETTENS *et al.* (1993a)   Gettens, R.J.; FitzHugh, E.W.; Feller, R.L. 'Calcium Carbonate Whites' *Artists' Pigments. A Handbook of their History and Characteristics* **2** Roy, A. (ed.) National Gallery of Art, Washington & Oxford University Press, Oxford (1993) 203–226

GETTENS *et al.* (1993b)   Gettens, R.J.; Kühn, H.; Chase, W.T. 'Lead White' *Artists' Pigments. A Handbook of their History and Characteristics* **2** Roy, A. (ed.) National Gallery of Art, Washington & Oxford University Press, Oxford (1993) 67–82

GETTENS *et al.* (1993c)   Gettens, R.J.; Feller, R.L.; Chase, W.T. 'Vermilion and Cinnabar' *Artists' Pigments. A Handbook of their History and Characteristics* **2** Roy, A. (ed.) National Gallery of Art, Washington & Oxford University Press, Oxford (1993) 159–182

GIESTER & RIECK (1994)   Giester, G.; Rieck, B. 'Effenbergerite, $BaCuSi_4O_{10}$, a New Mineral from the Kalahari Manganese Field, South Africa. Description and Crystal Structure' *Mineralogical Magazine* **58** #393 (1994) 663–670

GLOGER & HURLEY (1973)   Gloger, W.A.; Hurley, D.W. 'Antimony Oxide' *Pigment Handbook* **1** Patton, T.C. (ed.) John Wiley, New York (1973) 85–93

GRASSMANN & CLAUSEN (1973)   Grassman, W.; Clausen, H. 'Zinc Sulfide' *Pigment Handbook* **1** Patton, T.C. (ed.) John Wiley, New York (1973) 53–58

GRIBBLE & HALL (1992)   Gribble, C.D.; Hall, A.J. *Optical Mineralogy: Principles and Practice* UCL Press, London (1992)

GRISSOM (1970) Grissom, C.A. *A Study of Green Earth* Unpublished Master's Thesis, Department of Art, Oberlin College, USA (1970)

GRISSOM (1986) Grissom, C.A. 'Green Earth' *Artists' Pigments. A Handbook of their History and Characteristics* **1** Feller, Robert L. (ed.) National Gallery of Art, Washington & Cambridge University Press (1986) 141–169

HADEN (1973) Haden, W.L. 'Hydrated Magnesium Aluminium Silicate. Fuller's Earth' *Pigment Handbook* **1** Patton, T.C. (ed.) John Wiley, New York (1973) 269–273

HALL (1973) Hall, R.F. 'Calcium Carbonate, Natural' *Pigment Handbook* **1** Patton, T.C. (ed.) John Wiley, New York (1973) 109–117

HEINRICH (1965) Heinrich, E.W. *Microscopic Identification of Minerals* New York, McGraw-Hill Book Co. (1965)

HELWIG (1997) Helwig, K. 'A note on burnt yellow earth pigments' *Studies in Conservation* **42** (1997) 181–188

HERBST & HUNGER (1997) Herbst, W.; Hunger, K. *Industrial Organic Pigments: Production, Properties, Applications* 2nd ed., VCH, Weinheim (1997)

HERMENS (2002) Hermens, E. *Memories of Beautiful Colours. The Mariani Treatise and the Practice of Miniature Painting, Landscape Drawing and Botanical Illustration at the Pesaro Court in Early-Seventeenth Century Italy.* Unpublished Ph.D. thesis, Universiteit Leiden (2002)

HIBBARD (1995) Hibbard, M.J. *Petrography to Petrogenesis* Prentice Hall (1995)

HILL (2001) Hill, R. 'Traditional Paint from Papua New Guinea: Context, Materials and Techniques, and their Implications for Conservation' *The Conservator* **25** (2001) 49–61

HOWARD (1995) Howard, H.C. 'Techniques of the Romanesque and Gothic Wall Paintings in the Holy Sepulchre Chapel, Winchester Cathedral' *Historical Painting Techniques, Materials, and Studio Practice: Preprints of a Symposium, University of Leiden, the Netherlands, 26–29 June 1995* Wallert, A.; Hermens, E.; Peek, M. (eds.) The Getty Conservation Institute, Los Angeles (1995) 91–104

HOWARD (2003) Howard, H.C. *Pigments of English Medieval Wall Painting* Archetype Publications Ltd (2003)

HUCKLE et al. (1966) Huckle, W.G.; Swigert, G.F.; Wiberley, S.E. 'Cadmium Pigments – Structure and Composition' *Industrial and Engineering Chemistry Product Research and Development* **5** #4 (1966) 362–366

JAMBOR et al. (1996) Jambor, J.L.; Dutrizac, J.E.; Roberts, A.C.; Grice, J.D.; Szymanski, J.T. 'Clinoatacamite, a New Polymorph of $Cu_2(OH)_3Cl$, and its Relationship to Paratacamite and "Anarakite"' *Canadian Mineralogist* **34** (1996) 61–72

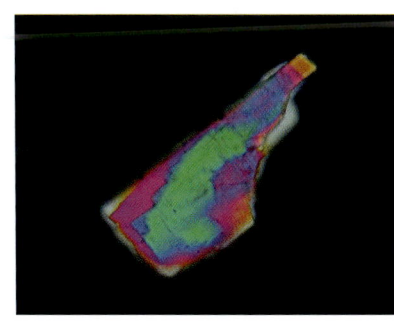

JANCZAK & KUBIAK (1992) Janczak, J.; Kubiak, R. 'Structure of the cyclic barium copper silicate $Ba_2Cu_2[Si_4O_{12}]$ at 300 K' *Acta Crystallographica* C**48** (1992) 8–10

JOHNSTON-FELLER (1986) Johnston-Feller, R. 'Standard Specification of Pigment Composition and Color' *Artists' Pigments. A Handbook of their History and Characteristics* **1** Feller, Robert L. (ed.) National Gallery of Art, Washington & Cambridge University Press (1986) 299–300

KAMPFER (1973) Kampfer, W.A. 'Titanium Dioxide' *Pigment Handbook* **1** Patton, T.C. (ed.) John Wiley, New York (1973) 1–36

KATIC (1973) Katic, J.M. 'Silica, Crystalline' *Pigment Handbook* **1** Patton, T.C. (ed.) John Wiley, New York (1973) 135–139

KITTEL (1960) *Pigmente. Herstellung, Eigenschaften, Anwendung [Pigments. Preparation, Properties, Use]* **3 vols** Kittel, H. (ed.) Wissenschaftliche verlagsgesellschaft, Stuttgart (1960)

KIRBY (1977) Kirby, J. 'A Spectrophotometric Method for the Identification of Lake Pigment Dyestuffs' *National Gallery Technical Bulletin* **1** (1977) 35–45

KOCKAERT (1979) Kockaert, L. 'Notes on the Green and Brown Glazes of Old Paintings' *Studies in Conservation* **24** (1979) 69–74

KRANICH (1973) Kranich, H. 'Silica, Diatomaceous' *Pigment Handbook* **1** Patton, T.C. (ed.) John Wiley, New York (1973) 141–155

KÜHN & CURRAN (1986) Kühn, H.; Curran, M. 'Chrome Yellow and Other Chromate Pigments' *Artists' Pigments. A Handbook of their History and Characteristics* **1** Feller, Robert L. (ed.) National Gallery of Art, Washington & Cambridge University Press (1986) 187–218

KÜHN (1986) Kühn, H. 'Zinc White' *Artists' Pigments. A Handbook of their History and Characteristics* **1** Feller, R.L. (ed.) National Gallery of Art, Washington & Cambridge University Press. Cambridge (1986) 169–186

KÜHN (1993a) Kühn, H. 'Lead-tin Yellow' *Artists' Pigments. A Handbook of their History and Characteristics* **2** Roy, A. (ed.) National Gallery of Art, Washington & Oxford University Press, Oxford (1993) 83–112

KÜHN (1993b) Kühn, H. 'Verdigris and Copper Resinate' *Artists' Pigments. A Handbook of their History and Characteristics* **2** Roy, A. (ed.) National Gallery of Art, Washington & Oxford University Press, Oxford (1993) 131–158

KUNCKEL (1689) Kunckel, J. *Ars Vitraria Experimentalis oder vollkommene Glasmacker-Kunst* Frankfurt and Leipzig (1689)

LAURIE (1914) Laurie, A.P. *The Pigments and Mediums of the Old Masters* Macmillan, London (1914)

LAVER (1997) Laver, M. 'Titanium Dioxide Whites' *Artists' Pigments. A Handbook of their History and Characteristics* **3** FitzHugh, E.W. (ed.) National Gallery of Art, Washington & Oxford University Press, Oxford (1997) 295–355

LEE *et al.* (1997) Lee, M.-Y.; Parkinson, G.M.; Smith, P.G.; Lincoln, F.J.; Reyhani, M.M. 'Characterization of Aluminium Trihydroxide Crystals Precipitated from Caustic Solutions' *Separation and Purification by Crystallization* (American Chemical Society Symposium Series 667) G.D. Botsaris and K. Toyokura (eds.) (1997) 123–133

LENGKE & TEMPEL (2003) Lengke, M.F.; Tempel, R.N. 'Natural Realgar and Amorphous AsS Oxidation Kinetics' *Geochimica et Cosmochimica Acta* **67** #5 (2003) 859–871

LEONA & WINTER (2001) Leona, M.; Winter, J. 'Fiber optics reflectance spectroscopy: a unique tool for the investigation of Japanese paintings' *Studies in Conservation* **46** #3 (2001) 153–162

LIDDICOAT (1993) Liddicoat, R.T. *Handbook of Gem Identification. Chapter IX; Colour Filters and Fluorescence* 12th ed., Gemmological Institute of America, Santa Monica, California (1993) 81–84

LOF (1983) Lof, P. *Elsevier's Mineral and Rock Table* Elsevier Science Publishers B.V., Amsterdam (1983)

LOVE (1973) Love, C.H. 'Calcium Sulfate, Anhydrous' *Pigment Handbook* **1** Patton, T.C. (ed.) John Wiley, New York (1973) 289–292

MACKENZIE & GUILFORD (1980) MacKenzie, W.S.; Guilford, C. *Atlas of Rock Forming Minerals in Thin Section* Longman (1980)

MACKENZIE *et al.* (1984) MacKenzie, W.S.; Donaldson, C.H.; Guilford, C. *Atlas of Igneous Rocks and their Textures* Longman (1984)

MACTAGGART & MACTAGGART (1980) Mactaggart, P.; Mactaggart, A. 'Refiners' Verditer' *Studies in Conservation* **21** (1980) 37–45

MACTAGGART (2002) Mactaggart, P. *Pigment ID. A Suite of Files to Help Those Who Want to Look at, or Identify Pigments with a Polarising Microscope* (interactive CD-ROM) Mac & Me Ltd, Somerset (2002)

MANGE & MAURER (1991) Mange, M.A.; Maurer, H.F.W. *Heavy Minerals in Colour* Harper Collins (1991)

MASON & BERRY (1968) Mason, B.; Berry, L.G. *Elements of Mineralogy* W.H. Freeman and Company, San Francisco (1968)

MASSON (1986) Masson, A. 'Les ocres et la pétroarchéologie: l'aspect taphonomique' *Revue d'Archéométrie* **10** (1986) 87–93

MCCORMACK (2000) McCormack, J.K. 'The Darkening of Cinnabar in Sunlight' *Mineralium Deposita* **35** #8 (2000) 796–798

MCCRONE *et al.* (1979) McCrone, W.; McCrone, L.; Delly, J.G. *Polarised Light Microscopy* Ann Arbor Science Publishers, Ann Arbor, Michigan (1979)

MCCRONE *et al.* (1973–80) McCrone, W., *et al. The Particle Atlas: An Encyclopedia of Techniques for Small Particle Identification* 2nd ed., 6 Vols., Ann Arbor Science, Ann Arbor (1973–80)

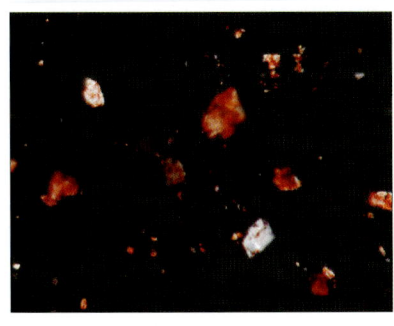

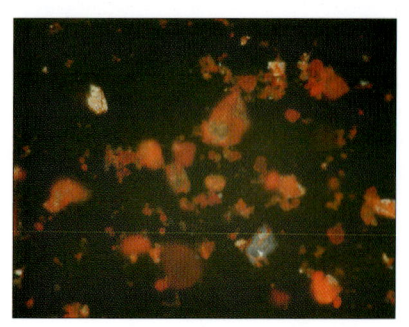

Merwin (1917) Merwin, H.E. 'Optical Properties and Theory of Color of Pigments and Paints' *Proceedings of the American Society for Testing Materials* **17** (1917) 494–520

Millette *et al.* (1993) Millette, J.R.; Hopen, T.J.; Bradley, J.P. 'Titanium Dioxide: Anatase or Rutile?' *Microscope* **41** (1993) 147–153

Moser (1973) Moser, F.H. 'Barium Sulfate, Synthetic (Blanc Fixe)' *Pigment Handbook* **1** Patton, T.C. (ed.) John Wiley, New York (1973) 281–288

Mühlethaler & Thissen (1993) Mühlethaler, B.; Thissen, J. 'Smalt' *Artists' Pigments. A Handbook of their History and Characteristics* **2** Roy, A. (ed.) National Gallery of Art, Washington & Oxford University Press, Oxford (1993) 113–130

Mullin (2001) Mullin, J.W. *Crystallization* 4th ed., Butterworth-Heinemann, Oxford (2001)

Muñoz Viñas & Farrell (1999) Muñoz Viñas, S.; Farrell, E.F. *The Technical Analysis of Renaissance Illuminated Manuscripts from the Historical Library of the University of Valencia* Harvard University Art Museums & Universidad Politécnica de Valencia (1999)

Nassau (1983) Nassau, K. *The Physics and Chemistry of Color: The Fifteen Causes of Color* John Wiley & Sons (1983)

Naumova & Pisareva (1994) Naumova, M.N.; Pisareva, S.A. 'A note on the use of blue and green copper compounds in paintings' *Studies in Conservation* **39** (1994) 277–283

Nelson (1940) Nelson, H.A. *The Versatile Paint-making Properties of Zinc Oxide* New York (1940)

Newman (1997) Newman, R. 'Chromium Oxide Greens. Chromium Oxide and Hydrated Chromium Oxide' *Artists' Pigments. A Handbook of their History and Characteristics* **3** FitzHugh, E.W. (ed.) National Gallery of Art, Washington & Oxford University Press, Oxford (1997) 273–293

Noble & Wadum (1998) Noble, P.; Wadum, J. 'The Restoration of the 'Anatomy Lesson of Dr Nicolaes Tulp' *Rembrandt under the scalpel. The Anatomy Lesson of Dr Nicolaes Tulp Dissected* Mauritshuis, Den Haag (1998) 51–72

O'Donoghue & Joyner (2003) O'Donoghue, M.; Joyner, L. *Identification of Gemstones* Butterworth Heinemann (2003)

Pabst (1959) Pabst, A. 'Structures of Some Tetragonal Sheet Silicates' *Acta Crystallographica* **12** (1959) 733–739

Pagès-Camagna *et al.* (1999) Pagès-Camagna, S.; Colinart, S.; Coupry, C. 'Fabrication Processes of Archaeological Egyptian Blue and Green Pigments Enlightened by Raman Microscopy and Scanning Electron Microscopy' *Journal of Raman Spectroscopy* **30** (1999) 313–317

Patton (1973a) Patton, T.C. 'Silica, Amorphous' *Pigment Handbook* **1** Patton, T.C. (ed.) John Wiley, New York (1973) 129–134

Patton (1973b) Patton, T.C. 'Magnesium Silicate (Talc)' *Pigment Handbook* **1** Patton, T.C. (ed.) John Wiley, New York (1973) 233–242

Patzelt (1974) Patzelt, W.J. *Polarized-light Microscopy: Principles, Instruments, Applications* Ernst Leitz Wetzlar GMBH (1974)

Pey (1987a) Pey, E.B.F. 'The Hafkenscheid Collection' *Maltechnik-Restauro* **93** #2 (1987) 23–33

Pey (1987b) Pey, E.B.F. 'De Firma Michiel Hafkenscheid en Zoon. Een negentiende-eeuwse handel in schilder-materialen te Amsterdam' ('The firm Michiel Hafkenscheid and Son. A nineteenth-century business in painting materials in Amsterdam') *Bulletin Koninklijke Nederlandse Oudheidkundige* Bond **86** #2 (1987) 49–70

Pey (1989) Pey, E.B.F. 'The Organic Pigments of the Hafkenscheid Collection' *Maltechnik-Restauro* **95** #2 (1989) 146–150

Pey (1998) Pey, I. 'The Hafkenscheid Collection' *Looking through Paintings* E. Hermens, A. Ouwerkerk & N. Costaras (eds.) Uitgeverij de Prom, Baarn & Archetype Publications, London (1998) 465–500

Plesters (1993) Plesters, J. 'Ultramarine Blue, Natural and Artificial' *Artists' Pigments. A Handbook of their History and Characteristics* **2** Roy, A. (ed.) National Gallery of Art, Washington & Oxford University Press, Oxford (1993) 37–66

Powers (1953) Powers, M.C. 'A New Roundness Scale for Sedimentary Particles' *Journal of Sedimentary Petrology* **23** (1953) 117–119

Putnis (1992) Putnis, A. *Introduction to the Mineral Sciences* Cambridge University Press (1992)

Read (1957) Read, H.H. *Rutley's Elements of Mineralogy* 24th ed., George Allen & Unwin Ltd (1957)

Richter et al. (2001) Richter, M.; Hahn, O.; Fuchs, R. 'Purple fluorite: a little known artists' pigment and its use in late Gothic and early Renaissance painting in Northern Europe' *Studies in Conservation* **46** (2001) 1–13

Riederer (1997) Riederer, J. 'Egyptian Blue' *Artists' Pigments. A Handbook of their History and Characteristics* **3** FitzHugh, E.W. (ed.) National Gallery of Art, Washington & Oxford University Press, Oxford (1996) 23–45

Roberts et al. (1980) Roberts, A.C.; Ansell, H.G.; Bonardi, M. 'Pararealgar, a New Polymorph of AsS, from British Colombia' *Canadian Mineralogist* **18** (1980) 525–527

Robinson (1973) Robinson, D.J. 'Chromium Oxide Green' *Pigment Handbook* **1** Patton, T.C. (ed.) John Wiley, New York (1973) 351–354

Robinson & Bradbury (1992) Robinson, P.C.; Bradbury, S. *Qualitative Polarized-light Microscopy. Royal Microscopical Society Microscopy Handbooks* **9** Oxford Science Publications (1992)

Rost (1992, 1995) Rost, F.W.D. *Fluorescence Microscopy* 2 Vols., Cambridge University Press (1992, 1995)

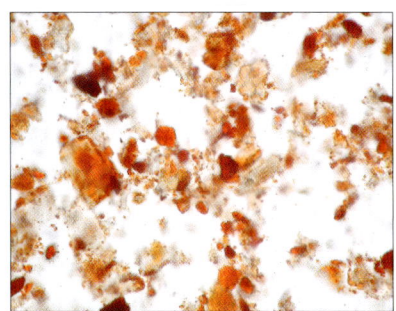

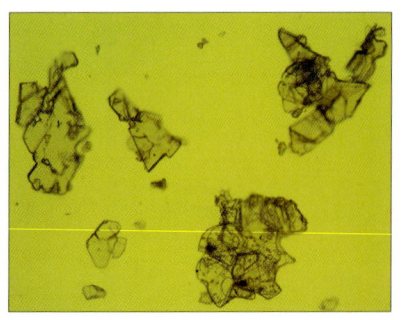

ROY & BERRIE (1998) Roy, A.; Berrie, B. 'A New Lead-based Yellow in the Seventeenth Century' *Painting Techniques. History, Materials and Studio Practice. Contributions to the IIC Dublin Congress, 7–11 September 1998* (1998) 160–165

RUTLEY (1988) Rutley, F. *Rutley's Elements of Mineralogy* 27th ed., Revised by C.D. Gribble Unwin Hyman (1988)

SALTER (1869) Salter, T.W. *Field's Chromatography; or, Treatise on Colours and Pigments as used by Artists* Winsor and Newton, London (1869)

SATTLER (1888) Sattler, H. 'Die mikroscopische Untersuchung des Schweinfurter Grüns und seiner Bildung' *Zeitschrift für angewandte Chemie* no. 235–247 (1888)

SCHWEPPE & ROOSEN-RUNGE (1986) Schweppe, H.; Roosen-Runge, H. 'Carmine' *Artists' Pigments. A Handbook of their History and Characteristics* **1** Feller, R.L. (ed.) National Gallery of Art, Washington & Cambridge University Press (1986) 255–284

SCHWEPPE (1997) Schweppe, H. 'Indigo and Woad' *Artists' Pigments. A Handbook of their History and Characteristics* **3** FitzHugh, E.W. (ed.) National Gallery of Art, Washington & Oxford University Press, Oxford (1997) 80–107

SCHWEPPE & WINTER (1997) Schweppe, H.; Winter, J. 'Madder and Alizarin' *Artists' Pigments. A Handbook of their History and Characteristics* **3** FitzHugh, E.W. (ed.) National Gallery of Art, Washington & Oxford University Press, Oxford (1997) 109–142

SCOTT (2002) Scott, D.A. *Copper and Bronze in Art: Corrosion, Colorants, Conservation* Getty Conservation Institute, Los Angeles (2002)

SEYLER (1929) Seyler, C.A. *The Microscopical Examination of Coal* Fuel Research: Physical and Chemical Survey of the National Coal Resources, No. 16. HMSO, London (1929)

SHELLEY (1992) Shelley, D. *Igneous and Metamorphic Rocks under the Microscope* Chapman and Hall (1992)

SHORT (1940) Short, M.N. *Microscopic Determination of the Ore Minerals* 2nd ed. USGS Bulletin 914 (1940)

SHORT (1940) Short, M.N. *Microscopic Determination of the Ore Minerals* US Department of Interior, bulletin 914, 2nd ed. (1940). [Reprinted 1964]

SINGH *et al.* (2003) Singh, N.; Singh, J.; Kaur, L.; Singh Sodhi, N.; Singh Gill, B. 'Morphological, Thermal and Rheological Properties of Starches from Different Botanical Sources' *Food Chemistry* **81** (2003) 219–231

SINKANKAS (1966) Sinkankas, J. *Mineralogy. A First Course* D. Van Nostrand Co., Princeton, New Jersey (1966)

SPRING (2000) Spring, M. 'Occurrences of the Purple Pigment Fluorite on Paintings in the National Gallery' *National Gallery Technical Bulletin* **21** (2000) 20–27

TALSKY & RISTIC-SOLAJIC (1987) Talsky, G.; Ristic-Solajic, M. 'High resolution/higher order derivative spectrophotometry for identification and estimation of synthetic organic pigments in artists' paints' *Anal. Chim. Acta* **196** (1987) 123

TOWNSEND (1993) Townsend, J.H. 'The Materials of J.M.W. Turner: Pigments' *Studies in Conservation* **38** (1993) 231–254

TRÖGER (1952) Tröger, W.E. *Tabellen zur optischen Bestimmung der gesteinsbildenden Minerale* Schweizerbart'sche, Stuttgart (1952)

TRÖGER (1979) Tröger, W.E. *Optical Determination of Rock-Forming Minerals. Part 1. Determinative Tables.* English Edition of the Fourth German Edition, Bambauer, H.U.; Taborszky, F.; Trochim, H.D.E. (tr.) Schweizerbart'sche Verlagsbuchhandlung (Nägele u. Obermiller), Stuttgart (1979)

TUCKER (1991) Tucker, M.E. *Sedimentary Petrology* 2nd ed., Blackwell Scientific Publications (1991)

VANDEMAELE (1965) Vandemaele, J. 'Les oxydes de zinc' *Double-Liaison* **119** (1965) 47–63

VERBITSKAYA & BURAKOVA (1965a) Verbitskaya, T.N.; Burakova, T.N. 'Crystal-optical study of the $PbTiO_3$-$PbZrO_3$-$PbSnO_3$ system' *Izv. Akad. Nauk. SSR, Ser. Fiz.* **29** (1965) 2059–2063

VERBITSKAYA & BURAKOVA (1965b) Verbitskaya, T.N.; Burakova, T.N. 'Crystal-optical study of the $PbTiO_3$-$PbZrO_3$-$PbSnO_3$ system' *Bull. Acad. Sci. USSR, Phys. Ser.* **29** (1965) 1893–1897

WAGNER *et al.* (1933) Wagner, H.; Zipfel, M.; Heintz, G.; Haug, R. 'Ursachen und Behebung der Cromgelb-Lichtunechtheit' *Farben Zeitung* **38** (1933) 932–934

WAINWRIGHT *et al.* (1986) Wainwright, I.N.M.; Taylor, J.; Harley, R. 'Lead Antimonate Yellow' *Artists' Pigments. A Handbook of their History and Characteristics* **1** Feller, R.L. (ed.) National Gallery of Art, Washington & Cambridge University Press (1986) 219–254

WALLERT (1984) Wallert, A. 'Orpiment and Realgar' *Maltechnik-Restauro* **90** 4 (1984) 45–57

WEBMINERAL (2003) www.webmineral.com (accessed August 2003)

WELSH (1988) Welsh, F.S. 'Particle Characteristics of Prussian Blue in an Historical Oil Paint' *Journal of the American Institute for Conservation* **27** (1988) 55–63

WHITBREAD (1989) Whitbread, I.K. 'A Proposal for the Systematic Description of Thin Sections Towards the Study of Ancient Ceramic Technology' *Archaeometry* **31** (1989) 127–138

WINCHELL (1929) Winchell, A.N. *Elements of Optical Mineralogy: Part III Determinative Tables* 2nd ed., John Wiley & Sons, New York (1929)

WINCHELL (1931) Winchell, A.N. *The Microscopic Characters of Artificial Inorganic Solid Substances or Artificial Minerals* 2nd ed., John Wiley & Sons, New York (1931)

WINTER (1997)   Winter, J. 'Gamboge' *Artists' Pigments. A Handbook of their History and Characteristics* **3** FitzHugh, E.W. (ed.) National Gallery of Art, Washington & Oxford University Press, Oxford (1997) 143–155

WINTER (1981)   Winter, J. ' "Lead white" in Japanese paintings' *Studies in Conservation* **26** (1981) 89–101

WINTER (1983)   Winter, J. 'The characterization of pigments based on carbon' *Studies in Conservation* **28** (1983) 49–66

WINTON (1906)   Winton, A.L. *The Microscopy of Vegetable Foods* John Wiley & Sons, New York (1906)

WOERNER (1973)   Woerner, P.F. 'Calcium Carbonate Synthetic (Precipitated)' *Pigment Handbook* **1** Patton, T.C. (eds.) John Wiley, New York (1973) 119–128

YANG & PREWITT (1999)   Yang, H.; Prewitt, C.T. 'On the Crystal Structure of Pseudowollastonite (CaSiO$_3$)' *American Mineralogist* **84** (1999) 929–932

YARDLEY *et al.* (1990)   Yardley, B.W.D.; MacKenzie, W.S.; Guilford, C. *Atlas of Metamorphic Rocks and their Textures* Longman Scientific and Technical (1990)

YOUNG *et al.* (1997)   Young, J.R.; Bergen, J.A.; Bown, P.R.; Burnett, J.A.; Fiorentino, A.; Jordan, R.W.; Kleijne, A.; Niel, B.E. van; Romein A.J.T.; Salis, K. von 'Guidelines for Coccolith and Calcareous Nannofossil Terminology' *Palaeontology* **40** #4 (1997) 875–912

ZIOBROWSKI (1973)   Ziobrowski, B.G. 'Molybdate Orange' *Pigment Handbook* **1** Patton, T.C. (eds.) John Wiley, New York (1973) 375–383

ZONA (1996)   Zona, C.A. 'The Microscopical Examination of Organic Pigments: The Hansa Yellows' *Microscope* **44** #4 (1996) 195–202

Index

(I) = illustration
(T) = text